Daring Dreams of the Future

THOUGHT, SOCIETY, CULTURE
Slovenian and South Eastern European Perspectives

Edited by the Research Centre of the Slovenian Academy of Sciences and Arts

Vol. 5

Notes on the quality assurance and peer review of this publication

Prior to publication, the quality of the work published in this series is reviewed by two external referees appointed by the editorship.

Aleksej Kalc / Mirjam Milharčič Hladnik /
Janja Žitnik Serafin

Daring Dreams of the Future

Slovenian Mass Migrations 1870–1945

Translated by Barbara Skubic

PETER LANG
Berlin - Bruxelles - Chennai - Lausanne - New York - Oxford

Bibliographic Information published by the Deutsche Nationalbibliothek
The Deutsche Nationalbibliothek lists this publication in the Deutsche Nationalbibliografie; detailed bibliographic data is available in the internet at http://dnb.d-nb.de.

ISSN 2195-2191
ISBN 978-3-631-89897-0 (Print)
E-ISBN 978-3-631-89904-5 (E-PDF)
E-ISBN 978-3-631-91003-0 (EPUB)
DOI 10.3726/b21284

Published by:
Peter Lang GmbH, Berlin, Deutschland

info@peterlang.com - www.peterlang.com

And the eye wandered through bright regions,

It dreamed daring dreams of the future,

Followed an orator's powerful gestures,

And kept faith for a generation which is not yet born ...

Oton Župančič, Duma

(translated by Henry R. Cooper, Jr.)

This book is dedicated to all the people on the move who dare to dream their future.

Contents

Translator's Note

As with any translation, the team working on the English language version of this book found ourselves in front of a series of challenges that we had to solve to make it faithful to the Slovenian original, as well as reader-friendly for those interested in (Slovenian) migration who might not be deeply familiar with the Slovenian ethnic territory, its people, and its history.

The book is complex: it approaches Slovenian migration through historiography, literary studies, and sociology, so translating it at times felt like cramming for comprehensives in three academic fields. My first task as a translator was thus to find solutions that would work throughout the book, as not every field uses the same vocabulary to describe the same thing. The book covers almost a century, including World War I, with all the breaks in continuity and thinking that it brought. This fact required linguistic solutions that would make sense for 1839 as well as for 1939.

The first such choice was what to call people who moved, either within the state or internationally. We reserved the use of the more traditional terms of emigrant or immigrant for situations in which a person's legal status is discussed, that is, when migrations are seen from the perspective of the country of origin and/or the receiving countries. When talking about the people themselves, we use migrants.

The territory and the identities of its inhabitants presented a further, and perhaps the most complex, challenge: what to call the territory that, over the span of a century, belonged to different states, not to mention its people, who lived under different regimes and adopted different identities, sometimes changing them, whether voluntarily or not? In the end, we decided on "Slovenian ethnic territory" to describe the regions that, throughout history, have had a contiguous Slovenian population. This territory has never been monolingual or monocultural; its inhabitants may have adopted different national or (particularly in earlier times) regional—as well as cultural and linguistic—identities, even within the same family and a single generation.

Some parts of this territory have long-established names in English, such as Carniola, Styria, or Carinthia, and we have used them throughout. Some parts do not; for those, we used current Slovenian names instead of their Italian or German "variants" or calques that sometimes appear in academic texts. We therefore use Prekmurje, Bela Krajina, Goriška, Primorska. Another deviation from the standard use is Benechia (the slightly anglicized version of Beneccia/

Benečija). Such use may sometimes create unusual linguistic combinations, for example, when we discuss Styria and Prekmurje or Carniola and Goriška, but since administrative borders often cannot (and do not) correlate with the ethnic and/or linguistic borders, we felt such choices were perfectly justifiable.

At the beginning of the period covered, the Slovenian ethnic territory was a part of the Habsburg Empire and, at its end, a part of the Kingdom of Yugoslavia, the Kingdom of Italy, and the interwar Kingdom of Hungary. Within the Austro-Hungarian Monarchy, it was divided between the Austrian and Hungarian parts, which we call Imperial Austria and the Kingdom of Hungary, rather than the perhaps more traditional names Cisleithania and Transleithania. Slovenian-speaking inhabitants of the Slovenian ethnic territory are sometimes simply called Slovenians (particularly in the diaspora), even when discussing the time when regional identities were perhaps more prevalent. The text often refers to Yugoslavia or Yugoslav territories, even when speaking about the time before 1929. We did so consciously to make it easier to read.

Another segment that required a decision on what to translate (if at all) and how to translate it were the names of (diaspora) publications and organizations. Like other migrants, Slovenians organized in clubs and associations, published papers, magazines, and books, entered political alliances with like-minded people from their own and other communities. In different places, and in different times, they made different choices to name their papers, clubs, or parties. Sometimes, the names were in Slovenian only, sometimes they were bi- or multilingual. Often, associations and papers merged and combined, and consequently dropped parts of their names or invented new ones. We therefore had to decide which names to translate and which ones to leave in Slovenian only. It was not an easy decision to make—and it was even harder to stick to it. In general, anything that was bilingual in real life, is bilingual in the book. We added the English in parentheses after the Slovenian one if we felt it would benefit the reader: for the organizations active for a long time, or in a large area, for papers that were very influential or were published for a long time, for certain ad hoc political associations that were established around a vital cause. We were also more inclined to translate the names of those publications that have already been digitalized and are accessible to readers. We avoided translating the names of associations that were short-lived and papers that only had a couple of issues published and are difficult to access.

It is worth noting, too, that certain names repeat throughout the diaspora, regardless of the country they are in and the period in which they appear. They often included (a variant of) *narod* (nation), *dom* (home), *zarja* or *zora* (dawn), *danica* (morning star), *delavec* or *proletarec* (worker), *zvon* (bell), *matica* (center, base, or headquarters).

The names of cultural and social associations, in particular, often reflected their political orientation and the time in which they were created (frequently noticeable in the choice of words, such as *jednota* for unit or *odbrana* for defense). Cultural associations, especially drama clubs, were often named after the famous Slovenian playwright Ivan Cankar. *Prosveta* indicated that a particular organization took care of the community's cultural and educational needs. More than one association was called Triglav, after Slovenia's highest peak. Choral societies were often named after instruments or birds, with *slavec*—or the diminutive *slavček*, if it was for children—being a perennial favorite. And then there were certain associations whose names were never translated: Prosvetna matica and Glasbena matica are such examples.

Finally, a text of this length will always present a challenge for the translator.

I would like to thank the authors, Mirjam Milharčič Hladnik, Janja Žitnik Serafin, and Aleksej Kalc, who patiently answered my incessant questions and worked with me through several drafts of the translation, friends—historians and linguists—whom I pestered to help me with tidbits about the countries and languages with which I am not particularly familiar, and all the librarians who helped me locate the sources that enabled me to make translation choices that would make the most sense for international readers. Professor Annemarie Steidl read an early draft of the translation; her comments on place names were invaluable, as were her comments on the parts of the text where the translation was not yet clear. Jana Renée Wilcoxen worked her editing magic to turn the final draft into a book, and using her fresh eye Ana Čavić made the final touches; it is difficult to explain just how much thought and work went into those two tasks. All mistakes that may have remained in the translation, of course, are mine.

Acknowledgements

The authors would like to thank the many colleagues who helped to bring this book to form through their time, insight, clarifications, and other contributions: Annemarie Steidl, Dirk Hoerder, Miha Zobec, Marina Lukšič Hacin, Majda Kodrič, Jernej Mlekuž, Danijel Grafenauer, Katalin Munda Hirnök, Attila Kovács, Matjaž Klemenčič, Žarko Lazarević, Aleksander Panjek, Mateja Rihtaršič, Deja Gliha, Tadej Turnšek, Oto Luthar, Martin Pogačar, Daniel Necas, Joseph Valencic, Marko Zajc, and Peter Vodopivec. Our deepest gratitude goes to Barbara Skubic for her translation, Jana Renée Wilcoxen for her copy editing, and Ana Čavić for her proofreading.

Introduction

America is our problem: it has been our economic, social, national, cultural problem ever since our people started moving across the sea. Adamič claims this happened in 1870. For Bela Krajina, however, one must go at least twenty years back, the first among our men emigrated around 1850. I remember from my young age how, in every tavern, side by side with religious icons on glass and color prints of the "Stages of Human Life," there were giant adverts for massive steamships with signs that said, "Red Star Line," "Hamburg-Amerika Linie," "Bremen-New York," "Trst-Amerika." People emigrated in droves, so more than one house was left without a master or even people altogether. Younglings at school dreamt about Bremen and Hamburg; New York, Chicago, Cleveland, Montana, Minnesota, Pennsylvania, Pittsburgh, Dakota, Nebraska, California were familiar names, as they were where their fathers, older brothers, uncles toiled in mines, factories, saloons, and farms, and from where they "sent back" to pay off debts, to fix the house, to purchase a bit more, to pay taxes, to buy some treats. Many a thatched roof disappeared, and a log cabin turned into a brick house. My five aunts—because five were all the sisters that my mother had—were in America. When my father sold the house—to an American, of course—he took me, because I did well in fourth grade, to Bosnia with him so we could buy a house there. Had we indeed bought it, what would have become of my Slovenehood? Bela Krajina leans south. We didn't buy it. My mother yearned to follow her sisters to America. Had she had her way—what would have become of me?
Oton Župančič (1932)

This is how Oton Župančič (1878–1949), the renowned Slovenian poet, playwright, translator, and influential editor of the literary magazine *Ljubljanski zvon*, reflected on mass emigration from the Slovenian ethnic territory, a phenomenon he had witnessed since childhood and which, at the time of writing these lines in the early 1930s, remained a burning societal and national question. From the mid-19th century until World War II, emigration shortchanged the demographic growth within the territory of today's Slovenia by 440,000 inhabitants. We can only fathom the impact and general importance of this emigration once we remember that the entire territory we are discussing had a mere total of one and a half million inhabitants at the end of the aforementioned period. Migration processes, fueled mainly by economic constraints and social restructurings, were largely responsible for such a deficit. During and after World War I, the war and the newly created political circumstances, coupled with the existing economic and social causes and motivations, added to migration a character of evacuation and political exile. Slovenian Lands in the Austro-Hungarian Monarchy and, later, in interwar Yugoslavia topped all the lists of territories by the rate of emigration. They were also at the top in specific periods when considering the European continent.

Emigration was a characteristic trait of the accelerated geographic mobility and different migration processes that, in the 19th and 20th centuries, also impacted the Slovenian ethnic territory and left a deep mark on its social and political history. This book tries to paint a picture of the "great" migrations in the Slovenian ethnic territory in their general and more detailed traits, show them in the light of the local and international socioeconomic and political contexts in which they emerged, and point out how migrations were a constitutive part of the everyday life of communities, families, and individuals. The book also wishes to show how, through mass migrations, the Slovenian "ethnic and cultural space" expanded well beyond its historical borders and how, with the national idea and national identifications, alongside the Slovenia "at home," other "Slovenias" around the world also appeared.

Beginning our research with the thesis that geographic mobility and migrations have been constitutive and structural elements of social processes throughout history, we sought to integrate migration events into a historical presentation that reveals the most common problems and themes. We wanted to draw attention to the great diversity of these movements and particularly show those processes that do not usually get attention and are absent from the mainstream perception of migration in Slovenia. At the same time, we tried to highlight the complexity of migration phenomena, their causes and consequences, and their changeability in accordance with the situation in time and space. Despite all this, the book is not and does not wish to be a comprehensive history of migration in the Slovenian ethnic territory. The emphasis of our research remains on the prevalent migration phenomena, that is, those that led people to leave the Slovenian ethnic territory.

Studying migrations in the Slovenian ethnic territory is particularly complex and challenging due to its political-administrative conformation throughout history. In our research, the Slovenian ethnic territory describes a geographical area of historical and contiguous settlement of the Slovenian-speaking population. This territory stretches beyond the borders of today's Republic of Slovenia (and its predecessors within the Socialist Federative Republic of Yugoslavia, 1945–1991; the Kingdom of Yugoslavia, 1929–1945; and the Kingdom of Serbs, Croats and Slovenes, 1918–1929), which were only defined after World War II. During the period under study in this book, the Slovenian ethnic territory was a part of different state entities and linked to the history of different empires and political spaces. Until the end of World War I, it was mostly a part of the Austrian half of the Habsburg Monarchy, with Prekmurje, the eastern region across the Mura River, being a part of its Hungarian half. Benechia and Resia in the west were a part of the Kingdom of Italy. The area under Austria was a part of the historical hereditary and crown lands, of which only Carniola had a Slovenian-speaking majority. In Styria, the Slovenian ethnic territory overlapped with the lower part of the land, that is, Lower Styria, in Carinthia with its southern part, while in the Austrian Littoral, the Slovenian ethnic territory mostly comprised Goriška (the eastern and northern part of the Princely County of Gorizia and Gradisca), Trieste and its environs, and the northern and eastern parts of Istria. Between the two world wars, the majority of the Slovenian ethnic territory (Carniola, Lower Styria, and Prekmurje) passed to the newly formed Kingdom of Serbs, Croats and Slovenes (from 1929 onward, the Kingdom of Yugoslavia). Almost the entire southern part of Carinthia became a part of the Republic of Austria, and the entire western territory (including the former Austrian Littoral and the fringes of Carniola) was annexed to the Kingdom of Italy.

Neither was the Slovenian ethnic territory a homogenous ethnic space: members of other linguistic and ethnic groups lived in it, either among the Slovenian-speaking population or in isolated linguistic islands, so that the territory was—at its edges, but partly also inland—a part of the wider central and eastern European mosaic and a crossroads of nations, languages, and cultures. Its belonging to and dependence on different centers of power, as well as internal administrative fragmentation into historical lands, influenced the socioeconomic and cultural differentiation/differences between individual parts of the Slovenian ethnic territory. Its geographical position, morphological/natural characteristics, and historical events also left a mark on migration phenomena. Migrations, in turn, impacted upon its development and contributed to forming the social, economic, and cultural image of the Slovenian ethnic territory. The geographical space that is at the center of our research allows us to observe the effects of the great migrations of the 19th and 20th centuries between the Northern Adriatic, the Alps, the Pannonian Basin, and the Balkans in the light of administrative and geopolitical fragmentation, under the influence of historical breaks and the shifting of state borders. To this, we should add interethnic relationships that, in the period under scrutiny, sometimes grew into national conflicts and were in different ways intertwined with migrations.

Map 0.1: The Habsburg Monarchy and the Slovenian ethnic territory 1866–1918.

Studying migrations within a territory thus defined, of course, runs into technical and methodological problems because this space, for the most part, does not correspond to the political-administrative division to which statistics and administrative and historical sources refer in different historical periods. An additional problem for the quantitative study is that Imperial Austria did not keep direct statistics of internal migrations prevalent in specific periods and specific places. At the same time, the data on migration abroad are insufficient and unreliable: the Austrian authorities themselves monitored transatlantic emigration using the data of emigration ports and the statistics of receiving countries. To study the longitudinal phenomenon of migrations from the Slovenian ethnic territory abroad, we only have statistics on the transatlantic migration from Carniola. For other Slovenian Lands, all that is available are fragments or incomplete data for smaller districts. We have more data—statistical series and administrative sources—for the period of interwar Yugoslavia when migration fell under the competency of a dedicated central state service with regional branches. Unfortunately, these sources do not allow for a comprehensive qualitative and quantitative insight into movements, particularly not into the extensive migrations within the Yugoslav space. Migrations in the parts of the Slovenian ethnic territory that belonged to other countries constitute a particular problem.

We encounter similar incongruencies in the receiving countries whose sources record Slovenian immigrants as citizens of Austria, Hungary, or Italy until the end of World War I. In the United States—the primary destination for Slovenian migrants abroad—the immigration statistics classify Slovenian migrants together with Croatians when displaying migrants' origins and other characteristics. The census only started using the mother tongue as a criterion in 1910. However, such a division misses all those who did not declare themselves Slovenians and those who belonged to other ethnic/linguistic groups. After World War I and the creation of the new countries, which changed the geographical notions of "internally" and "internationally," the difficulties do not cease. In the statistics of the receiving countries, Slovenian emigrants from Yugoslavia are registered as Yugoslavs; those from the territories outside of the Yugoslav borders are recorded as Italians, Austrians, or Hungarians. Statistics are just the most direct proof of the conundrums and difficulties that are characteristic of all sources from administrative institutions, difficulties that any scholar of Slovenian migrations—be it in the country of origin or in the receiving countries—regularly faces.

As we have mentioned, this book aims to offer a general and yet as complex as possible historical overview of mobility and migration processes in the Slovenian ethnic territory from the second half of the 19th century until World War II. We wished to present migration phenomena and their structural and dynamic characteristics across a long temporal arc, from the points of view of causes and effects, forms, and geographical trajectories, and the contexts of the places of origin and the receiving countries. At the same time, we wanted to provide a human-centric insight into the events and complete the explanations based on the schematic analytical approaches with migrants' concrete experiences. For this reason, we conceived this presentation in three parts.

The first part, written by Aleksej Kalc, places different migration phenomena in the Slovenian ethnic territory into a spatial and historical context, showing their general and detailed dynamics, their structural and typological characteristics, and their links to economic, social, and political events in Slovenia and the destination countries. From the demographic aspect, we focus on dynamics, morphological traits, original geography, and destinations of migrations. The emphasis is on regional differences and socioeconomic contexts and factors. At the forefront of the book's first part are migration systems that used individual migration patterns and their functional origins to connect emigration and immigration environments to economic, social, and cultural transfers. We also look into other factors that influenced migrations, for example, gender differentiations, structural aspects, and legal frameworks. We particularly emphasize continuities and breaks that marked migrations, shifted their routes, or altered their scope, character, and impact in the period researched. We study them from the socioeconomic aspect and from the national, political, and organizational aspects.

The second part, by Janja Žitnik Serafin, in which the aspects of cultural and general history complement each other, focuses on Slovenian migrant communities in receiving countries, emphasizing their organizational and creative endeavors. This part illustrates how the organized life of Slovenian migrant communities started and developed, presents the types of activities they engaged in, and discusses the worldview and political and ideological rifts and conflicts that have continuously stirred the Slovenian diaspora. One section focuses on the Slovenian diaspora press as one of the most important sources for understanding the events and internal relationships that shaped the Slovenian migrant communities. In continuation, it shows how the migrant communities dealt with the social and cultural environment in the receiving countries and how Slovenian "national activists" (emigrant teachers, priests and school sisters, and others) aided the transfer of the Slovenian language and culture to the descendants of migrants, in everyday life or via cultural production. It shows the differences in intergenerational relationships in the diaspora regarding organizations and the cultural position of immigrants in different migrant environments and periods. We dedicate special attention to the question of how the cultural position of Slovenian migrants is directly reflected in the level of linguistic, cultural, and—in some places—religious assimilation of its members in the circumstances wherein the receiving societies did not encourage the preservation and use of the immigrant/minority languages and cultures.

In the book's third part, by Mirjam Milharčič Hladnik, the selected life stories illustrate periods, patterns, and trajectories within the era of Slovenian great migration "from below." The narratives and memoirs, letters, diaries, stories, testimonies, and other "documents of life" explain migration from a human perspective. Each story describes migration as "a total social act," in which human practices and experiences are involved in myriad interdependent relationships and interactions with social, economic, cultural, political, and religious spheres and with social representations of the world. Subjective perceptions of the migration process reveal the full complexity of migratory experiences, which are determined by family, politics, transnational community, moral norms, prescribed gender roles, religion, ethnicity, and nationality. The stories reveal the richness of the reflections, decisions, and identity transformations, and the transnational networks that marked the lives of all involved, both those who left and those who stayed. The focus is on individuals as agents of the migratory phenomenon, interpreting both their own mobility and that of the societies in and between which they move.

The book covers migrations in the Slovenian ethnic territory in two particularly distinct historical phases, very different from each other: the time of relatively free movement, within the state borders and beyond them, in the 19th and the early 20th centuries, and the time between the two world wars, nicknamed "the tyranny of the national" in the history of migration, with its regulations and restrictive migration policies. World War II had an even more dramatic and violent impact on the movements of the Slovenian population. It was followed by a new era, marked by the new political geography and the new Cold War division of Europe into two blocs. The purpose of this book is not to draw a complete panoramic picture of all the documented migration activity but, instead, to highlight migration dynamics and their prevalent or original patterns and traits from the perspective of continuities and changes in the context of the historical development in the Slovenian ethnic territory and the world at large. The research we undertook for this book was done to partially close the significant gaps in this under-researched area in the historiography of Slovenia, despite the advancements of recent decades. The English version additionally hopes to contribute to the historical atlas of European migration and shed light on the territory that only appears as a focus of international research episodically and in "special cases."

1 Historical, Social, Economic, and Political Aspects of Migrations

From Tradition to Modernity

Mass migrations and population shifts rank among the most remarkable social phenomena of the 19th and 20th centuries. Historiography and other disciplines have long studied them through the lens of modernization and its landmark societal, economic, political, and cultural changes that characterized this period. The paradigm of modernization influenced how migration and intense spatial mobility were presented. Migration was understood as a typical trait of modern society, linked to industrialization, the growth of the middle class and the proletariat, the technological and transport revolution, urbanization, as well as to the implementation of personal liberties, expansion of new needs, consumerism, and new ways of life. On the other hand, traditional rural pre-industrial society was considered sedentary due to its dependence on the land and its entrapment in feudal structures. In it, people were presumed to live within their local communities and only move—if they moved at all—within the limits of their neighboring regions. Official records seemingly support such an image: unlike previously, migrations and spatial mobility feature prominently in 19th- and 20th-century sources, be it as part of major historical events or events that occurred at the local level, in the lives of families and individuals. Their scope and their social and economic—as well as legal, cultural, and political—implications made migrations subject to increasingly precise observations, definitions, and classifications[1] and consequently more formally visible and recognizable. Likewise, the rise of nation-states and nation-building ideologies shaped both the definition of the migration phenomenon and its research. Receiving countries and countries of origin focused their research interest conceptually on the phenomenon of immigration or emigration by positioning the issue within national paradigms and ethnic categories. The focus on modernization and economic theories about mobility highlighted the importance of modern migrations and their relation to

1 Such developments were influenced by the development of citizenship and civic rights, social care systems, public education, universal conscription, and other socio-administrative systems for establishing evidence of the holders of rights or duties (Green and Weil, 2007; Kalc, 2016c, 24).

the changes brought by the 19th and 20th centuries but downplayed the role of mobility and migration patterns in pre-modern times.

This historical perception (and consequently, interpretation) of migrations has changed over the decades. The images of the moves turned out to be deceptive: they were constructed upon official definitions and categories created for administrative purposes to align with specific interests in the era of industrialization and the modern nation-state. The inappropriate nature of unified analytical models and one-sided and generalized explanations became apparent. Case studies, reconsiderations of methodological approaches, and epistemological questions revealed the traps of terminological and conceptual formulations that overshadowed the phenomenon's characteristics, nature, and dimensions. New ways of seeing, interpreting, and understanding migrations overcame the classical typological divisions and categories. Emigration and immigration, emigrants and immigrants, were replaced by migration and migrants. The focus of research expanded from the phenomenon of migration to include people and their practices and experiences. Migration studies thus replaced the history of emigration and immigration. These studies do not read the phenomenon through deterministic patterns but rather observe it as a complex interplay of material and non-material factors while treating migrants as conscious actors moving within it.

New trends in migration research understand migrations as passing between social and cultural spaces rather than merely crossing state and administrative borders and geographical mobility as "a social process and [...] a basic condition of human societies" (Hoerder, 2002, xix). This research has revealed that pre-modern society was far more mobile than the followers of modernist theories believed (Hochstadt, 1999; Lucassen and Lucassen, 2015). It has also deconstructed the thesis that migrations were the effect of times of crisis and watershed moments in history, showing them instead as a continuous phenomenon structurally built into everyday social, economic, and cultural organizational frames and dynamics. This scholarship highlighted the inconsistency of the dichotomous differentiation between the pre-industrial and industrial eras and the theoretical model of the transition from a society that was geographically immobile or barely mobile to a mobile one (Zelinsky, 1971; Moch, 1992; Rosental, 1999; Lucassen and Lucassen, 2009). The role of migrations and mobility in the long course of history became clear. Thus, the concept of *homo migrans* entered migration studies to draw attention to human mobility as a historical, social, and cultural practice (Bade, 2001).

The new scholarship also recognized that sedentariness and stability were idealized social behaviors promoted as "natural" by authorities and political and public opinion makers, which tended to ascribe negative meanings to mobility and foster mistrust and fear toward migrants (cf. Becker, 2019, 218). Migrations (have always) had a certain "reputation": on the one hand, they cause "losses" to their nation and country of origin; on the other, the "invasion" of foreigners and undesirable others into the imaginary homogenous communities triggers a sense of endangerment (Corti, 2015). Modernization did profoundly change the scope and intensity and the directions and distances of migrations, and the mid-19th century was a turning point in this sense. However, we need to observe these mobilities parallel to the revolution in transportation, which simplified and increased movements, rather than parallel to the idea of modernization per se (cf. Lucassen and Lucassen, 2015; Moch, 1992). Migration did not emerge as a new phenomenon with modernization. Instead, it adjusted—in form, methods, and extent of movement—to the times and means and started happening with an increasing speed in an expanding international and transcontinental realm.

The "Slovenian case" was no exception. It displays the continuities and socially constitutive role of the migration phenomenon throughout the centuries and its expansion and changes in the context of 19th-century modernization. In the Middle Ages and the early modern period, the Slovenian ethnic territory[2] provided a stage for dynamic spatial mobility, the comings and goings that, over centuries, helped form its social fabric and cultural character (Štih, 2010). The remnants of these movements can even be traced in common Slovenian surnames with geographical or ethnolinguistic etymology, such as Lah (Italian), Nemec (German), Čeh (Bohemian), Turk (Turk or Ottoman), Korošec (Carinthian), Kranjec (Carniolan), Kočevar (from Kočevje), Furlan (Friulian), Horvat (Croat), and similar (Melik, 1963). Historical research shows that the predominantly rural population in Slovenia was all but settled. The level of mobility varied, but in many places only a narrow segment of families remained in the same homesteads or settlements for longer periods. Research also shows that short- or long-distance moves and resettlements were a constitutive part of

2 That is, the Crown Lands of Imperial Austria or their parts with a predominantly or substantially Slovenian population—Carniola/Kranjska/Krain, Lower Styria/Spodnja Štajerska/Untersteiermark, Carinthia/Koroška/Kärnten, and the Austrian Littoral/Avstrijsko Primorje/Österreichisches Küstenland/Litorale Austriaco comprising Goriška (the eastern part of Gorizia and Gradisca), Istria, and Trieste—as well as Prekmurje/Muravidék/Übermurgebiet in the Kingdom of Hungary and Benechia in the Kingdom of Italy.

everyday life, whether conditioned by economic or other motives (Melik, 1963, 331; Voje, 1978, 149; Gestrin, 1991, 1998; Kosi, 1998).

Indeed, mobility and migrations were part of economic practices and social strategies. The images of the Slovenian peasant living off the land, engaging in animal husbandry, earning a living by selling produce, or even striving toward agrarian self-sufficiency, are deceptive. They correspond to the romantic imagery of the Slovenian population's intimate attachment to the land and the idealization of "home turf" and the rural world as a source of life's happiness, an image reinforced in cultural and political thought as the Slovenian national idea developed in the 19th century. The agrarian economy often only partially met the needs of the rural population. Thus, non-agrarian activities, from commerce to transportation, craft trades to mining and ironworks, supplemented the economy and often became predominant components of an integrated (or mixed) economic system. This course of development was particularly noticeable from the 15th century onward. In the 18th century, an average of one-third of earnings in rural Slovenia came from non-agrarian economic activities (Blaznik et al., 1970, 627–631; Valenčič, 1981; cf. Panjek, 2018; Beguš, 2018). However, spatial mobility and migration were not equally distributed across the territory. Migrations were more widespread in places where local agrarian sources were meager (for example, in mountainous and karstic regions) or where additional opportunities to make a living were not readily available within the range of day mobility. People from such environments were likelier to participate in seasonal or extended migration to rural and urban areas. In the areas that did provide easier access to either daily or itinerant work in agrarian or urban economies, the option to plan work-related activities based on mobility or circular migration and integrate agrarian and non-agrarian income often accelerated farm fragmentation (Voje, 1978, 149; Panjek, 2022).

Specific categories of workers, such as merchants and artisans, were characteristically mobile. Merchants—locals and foreigners, often from Italian regions—moved in search of opportunities and a suitable environment for their business. Many were peddlers, a line of work characterized by circular migration. In their constitutive and institutionalized practices, trades depended on migrations: artisans started with apprenticeships and continued with wandering, structured to help apprentices gain professional experience before they could qualify as journeymen and later settle and become masters of the trade. Not only journeymen but all precarious workers were highly mobile and gravitated toward urban centers. Their mobility contributed to the ever-changing dynamics of cities, which, because of high mortality rates, depended on newcomers to replace and maintain their population (cf. Gestrin, 1991, 150–155; Golec, 1999, 775, 786–788; Kalc, 2008, 2010).

The Slovenian ethnic territory experienced increased local and interregional migration throughout the 18th century. It coincided with the state's relaxation of legal limitations to personal mobility as a part of reforms and mercantilist policies to stimulate population growth, social and occupational mobility, and economic productivity. Such moves did not significantly influence the demographic growth in the Slovenian ethnic territory, which continued to be driven by natural factors throughout the early modern period. Nevertheless, their demographic effects were visible locally, for example, in clusters of cottagers on common land in areas where transportation infrastructure or other sources enabled the development of non-agrarian economic activities, such as the area around Logatec on the Ljubljana–Trieste route, where the number of cottagers rapidly increased (Voje, 1952/53; Gestrin, 1991, 20).

Demographic changes were more pronounced in Trieste where, due to mass immigration, the number of inhabitants in the 18th century increased from 4,000 to 20,000 (Kalc, 2008). This increase was a consequence of state policy. Vienna awarded Trieste the status of a free port in 1719, thus creating a maritime emporium in the North Adriatic to promote domestic and international trade. Immigration from the immediate hinterland and more remote European and Mediterranean areas became the driving force behind the Triestine demographic expansion. Newcomers were attracted by the free port liberties and the policies the authorities in Vienna employed to encourage merchants, artisans, and all the economically "useful" population to settle in the city. In a century, the stagnant, communal Trieste, founded on medieval administrative autonomy, transformed itself into a vibrant, multi-ethnic urban hub, which—even before Joseph II's Patent of Tolerance—guaranteed religious freedom and special status for its non-Catholic communities (Finzi and Panjek, 2001; Kalc, 2008, 2012). Although Slovenian migrants to Trieste were primarily members of the lower social strata, merchants, artisans, and other successful professionals were also among them. These migrants contributed to developing the urban social fabric and the rural areas around the city where the Slovenian settlers had purchased or rented the land from its city-dwelling owners and created suburban agrarian communities. Some Italian historians from Trieste named this process the "slavicization" of the *ager* of Trieste (Kalc, 2009, 2018b). Another place where remarkable demographic changes occurred in the 18th century because of migrations was Idrija, the site of Europe's second-richest mercury ore deposit. The influx of miners made it the one of the most populated towns in the Slovenian ethnic territory, after Trieste and Ljubljana (Arko, 1931).

Migrations influenced the structure of the inhabitants, and not just in Trieste and other cities. In the era of mercantilism, the absolutist state banned the drain of skilled and other specialized workers and economic resources while attracting them in various ways from abroad. For example, La Serenissima (Republic of Venice) proclaimed in the 1770s that emigration was a "delict against the state" and imposed hefty fines for those breaking the law. Numerous master artisans and workers (and their families) moved from the Venetian Republic to work in the paper mills and textile factories around Gorizia/Gorica/Görz because the local specialized workforce was scarce. In time, more than a thousand of these Venetian textile workers moved from Ajdovščina in the Vipava Valley to other Austrian textile centers (Iancis, 2006). Similarly, artisans and those with special skills moved to the Slovenian territory from other places.

The "Weight" of the Population and the Beginning of Mass Emigration

The demographic picture and the impact of migrations on the whole Slovenian ethnic territory started changing in the Pre-March Period (Vormärz, i.e. 1815–1848). The demographic transition began then, and the precedent of predominantly stationary long-term demographic dynamics gave way to population expansion. Available enumerations and vital series show that the growth in the first half of the 19th century, a result of declining mortality and a rise in fertility, as well as immigration from non-Slovenian territories to the towns and areas with early factories, particularly the ironwork industry, was the fastest in the Slovenian ethnic territory in centuries. However, from the end of the Pre-March Period, the backwardness of the pre-modern rural economic structures and the inconsistency of industrialization became apparent. In these circumstances, migration toward the areas of economic expansion in other parts of the monarchy and abroad increased exponentially. It became a structural factor of the continuing demographic and social development.

Statistical data and estimates clearly show the magnitude and consequences of this remarkable phenomenon. Between the mid-19th century and World War I, a net of 310,000 people emigrated from the territory of today's Slovenia (Blaznik et al., 1970, 103–105). As a result of emigration, natural demographic growth was reduced by almost 58 percent. Consequently, the population in the territory of today's Slovenia only grew by about 23 percent (from 1,072,000 to 1,321,000) in the period between the censuses of 1857 and 1910, while in the entire Slovenian ethnic territory (i.e., including southern Carinthia, parts of Goriška, and Trieste), the population grew by 33 percent. By comparison, the entire Imperial

Austria, which was no stranger to mass emigration, saw its population increase by 52 percent in the same period (Šifrer, 1962, 17). The Slovenian ethnic territory, particularly its predominantly rural central part, was among the areas with the highest emigration rates in the Austria-Hungary and proportionally one of the most prominent European population and workforce exporters before World War I (Hoerder, 1996, 37).

We can only establish the process and the scope of migration from the Slovenian ethnic territory roughly and indirectly, using population census data and net migration rates. The migration dynamics had several phases. In the first, the politically turbulent period from the mid-1840s to the census in 1857, when the Austrian Empire recorded significant internal movements and numerous departures abroad, migration from the Slovenian ethnic territory was already quite pronounced. Structural changes that appeared after the abolition of feudalism corroded the traditional social and economic relationships. Along with modernization processes, such as the construction of the railway and the emergence of the first industrial plants, these changes spurred vivacious emigration and immigration movements.

Emigration (and years with increased mortality) decreased the number of inhabitants in Carniola and the Littoral during this politically and economically critical time (Blaznik et al., 1970, 103–106). According to the first modern census in 1857, almost 25,000 people with a right of domicile in Slovenian Lands were living in other parts of the monarchy (Valenčič, 1990a, 49). In the 1860s and 1870s, emigration slowed down slightly but accelerated again in the 1880s, when it entered its most intensive phase, which lasted until World War I.

As explained above, the dynamics and demographic impact of the emigration phenomenon are most pronounced if we exclude the urbanized areas along the western and northern edge of the Slovenian ethnic territory (where the urban population was increasing with the immigration from inland and, in Trieste, also from abroad) and focus on the territory of today's Slovenia. Graph 1.1 and Table 1.1 reveal how, since the 1870s, despite the increasing natural growth, the demography of the central Slovenian territory was shaped by emigration. For example, in the 1860s and 1870s, emigration subtracted 33 and 41 percent of the natural increase, respectively, the percentage in the 1880s was 47 percent, and in the 1890s, 68 percent. In the first decade of the 20th century, despite the exceptional contribution of the natural growth, emigration reduced its effect by 61 percent.

Table 1.1: Population growth and migration in the territory of today's Slovenia, 1857–1910 (in thousands). Source: Šifrer, 1962.

	1857–1869	1869–1880	1880–1890	1890–1900	1900–1910
Natural increase	92	81	98	108	135
Population increase	62	48	52	34	53
Migration (net)	–30	–33	–46	–74	–82
Final population	1,134	1,182	1,234	1,268	1,321
Population growth index (1869=100)	*100.0*	*104.2*	*108.8*	*111.8*	*116.5*
Net migration x 100 / Natural increase	*32.6*	*40.7*	*46.9*	*68.5*	*60.7*

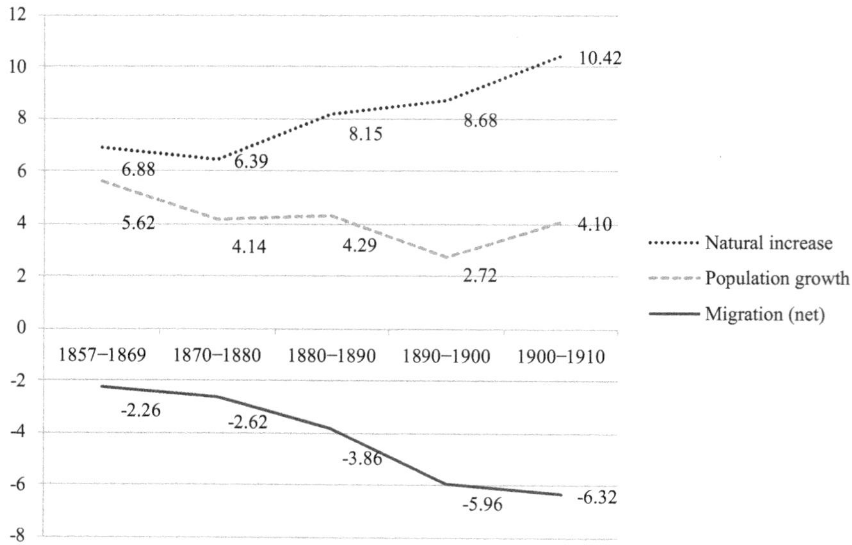

Graph 1.1: Rates of natural increase, population growth, and net migration per 1,000 inhabitants in the territory of today's Slovenia, 1857–1910. Source: Šifrer, 1962.

Regional Differences and Specifics

Different Slovenian regions—and even parts of regions—experienced migration and its effects differently. These variances resulted from the territorial population distribution, the tendencies of natural demographic growth, and the social and economic circumstances that shaped the population's different migration patterns and rearrangements. Carniola, the largest Slovenian Land, where the rate of natural increase was higher than in Styria or Carinthia, also experienced the most prominent migration. Between the 1869 and 1910 censuses, which registered 463,273 and 525,995 inhabitants, respectively, the Carniolan net emigration amounted to just over 100,000 people, or 21.6 percent, of the initial population.[3] The net emigration of Styria (17.6 percent) and Goriška (16.6 percent) was lower; in absolute numbers, that meant 71,000 and 33,000 people, respectively. The emigration rate in Prekmurje was slightly lower still (14.6 percent) while in Istria it was the lowest (9.2 percent). The population deficit in the last two regions amounted to just over 9,000 people. Some data must be considered with caution, particularly when we talk about the regions that include large or quickly developing cities: in the part of Carinthia that Slovenians populated, the recorded net emigration was 1 percent. However, this information is misleading, as it is severely skewed by the growth of the town of Klagenfurt/Celovec, which received populations from the region and beyond. For the same reason, this analysis excludes the city of Trieste. Although Trieste was part of the Slovenian Lands and a historic settlement territory, its size and substantial immigration overshadowed the migration phenomena in the Littoral and the migration ratios in the wider Slovenian space (see Table 1.2).

Carniola and Styria experienced consistent emigration from the 1870s onward (see Table 1.3). Carniola's emigration showed a linear increase throughout the decades, and—with the exception of Ljubljana-environs and Radovljica/Radmannsdorf—all the districts contributed high absolute and relative shares. However, as far as the intensity is concerned, the true epicenter of emigration (also concerning the entire Slovenian ethnic territory) was Lower Carniola (Dolenjska and Bela Krajina). Here, mass emigration started earlier than elsewhere and became more intense. In the most affected districts, Kočevje/Gottschee and Črnomelj, the net emigration rate was 42 and 41 percent of the

3 Studies display different numbers, depending on the territory considered and the method of calculating coefficients. Here, we amalgamated the data from Valenčič, Blaznik, and Šifrer.

Table 1.2: Migration, natural increase, and population increase in the Slovenian ethnic territory by Crown Land, 1869–1910. Source: Valenčič, n.d., 161–172.

	Migration (net)						**Natural increase**	**Population increase**	**Mig(net)/ natural increase**
	%					N	%		
Lands	1869–1880	1880–1890	1890–1900	1900–1910	1870–1910	1870–1910	1870–1910	1870–1910	1870–1910
Lower Styria	−4.1	−4.6	−7.6	−7.4	−17.6	−71,376	35.0	17.4	−50.3
South Carinthia	−0.6	−1.9	−5.7	−0.5	−1.0	−1,922	25.6	24.7	−3.8
Carniola	−2.8	−4.3	−6.5	−6.7	−21.6	−100,046	35.1	13.5	−61.5
Trieste	15.0	5.6	10.7	20.5	65.0	80,035	21.4	86.4	303.4
Goriška	−4.6	−6.0	−3.7	−1.6	−16.6	−33,825	44.3	27.8	−37.4
Istria	0.5	−2.4	−3.0	−2.7	−9.2	−9,172	53.9	44.7	−17.1
Prekmurje	6.2	−1.5	−4.9	−9.5	−14.6	−9,426	54.5	40.0	−26.7

Table 1.3: Migration, natural increase, and population increase in the Slovenian ethnic territory by district, town/city, and land, 1869–1910. Source: Valenčič, n.d., 161–172.

	Migration (net)						**Natural increase**	**Population increase**	**Mig(net)/ natural increase**
	%					N	%		
Districts or towns/ cities **Lands**	1869–1880	1880–1890	1890–1900	1900–1910	1870–1910	1870–1910	1870–1910	1870–1910	1870–1910
Maribor-city	36.1	15.4	22.7	13.5	118.4	15,193	–0.2	118.2	
Celje-city	37.0	26.9	24.2	20.0	139.0	5,873	–75.2	63.8	
Ptuj-city	80.3	–7.8	40.6	31.0	189.2	4,468	–93.4	95.9	
Celje-environs	–2.2	–4.7	–7.1	–7.4	–23.3	–27,559	41.5	18.2	–56.2
Ljutomer	–5.7	–2.4	–6.1	–6.0	–21.1	–5,275	31.0	9.9	–68.1
Maribor-environs	–5.0	–3.9	–7.4	–5.2	–22.4	–18,740	35.5	13.0	–63.2
Ptuj-environs	–4.5	–4.2	–8.7	–8.7	–27.3	–20,337	37.4	10.1	–73.0
Brežice	–5.9	–7.7	–10.2	–7.2	–31.9	–14,677	41.4	9.5	–77.1
Slovenj Gradec	–3.7	–4.1	–5.5	–11.0	–26.1	–10,322	35.7	9.6	–73.1

(continued)

Table 1.3: Continued

	Migration (net)						Natural increase	Population increase	Mig(net)/ natural increase
Lower Styria	**−1.9**	**−3.4**	**−5.4**	**−5.5**	**−17.6**	**−71,376**	**35.0**	**17.4**	**−50.3**
Lower Styria without cities	*−4.1*	*−4.6*	*−7.6*	*−7.4*	*−25.1*	*−96,910*	*38.1*	*13.1*	*−65.7*
Klagenfurt/ Celovec-city	30.3	4.7	20.2	23.3	99.3	15,181	−10.2	89.1	
Hermagor/Šmohor	−5.2	−8.5	−9.0	−6.1	−29.4	−5,213	36.7	7.4	−80.0
Klagenfurt/ Celovec-environs	−0.2	1.3	−4.4	−1.0	−4.8	−2,826	26.4	21.6	−18.1
Villach/Beljak	1.6	−0.7	−2.3	5.8	5.2	2,797	36.3	41.4	14.2
Völkermarkt/ Velikovec	−1.7	−4.7	−10.1	−5.8	−22.6	−11,861	20.5	−2.1	−110.3
South Carinthia	**1.8**	**−1.3**	**−3.4**	**2.1**	**−1.0**	**−1,922**	**25.6**	**24.7**	**−3.8**
South Carinthia without Klagenfurt	*−0.6*	*−1.9*	*−5.7*	*−0.5*	*−9.3*	*−17,103*	*28.6*	*19.3*	*−32.5*

Ljubljana-city	25.6	25.4	21.5	18.0	113.3	25,603	−28.6	84.7	
Postojna	−7.0	−8.2	−8.1	−8.4	−31.9	−13,148	36.7	4.8	−86.9
Kočevje	−2.9	−9.0	−12.5	−13.8	−42.2	−16,078	51.6	9.5	−81.7
Krško	−3.9	−4.9	−7.6	−9.2	−26.9	−13,439	37.8	10.9	−71.0
Kranj	−6.6	−5.0	−5.4	−8.6	−25.2	−13,583	24.4	−0.8	−103.4
Ljubljana-environs	−2.9	−2.8	−7.0	−1.1	−15.2	−7,662	47.9	32.7	−31.7
Litija	−2.6	−4.6	−12.1	−8.9	−31.0	−10,117	43.1	12.1	−72.0
Logatec	−1.6	−4.4	−10.0	−10.8	−30.2	−10,633	46.9	16.7	−64.5
Radovljica	−8.6	−4.6	2.3	1.3	−9.5	−2,543	36.5	27.0	−26.0
Novo Mesto	−2.2	−7.5	−8.4	−12.5	−32.9	−14,671	39.6	6.6	−83.2
Kamnik	−3.5	−4.9	−9.8	−11.0	−30.4	−11,613	33.9	3.6	−89.5
Črnomelj	−5.9	−12.0	−12.3	−12.7	−41.0	−12,162	24.5	−16.5	−167.3
Carniola	**−2.8**	**−4.3**	**−6.5**	**−6.7**	**−21.6**	**−100,046**	**35.1**	**13.5**	**−61.5**
Carniola without Ljubljana	*−4.3*	*−6.0*	*−8.3*	*−8.6*	*−28.5*	*−125,649*	*38.4*	*9.9*	*−74.2*

(*continued*)

	Migration (net)						Natural increase	Popu-lation increase	Mig(net)/ natural increase
Trieste	**15.0**	**5.6**	**10.7**	**20.5**	**65.0**	**80,035**	**21.4**	**86.4**	**303.4**
Gorizia-city	27.2	8.5	23.5	25.6	107.7	17,934	−21.6	86.1	
Gorizia-environs	−1.7	−6.6	−5.2	−6.8	−23.0	−12,926	54.8	31.7	−42.1
Gradisca	−11.3	−10.3	−7.5	−1.8	−31.2	−20,793	62.1	30.9	−50.3
Sežana	−9.0	−5.7	−8.1	−9.4	−33.3	−9,029	45.5	12.2	−73.1
Tolmin	−7.8	−5.9	−6.4	−4.3	−24.0	−9,011	25.7	1.7	−93.3
Goriška	**−4.6**	**−6.0**	**−3.7**	**−1.6**	**−16.6**	**−33,825**	**44.3**	**27.8**	**−37.4**
Goriška without Gorizia	*−7.4*	*−7.6*	*−6.7*	*−4.9*	*−27.6*	*−51.759*	*50,2*	*22,6*	*−55,0*
Capodistria/Koper	2,9	−3,0	−3,5	−5,1	−11,3	−6.995	55,1	43,8	−20,4
Volosko	−3.6	−1.5	−2.1	1.4	−5.8	−2,177	52.0	46.2	−11.2
Istria	**0.5**	**−2.4**	**−3.0**	**−2.7**	**−9.2**	**−9,172**	**53.9**	**44.7**	**−17.1**
Prekmurje	**6.2**	**−1.5**	**−4.9**	**−9.5**	**−14.6**	**−9,426**	**54.5**	**40.0**	**−26.7**

initial population, respectively. In the Črnomelj district, emigration amounted to 167 percent of the natural growth, causing a population drop from 29,600 to 24,700 inhabitants, or by 16 percent, between 1870 and 1910. Kočevje, although demographically less impoverished by the phenomenon, was the district with the highest net emigration balance in the same period.

Ranked by emigration intensity, the districts of Postojna, Litija, Logatec, Novo Mesto, and Kamnik follow them with different time dynamics. The districts of Krško and Kranj registered a bit lower but still show very significant net emigration balances, while Ljubljana-environs and Radovljica districts recorded the lowest deficits. In Radovljica, the net migration rate in the 1890s turned positive due to the development of the industrial town Jesenice.

In Styria, the relative deficits ranged from 21.1 to 31.9 percent of the initial population, with the Brežice district in the top position and the Celje-environs district suffering the highest absolute loss (27,000 people), followed by Ptuj-environs (20,000) and Maribor-environs (19,000).

In the southern—Slovenian—part of Carinthia, emigration reached its crucial period in the 1890s. The highest ratios and absolute numbers were in the districts of Hermagor/Šmohor (29.4 percent) and Völkermarkt/Velikovec (22.6 percent), where net emigration numbers reached about 5,000 and 12,000, respectively. Prekmurje, the eastern part of the Slovenian ethnic territory in the Hungarian half of the monarchy, experienced significant immigration in the 1870s. However, a decade later, emigration started and increased over the next three decades.

At the other end of the Slovenian territory, Trieste's development strongly influenced the population movements in Goriška and Istria, though differently in each region. In the 18th century, these two regions registered the highest rates of natural growth (Blaznik et al., 1970, 100, 114), which remained at high levels also in the 19th century. In Goriška, a significant share of this growth was absorbed by Trieste ever since the city was given the status of a free port. The net emigration from Goriška gradually decreased from the 1890s onward. The decrease was a consequence of the growth of the town of Gorizia and industrial centers, the development of transportation infrastructure, the rise of market-led agriculture, and the immigration from Friuli (Kingdom of Italy) and the Austrian Crown Lands. Emigration was the most drastic in the Sežana district—a direct neighbor of Trieste—where the net emigration for the period 1869–1910 amounted to 33 percent of the initial population. In contrast, Istria's net emigration was more modest. However, Istria's Koper/Capodistria district felt it deeply, particularly at the beginning of the 20th century.

Map 1.1: The Slovenian ethnic territory: Crown Lands and political districts subdivision 1866–1918.

Net migration changes show the effects of emigration, but they reflect neither its more detailed dimensions nor, at times, the actual traits of movements. There were intensive migrations within the Slovenian ethnic territory itself and many districts witnessed mass emigration and immigration. The main migration destinations were towns. The urban population in Slovenia at that time grew significantly, not only in larger towns but also in settlements with more than 1,000 inhabitants, where commercial activities and artisan or industrial production increased. These settlements provided administrative or transportation functions and adopted socio-economic traits that could classify them as urban in the Slovenian context, even though their social structure and appearance intertwined urban and typically rural elements (Melik, 1956). In Ljubljana, Gorizia, and Klagenfurt, the population increased by 85 percent between 1870 and 1910, in Celje and Idrija by more than 60 percent, in Maribor by almost 120 percent, and in the coal-mining town of Trbovlje by 180 percent. In Trieste—the only large city—the population increased by 95 percent, but the increase in absolute numbers surpassed the increase of all other towns in the Slovenian ethnic territory combined. Furthermore, the urban population growth was largely due to migrations.

Migrations between rural districts were also dynamic. Rural families, agricultural laborers, and domestic workers moved to the predominantly agricultural districts in Styria, Carinthia, and Carniola from other parts of Slovenia and also Croatia (Valenčič, n.d., 47, 53). The Slovenian parts of Carinthia and Lower Styria lost much of their population to the industrially more developed Austrian Crown Lands. However, they received many migrants from other Slovenian and Austrian regions. In 1900, 7,500 persons born in predominantly German-speaking regions lived in Lower Styria (Mačkovšek, 1911, 250). Between 1890 and 1910, the Radovljica district had a positive net migration rate thanks to the expansion of the ironwork center Jesenice, but many of the district's natives moved abroad (Drnovšek, 1989, 205–206). In Istria, the district of Volosko had the highest absence of the native population as people moved in large numbers to Pula, Rijeka, and other towns in present-day Croatia. However, the same district was also the end destination for migrants from elsewhere because Opatija was developing into an important Austrian tourist center (Kavrečič, 2015). In the Kamnik district, from the 1890s onward, the population dropped because of emigration, except in the Domžale municipality, where the population increased drastically due to immigration from surrounding areas and abroad linked to the straw-hat-making industry (Valenčič, 1958, 64–65). Kočevje also saw a certain level of immigration, although its level of emigration and population decrease was among the most prominent in general.

Net migration rates in rural districts everywhere show that emigration surpassed immigration. Nevertheless, the amplitudes and intensity of exchange differed from district to district and within districts; relatively modest negative net migration rates could, in fact, be hiding high immigration and emigration. Towns, in particular, experienced high mobility and population turnover; they were subject to constant growth due to the population arriving from other regional districts. At the same time, specific segments of their population were departing. These exchanges were reflected in the *Heimatrecht* (right of domicile) statistics. In 1900, only 10 percent of the inhabitants of Maribor had the *Heimatrecht* in the town. For Celje and Ptuj, it was 20 percent. In Ljubljana and Klagenfurt, 23 and 24 percent of the inhabitants had the right of domicile, respectively. The comparison is even starker in the context of the monarchy: the 1910 population census reveals that in the decade prior, three out of ten towns in the monarchy that experienced the highest in-migration and out-migration rates, respectively, were in the Slovenian territory: Celje (50.5, 74.0), Ptuj (34.4, 71.6), and Klagenfurt (41.8, 71.1) (Steidl, 2021, 51). The fluidity in the population of these cities can also be seen from the data about the natural demographic growth, which records exceptionally negative rates.

Emigration as a selective phenomenon also left deep traces in the population's structure. From the 1870s to 1910, the population in the age cohorts between 15 and 64 dropped from 62 percent to 58.5 percent. Among the nations in Imperial Austria, Slovenian speakers in Slovenia had, in 1900, the lowest number of inhabitants aged 20 to 40 years old. The 21 to 30 age group deviated by 25 percent from the country average; the 31 to 40 age group, by 12 percent. The Serbo-Croatian speakers—second on this list—deviated from the state average of 12.5 percent. Slovenians also had the highest percentage of the population above 50 years old (Mačkovšek, 1911, 247). The emigration outflow left a mark on the gender structure as well. The entire Slovenian ethnic territory between 1857 and 1910 averaged 5 percent more females than males. In Carniola, women of childbearing age outnumbered the men in their cohort by 10 percent (Fischer, 2005a, 19–20).

The Socioeconomic Context and the Factors of Imbalance

Migration patterns diversified in the second half of the 19th century up to World War I. Along with internal migrations, moves that turned away from the Slovenian ethnic territory became more frequent and, from decade to decade, transformed into negative migration balances. Mass migrations appeared within the context of social and economic restructuring processes caused by modernization and the implementation of capitalist relationships. In the Slovenian ethnic territory, this shift from the pre-modern to a modern socio-economic model based on rationalizing agrarian production methods for market purposes, expansion of industrialization, and other non-agrarian economic sectors was very slow and precarious. Before the 1880s, more than 80 percent of the Slovenian population was rural, which increased pressure on traditional agrarian structures following demographic growth. The agrarian population remained largely trapped in the pre-modern production logic. Due to the low or non-existent accumulation of capital and inability to invest in modernization, it could not keep up with the new times. The transformation toward a market-oriented economy and increasing revenue from farms in itself anticipated the reduction of the rural population, whose numbers surpassed the economic sustainability of the agricultural sector, and its reorientation to other economic sources and industries. However, industrial development also lagged when compared to the available human potential and the needs and aspirations of the population. Agriculture's slow modernization, the high concentration of the population in the agrarian countryside, and difficulties in raising the standard of living for the rural population were—as economic historians emphasize—fundamental social and economic problems in Slovenia, closely linked to migration, all the way to World War I, and partly to World War II (cf. Lazarević, 2015, 62–73).

The circumstances, factors, and mechanisms from which migrations emerged were complex. While implementing capitalist logic in the agrarian economy, the rural population faced many problems and challenges. The abrogation of the feudal order and agrarian relationships and the abolishment of serfdom in 1848 provided the rural population with access to land ownership, but for the small and mid-sized farms—and they were the most numerous—the property reform came with paying off compensations for 20 years. The farming economy was individualized, the ability to use common land—traditionally an important economic source—decreased, and the tax pressure and the money necessary to modernize farms' operations increased. All this led to an increased dependence on the market economy.

One serious problem of the agrarian economy was the small property size, which hindered and, in many cases, prevented the rationalization of rural production. The partitioned structure of the agrarian property in the first half of the 19th century slowed down the pace of development of the agrarian sector, preventing it from reaching its full potential. The 1868 law allowing free farm division further increased its fragmentation. At the beginning of the 20th century, when the share of the rural population decreased to 67 percent of the entire population, 55 percent of all farms were smaller than 5 hectares, and 36 percent were smaller than 2 hectares. One-third of the farms were mid-sized, with 5 to 20 hectares of land. Only 1.5 percent of farms were bigger than 50 hectares (Fischer, 2005b, 73). Moreover, because common land was divided according to the size of the farms, richer landowners benefited the most. They also expanded their property, profiting from the failure of small and mid-sized farms, and strengthened their economic position with non-agrarian sources of income.

Even larger farms met obstacles in converting to modern farming and intensive market production because of the size of arable land and farms and inauspicious natural circumstances. On average, the revenue on grain products in Slovenia was only half of those in the lands with progressive agrarian production. At the same time, farms smaller than 10 hectares could not modernize. These farms, run by using kin as the workforce, could compete with neither local modernized and capitalist-run farming nor cheaper imported produce. The agrarian economy worsened in the 1870s with the global agrarian crisis, accompanied in the 1880s by phylloxera and other plant diseases (Granda, 1999; Vodopivec, 1999). These affected the most profitable crops, for example, vine. Agriculture only recovered in the 1890s. After that, the circumstances in the rural sector gradually improved through modernization and increased production.

Significant changes, most acutely felt by the rural population, occurred with the railway and the transportation revolution. Railway construction offered many opportunities for employment and other forms of income. However, the effect of economic progress it provided was only temporary (Studen, 2005a). The railway transportation and the influx of domestic and foreign industrial products caused rural populations to lose many non-agrarian sources of income, which until then, particularly in the strata owning no or very little land, either complemented the income or constituted its main part. The need for traditional transportation of goods and roadside inns diminished in areas with the new traffic network.

The role of—and income from—commercial activities previously available to the peasantry also decreased because Slovenian market centers transformed into mere transit stations, and retail activities professionalized. The new situation thus mostly pushed the farmers out (Blaznik et al., 1970, 633–634; Valenič, 1981; Gestrin, 1991, 245–298; Fischer, 2005b). The railway took over the transportation of timber, and, increasingly, the only thing left to the horse-drawn carriages was getting timber out of forests and transporting it to the sawmills. Although some still transported it long-distance using horse-drawn carts until World War I, the revenue from this activity dropped.

Because of the land tenure reform, peasants also lost revenue from wood, which they could previously gather based on easement rights and sell independently. Such trade was now exclusive to the forests' owners (Valenčič, 1989). Transportation and industrial production decreased or annihilated the income from rural handicrafts, the production of goods—both cottage industry and manufacturing forms—disappeared, and artisans who were unable to modernize and adjust to the market circumstances were outcompeted and collapsed. Compared to previous centuries, this period brought the "ruralization" of the countryside, meaning that the population gave up on many non-agrarian activities and income sources, and the countryside gained a stronger agrarian character (Blaznik et al., 1970, 634).

From the technical and intensity aspect, the agrarian economy advanced after the land tenure reform, but the rural income decreased. In 1873, the economic crisis and then the agrarian—which lasted into the 1890s when it peaked in the first third of the decade—worsened the situation. The crisis and the increasing monetary expenditure for paying taxes, other monetary dues, and utilities caused small and mid-sized farms to incur even more debt; it only took a couple of decades for debts to increase significantly. Between 1858 and 1881, the farm debt in Carniola increased by 84 percent, in Carinthia by 80 percent, in Styria by 74 percent, and in the Littoral by 45 percent. In many districts, only one-third of smallholders and cottagers were debt-free in 1884 (Vošnjak, 1884; Valenčič, 1989, 33). In the 1890s, the situation improved, but the indebtedness increased until the turn of the century (Lazarević, 1994, 16–25). Because the underdeveloped loan system only improved in the 1890s with the emergence of rural credit unions and other cooperatives and so many farms experienced economic insecurity, access to loans was difficult and burdened by often exorbitant interest rates (Lazarević, 1993). As a result, owners were forced to sell either part of their farms or their entire estates, but even more often, their property was foreclosed and auctioned off. The exact amount of the debt taken on by Slovenian farms is unknown, but sources state that between 1868 and 1892, 10,190 farms were foreclosed and sold at auctions, which is 13 percent of all the farms in Carniola; between 1880 and 1882, 1,272 farms were forcibly sold (Lazarević, 1994, 16–19; Gestrin and Melik, 1966, 167; Vodopivec, 1999, 28). During the crisis in the 1870s and 1880s, the countryside saw the proliferation of small farms that barely survived and had to supplement their income with other agrarian or non-agrarian income. Smallholders also had difficulties retaining their fragile level of existence, susceptible to both small-scale inconveniences and large-scale damage. The increasingly smaller and weaker rural estates fueled the growing social division in rural society. A share of the population was pushed to the social margin, exacerbating an already hopeless situation. In 1902, one-third of rural households in Slovenia were small farms; one-fifth were able to live off farming only in exceptional circumstances, one-third were proper, solid rural homesteads, and the rest were large estates. At the same time, the share of the land owned by non-agrarian landowners and petty bourgeoisie increased (Blaznik et al., 1970, 174–179, 636–637).

One of the factors in the described process and in undermining the position of small landowners was the system of inheritance, which stipulated that all descendants of a landowner had the right to an equal share. Paying out the shares was a heavy burden for those who took over the homesteads and an additional cause of debt (Blaznik et al., 1970, 635). Younger brothers and sisters did not have many life opportunities at home on the farm. They had to leave it or accept the subordinate role of being a member of the family workforce whose position did not significantly differ from that of a servant. They could only emancipate themselves by leaving the family economy and searching for independent sources of income. A small rural household squeezed between rationalizing production and balancing production and consumption could not save its descendants from social marginalization, let alone provide for its family's reproduction and auspicious future. The growth of the rural population, the structure of estates, lack of investments in market-oriented farming, low productivity and the low value of agricultural products in the market, the prevalence of the subsistence agrarian economy, competition of foreign produce, low monetarization, and the lack of loan systems—all of these factors multiplied the ranks of semi-proletarians, as well as the full proletarians, who depended solely on wage work. Likewise, in urban environments, many small artisans succumbed to dependent work and proletarianization as they could not compete against factory production or production in the larger workshops equipped with machinery and managed according to the tenets of capitalism (Fischer, 2005c). In short, the population expanded, but the position of the members of its largest social classes—who had no possibility of supporting themselves at home, let alone the possibility for social emancipation in accordance with the standard of living that came along with modernization—weakened. The small rural estate itself, just like in the past, required additional resources for its existence and adjustment to the circumstances; to acquire those resources, its members had to enter the labor market, particularly the international.

On the other hand, the Slovenian ethnic territory in the 19th century and the early 20th century up to World War I saw great progress in the steel and wood industries, mining, and other industrial fields that employed more and more workers. The number of factories increased, and the non-agrarian economy was restructured based on the requirements of capitalism. Between the 1880s and World War I, industrialization created some 40,000 jobs (Lazarević, 2015, 67). In this period, the professional structure of the Slovenian Lands visibly changed. The population making a living in industry and trade increased by 84 percent between 1869 and 1910. The number and share of those active in other non-agrarian fields, such as transportation, commerce, public service, and independent professions, also grew, as did the number of those living on annuities. The increase in all the mentioned fields was proportionally faster than in other nations within Imperial Austria. Between the 1880s and World War I, the non-agrarian population grew from 17 to almost 32 percent. The greatest increase and shift to industrial and other non-agrarian activities happened in the early 20th century. Despite that, before World War I, approximately 66 percent of the active population got most of its revenue from agrarian activities. Among Imperial Austria's German-speaking population, this share stood at 30 percent (Fischer, 1988, 27; Hočevar, 1979, 50).

These changes and the progressive economic modernization were unevenly distributed throughout the Slovenian territory. Some areas showed a greater ability and susceptibility to it; their circumstances were more favorable for developing a more intensive economy. Others were excluded from these processes and remained trapped in the economic and social structures of the pre-modern era. In the Littoral and Carniola, the number of employees in trades and industry grew faster than in Carinthia and Styria. Between the censuses in 1869 and 1880, the active population in the Littoral increased by 34 percent, in Carinthia by 14 percent, in Styria by 8 percent, and in Carniola by merely 2 percent. In the decades that followed, these dynamics changed. Trieste was the primary catalyst for investments, entrepreneurship, and workforce on the western edge of the Slovenian ethnic territory. In 1869, of those working in industry and trade throughout the territory, 28.2 percent were concentrated in Trieste. Because these two economic sectors also advanced elsewhere, this share decreased in the following decades, but in 1910 still amounted to 22.4 percent of all those employed in them (Fischer, 1988, 28–30).

However, compared to the countries of intensive industrialization abroad and Austria itself, the economic progress was too slow to sustain and retain the rapidly growing population. The segment of independent artisans increased significantly but employed very few dependent employees. Greater possibilities for employment only existed in manufacturing and other industrial sectors, so the number of workers grew by 60 percent. However, the Slovenian ethnic territory's areas and sectors of the most intensive economic development could only absorb a part of the population excluded by the underdeveloped agrarian structure of the countryside (Lazarević, 2015, 2017). Additionally, the poorly educated and insufficiently skilled available workers forced many sectors, at least in certain phases of development, to recruit the appropriate or missing specialized workforce from elsewhere.[4] Also part of the game were the production and life cycles of the industrial plants and the oscillating circumstances in the labor markets, precarious employment practices, the problems of wages, union disputes, and the very nature of work and related types of workforce mobilization. These circumstances and the problematic penetration of economic modernization caused the process of easing the discrepancy between the population and the available

4 A very illustrative case related to this is the construction of the Southern Railway connecting Vienna and Trieste. The project did not employ much of the local workforce, but it did attract 10,000 Italian workers, mostly rural folk for whom such working migration was an established practice and who responded in an organized manner to the possibilities of employment in land digging works and other public works. The construction of the railway also increased the work in the steel mills. After the construction was completed, with the decrease in production, the possibility of employment also decreased (Granda, 1999, 33).

economic resources and life prospects to increasingly focus outside the Slovenian ethnic territory, both toward the more developed areas of the Austro-Hungarian Monarchy as well as abroad. It was encouraged not only by the economic sphere and material needs but also by the factors of social modernization, which promoted new standards and ways of life and opened new spaces to fulfill personal potentials and aspirations.

The Directions of Migration Routes: Internal and International Migrations

Before World War I, mass migrations from the Slovenian ethnic territory went in different directions. The general (and even some parts of the academic) perception has often linked this phenomenon with foreign countries, mainly the Americas. Indeed, the United States of America was the most frequent end destination of Slovenian transnational migration at the time, so leaving to go "across the pond" became synonymous with emigration. The United States always triggered the greatest expectations and greatly influenced the collective imagination (cf. Drnovšek, 1998b). It entered the popular imagery and speech as a metaphor for wealth, the modern way of life, and the possibility to rise above the dominant social standards in the "old place" (Hoerder, 1991, 80).

However, overseas migrations were only one part of a larger process of geographic mobility that marked this historical period. Like all other lands of the empire, the leading and most immediate destination of migration from the Slovenian Lands was the Austro-Hungarian Monarchy itself. Although Austria-Hungary was one of the top European countries in terms of emigration in the two decades leading up to the Great War, the majority of spatial mobility nevertheless occurred within its borders: the 1910 census shows that only 2.2 percent of the population of Imperial Austria migrated abroad, while 8.6 percent crossed provincial borders, and 25 percent crossed those of political districts (cf. Steidl et al., 2017, 52, 2021, 23–24, 30; Rosental, 2006). Studies of internal migrations conducted using data about the place of birth show that in 1900 between 60,000 and 65,000 Slovenian migrants were living in other Austrian Crown Lands; 46,000 of them lived in Middle and Upper Styria (cf. Pfaundler, 1907; Hecke, 1913). Thus far, historical and other migration studies have paid little attention to these migrations, which had just as an important influence on Slovenia's demographic and social development as the migrations abroad did.

Regional Similarities and Differences

The geography of migration routes in Slovenia was very complex: intensive migrations occurred within the Slovenian ethnic territory, but migrants also sought end destinations elsewhere in the monarchy and abroad. When we look at the scope, directions, and distances, we can notice four distinct areas: Lower Styria and southern Carinthia, the Littoral (Goriška and Istria), Carniola, and Prekmurje. Although the shares that the first area—Lower Styria and southern Carinthia—contributed to the migration abroad were significant, the main migrations from both regions were directed toward other parts of the monarchy, primarily to the German-speaking lands in Imperial Austria. People from Lower Styria mostly moved north to the German-speaking parts of Styria. According to the 1900 census, 70 percent of those with the right of domicile (used in this analysis as an indicator of moves and provenances inside the monarchy) in Lower Styria living in other parts of Imperial Austria resided in these two regions. The most common points of origin for the moves from Lower Styria were the districts of Maribor-environs, Celje-environs, Ptuj-environs, and the district of Slovenj Gradec. Another destination for migrants from Lower Styria was Lower Austria/Niederösterreich, with an established route also leading toward the Kingdom of Hungary. Migration from the southern—Slovenian-speaking—part of Carinthia was primarily directed toward the more economically developed parts of Imperial Austria. However, this migration was more equally distributed between the German-speaking parts of Carinthia and Styria and, to a smaller extent, Lower Austria and the western regions of Imperial Austria.

In the second area, consisting of Goriška and Istria (the districts of Koper and Volosko), movements had different geographical traits. Migrations toward inland Imperial Austria were very modest because, in the immediate vicinity, Trieste was a favored and traditional destination. In 1900, 29,000 inhabitants in Trieste had the right of domicile in Goriška, and more than 12,000 had it in Istria. These numbers represented three-quarters and two-thirds, respectively, of all those from the two regions residing in other parts of Imperial Austria. Because of the strong flux of its population into Imperial Austria's main seaport, Gorizia and Gradisca/Goriško-Gradiška (the Princely County of Gorizia and Gradisca, of which Goriška was the part with a predominantly Slovenian-speaking population) was among those with the highest level of interregional migration in Imperial Austria. The migrants from this province were the most numerous among all Austrian immigrants in Trieste. The migration into Trieste from the Slovenian part of Istria only accelerated in the 20th century. Before then, Slovenian Istria witnessed in-migration from other parts of Istria, linked to the development of agriculture for the expanding Trieste food demand. The 1910 census revealed that of all the districts that provided migrants to Imperial Austria's main port of Trieste, Koper contributed the most: 8,000 of its citizens lived there (Cattaruzza, 1979, 20–25). The migration from Goriška and Slovenian Istria also gravitated strongly toward Pula and Rijeka. At the same time, the northern mountainous political district of Tolmin contributed an important share to the moves toward German-speaking crown lands. Many migrants from Goriška, in particular, also migrated abroad.

In Carniola, the third area, the migration geography was particularly diverse. Many from the region moved to the economic centers in German-speaking Middle and Upper Styria, Lower Austria, and beyond, some also southwards to the Kingdom of Hungary, mainly Croatia. Larger movements from Carniola were also directed toward other Slovenian Lands. Trieste, a traditional destination for migrants from Carniola since the 18th century, absorbed a significant share of migrants. In 1900, 15,000 persons with the right of domicile in Carniola lived in Trieste. Many Carniolan migrants were also attracted to (Slovenian-speaking) parts of Styria and Carinthia: the 1900 census registered 11,000 and 6,000 persons, respectively, living in those areas with the right of domicile in Carniola, with a further 12,000 and 2,000, respectively, residing in the German-speaking parts of those two regions. Compared to the rest of the Slovenian ethnic territory, however, Carniola was at the forefront of migration abroad.

The fourth region, Prekmurje, was similar in that respect, as migrations there also took different directions. One part was oriented toward Lower Styria. The primarily seasonal and occasionally permanent moves were directed toward the Kingdom of Hungary's agrarian regions, and a significant route crossed the Atlantic (cf. Kuzmič, 2001).

In terms of distances, migrations from Styria and Carinthia, and also from Goriška and Istria, tended to be shorter and, in their geographical configuration, more intra- or inter-regional. Styria and Carinthia alone contributed the majority share to the entire emigration from the Slovenian ethnic territory to the rest of Imperial Austria, which reflected the proximity of these two regions to the economically most developed regions in the Austrian half of the monarchy. Similar was true for the western Slovenian territory, where the direction of migration reflected the gravitational role of Trieste and, increasingly, Pula and Rijeka. In contrast, the geography of migrations from Carniola highlighted the decentralization of the land and the influence of distances and connections of its parts to different centers of intensive economic development. These factors fostered migration within the monarchy that was smaller in scope but more diverse in its directions. At the same time, there was a more prominent trend to emigrate abroad.

At the beginning of the 20th century, these regional particularities were even more visible. Migration from the Slovenian part of Styria and Carinthia toward the neighboring German-speaking lands increased. In contrast, the migrations from Carniola and Gorizia and Gradisca toward German-speaking lands stopped and turned more toward the destinations on the Adriatic Coast. With increased maritime traffic and industrialization since the 1890s, Trieste was experiencing its second economic revival. Between 1890 and 1910, the number of its inhabitants born in Carniola and Gorizia and Gradisca increased by a quarter, and the number of those born in Slovenian districts in Istria doubled. Overall, the Slovenian territory saw massive migration abroad during the two decades before World War I, with Carniola leading the way by far.

The Slovenian Regions in the Context of Imperial Austria

The Slovenian ethnic territory reflected the general direction of migration in Imperial Austria, where, in a complex intertwining of internal and international movements and local particularities, two main tendencies stood out. The first one was intense short- and mid-distance migration to—and within—the regions that were developing and urbanizing faster: Lower Austria, Upper and Middle Styria, parts of Bohemia and Moravia, Vorarlberg, Salzburg, Trieste, and some other centers. These regions, which generally overlapped with the German linguistic territory, had positive net-migration balances, and immigration contributed to demographic growth. However, when we move toward the eastern and southern periphery of the state, we see a second tendency, in which the rates of internal migration drop. The regions that experienced severe migration abroad were particularly peripheral ones with a Slavic population in Imperial Austria (Galicia, Bukovina, Dalmatia, and Carniola) and the Kingdom of Hungary (Croatia-Slavonia, and Slovakia). These regions lagged behind or were even excluded from larger industrialization processes and retained the overwhelmingly rural and agrarian structure. This configuration of migration tendencies was reflected in the data about the domicile statistics. On average, in Austria in 1869, more than 80 percent of the population resided in their municipalities of domicile. By 1890, this share in Lower Austria, Upper Austria, and Bohemia fell below 50 percent because of internal migrations. That year in Galicia and Bukovina, it still came up to 84 percent and only dropped below 80 percent in 1910. In Carniola, where the migration patterns were directed toward other countries rather than within the country, it was 87 percent (Fassmann, 1985, 71; Steidl et al., 2017, 50–54).

Comparing natural and migration net balances can help us understand the dimensions, or ratios, between migration abroad and migration to non-Slovenian areas within Austria. At least 5,400 people moved to the latter in the last decade of the 19th century, while 29,000 inhabitants migrated abroad (Österreichische Statistik, 1903, XXXII–XXXIII). In the first decade of the 20th century, when emigration from Carniola toward the northern regions of Austria decreased significantly while migration abroad increased significantly, Carniola's migration to foreign countries increased to at least 42,000 people (Österreichische Statistik, 1914, 28*–29*). Between 1890 and 1910, Carniola, contributing around 5 percent to the total emigration from Imperial Austria, was fourth among Austrian Lands in the absolute number of people who migrated abroad. Higher on the list were the incomparably more densely populated Galicia (which contributed more than 50 percent), Bohemia (20 percent), and Moravia (7 percent). However, when we look at the relative extent of emigration, Carniola comes out first—by far: in the 1890s, 5.7 percent of the Carniolan population migrated abroad, and in the following decade, 8.3 percent.[5] Due to this high share of emigrants abroad, Carniola was, within Imperial Austria, considered a land of emigration par excellence (Meinzinger, 1903, 146).

In the 1890s, at least 9,000 persons migrated abroad from Gorizia and Gradisca, which amounted to less than 1 percent of all Austrian emigration and 3.3 percent of the population of the province in 1900 (Österreichische Statistik, 1903, XXXIII, 1914, 29*). The drop in emigration from Gorizia and Gradisca abroad in the 20th century was connected to the development of some industrial centers within the province itself, primarily the shipbuilding center Monfalcone/Tržič/Falkenberg. These centers strengthened the moves within the country and attracted the population from other Austrian regions and from abroad, particularly from the neighboring Italian Friuli. The improved transportation infrastructure and circumstances for market-oriented agriculture and trades strengthened the local sources of income from the economy and decreased the centrifugal migration tendencies (cf. Kalc, 2013a).

Migration abroad from Lower Styria and Southern Carinthia cannot be quantified in the same way for Carniola and Gorizia and Gradisca. Generally, we can state that Styria as a whole was, thanks to the intense economic development

5 To compare: between 1900 and 1910, 6.75 percent of the Galician population emigrated, and 5.3 percent of the Dalmatian.

of its northern and central parts, a direct opposite to Carniola. It took the last place among the Austrian Crown Lands regarding the ratio between emigration abroad and the total population. In the 1890s, there were around 9,000 persons fewer in the entire land due to emigration, and in the next decade, 7,000 persons (Österreichische Statistik, 1903, XXXIII, 1914, 29*). We can assume that a substantial part of the emigration was from the Slovenian part of the region. However, because Styria, Carinthia, and Gorizia and Gradisca recorded more immigration from abroad than Carniola did, the de facto emigration rate was substantially higher than the net migration balance shows.

To Imperial Austria and the Kingdom of Hungary

The overview of the scope and the directions of migration from the Slovenian ethnic territory to other parts of the Austro-Hungarian Monarchy can only be sketched broadly and indirectly based on the population censuses for Imperial Austria and the Kingdom of Hungary. There are no specialized statistics on these movements. In the 1850s, migrations from the Slovenian ethnic territory turned almost equally toward the two parts of the monarchy. When the phenomenon gained momentum from the 1860s onward, the destinations in Imperial Austria prevailed and attracted about four-fifths of the migrants. The primary receiving regions were the German-speaking areas of Styria and Carinthia, as well as Lower Austria. Styria, where industrialization and modernization were the fastest, received half of all the migrants from the Slovenian territory (cf. Vodopivec, 1979). More than 75,000 persons with the right of domicile in the Slovenian Lands lived in Upper and Middle Styria in 1900; they represented a quarter of immigrants there and, a decade later, a half. Most were proletarianized farmers employed in mining and industrial centers and towns (Rozman, 1979, 18).

In 1900, there were 25,000 in Graz, 8,000 in its surroundings, while thousands more lived in the mining and industrial cities of Leoben (9,000), Leibnitz/Lipnica (6,000), and Judenburg (6,000). Immigrants from Lower Styria in 1900 represented 9 percent of the population of Graz, and those from Carinthia and Carniola were 7 percent (Hubbard, 1984, 23). Other important places of immigration were Bruck an der Mur, Deutschlandsberg/Lonč, and Voitsberg, with around 4,000 immigrants each, and Bad Radkersburg/Radgona, with 3,000 inhabitants that had the right of domicile in Slovenian Lands. That same year, 18,000 were registered in the German-speaking part of Carinthia, and a further 27,000 in Lower Austria, of which 19,000 were in Vienna. These territories offered possibilities for employment to ordinary and skilled workers, other employees in different industries and trades, and, in bigger cities, services. In

Graz in 1900, for example, immigrants from Lower Styria and Carinthia represented 13 percent of the working-class population; those from Carniola represented 9 percent (Hubbard, 1984, 24). In Carinthia and Styria, many immigrants from the Slovenian territory were employed in the countryside as an agrarian workforce (Valenčič, 1990a, 58).

An important region of immigration in the south was present-day Croatia, where the 1900 census showed 14,000 citizens had the right of domicile in Slovenian territories. The most attractive to migrants was Istria's Pula, with its industrial and shipbuilding plants and military facilities. Many migrants were attracted by textile and other industries in Imperial Austria's western provinces, particularly Salzburg and Vorarlberg. The number of people with the right of domicile in Slovenian territories living there at the end of the century increased to 12,000. Fewer migrants left for the eastern regions of Imperial Austria, and the majority of the 6,000 persons resided in different places around Bohemia and Moravia. Dalmatia was a prominent destination for Slovenian migrants until the 1890s when 3,000 inhabitants lived there with the right of domicile in Slovenia at the beginning of the decade.

Slovenian migration to the Kingdom of Hungary primarily went toward Slavonia and Croatia; many people living near the Croatian border migrated there as early as the first half of the 19th century. Their main destinations were Varaždin, Bjelovar, and especially the Zagreb district. These migrants belonged to all social classes and professions, from farmers and servants to public servants and intelligentsia. Zagreb was a true Slovenian center. In 1880, almost 4,000 Slovenian speakers lived there, and in 1910, over 5,000 (Drnovšek, 1999a, 34; cf. Trdina, 1951).

The migration to the Kingdom of Hungary was the most intense in the 1870s and, to a considerable extent, originated from the cities, that is, Ljubljana and Maribor as well as Celje and Ptuj. After 1880, migration mostly turned westwards, particularly to Rijeka, where the number of inhabitants from the Slovenian ethnic territory increased significantly by 1910. Other routes decreased in the 1890s, so Slovenian presence in the Kingdom of Hungary did not grow visibly. At the 1890 census, more than 30,000 inhabitants with the right of domicile in the Slovenian parts of the Austrian half of the monarchy were recorded in the Hungarian half. The same census listed around 20,000 Austrian citizens with Slovenian as their mother tongue (Szabo, 1995).

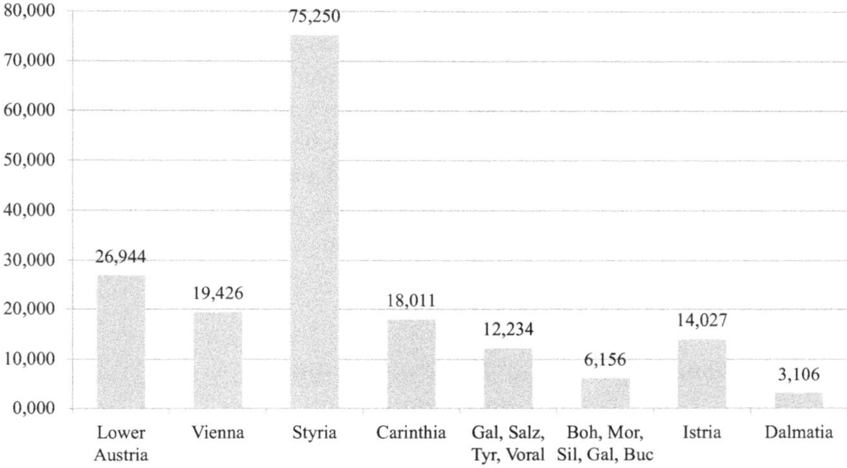

Graph 1.2: Population with the right of domicile (*Heimatrecht*) in the Slovenian ethnic territory (excluding Prekmurje and Benechia) residing in other parts of Imperial Austria in 1900. Sources: Valenčič, n.d., 221; Österreichische Statistik, 1902.

The Americas, Predominantly the United States

When spatial mobility began to grow noticeably in the mid-19th century, migrant paths from Slovenian regions also led to other countries. Some, mainly temporary, international moves were traditions held over from previous eras and specifically related to particular regions. In the first phase, migration beyond the borders of the monarchy was rare compared to internal migration. A visible change occurred in the 1880s when migration to the Americas became more commonplace: it gradually grew until World War I and became the most prominent migration phenomenon.

Slovenian transatlantic migration was part of a wider phenomenon that took over the entire Austro-Hungarian Monarchy and Eastern and South-Eastern Europe. Four-fifths of the transatlantic migrants from Austria-Hungary headed to the United States, the rest mainly to Argentina, Brazil, and Canada. Due to the structural characteristics of the migration, as well as for linguistic and cultural reasons, South American countries remained a relatively marginal goal destination for Austro-Hungarian emigrants. Between the mid-1870s and 1910, more than three and a half million people migrated from Austria-Hungary to the United States (Chmelar, 1974, 24). As late as the 1870s, only about 5,000 arrivals per year were registered (Hecke, 1916, 483). Before, intensive moves only appeared from Bohemia and some German-speaking parts of Austria. The 1880s were a turning point; transatlantic migration increased exponentially and peaked between the mid-1890s and the Great War. In the first decade of the 20th century alone, more than two million migrants left Austria-Hungary for the United States. If most migrants came from Bohemia and Vorarlberg before, this was the time when mass migrations took hold, primarily in the southern and eastern parts of the monarchy. From the 1870s on, more than 50 percent of the migrants came from the Austrian half, of those, 6 out of 10 were from its Slavic regions (Chmelar, 1974, 31; Klemenčič, 1987, 32–33; Willcox and Ferenczi, 1929, 384–393).

From the Pioneering Phase to "The Great Migration"

One characteristic of transatlantic migration from the Slovenian ethnic territory was its above-average orientation toward the United States. The Slovenian territory knew transatlantic migration even before the mid-century when individuals or smaller groups set off. These early migrants from Carniola, the Littoral, and Styria first came to attention at the 1857 Austrian population census, but their numbers were never precisely established (Valenčič, 1990b, 50). The pioneers included some typical characters that we can also find in other nations. They were some sort of "precursors" of the mass migration phenomenon. The peddlers who had been coming from Lower Carniola since the 1830s were one such early example of economic migrants. By virtue of their profession, peddlers were susceptible to discovering new ways to make money. By the 1860s, some had become established merchants or business owners in US towns and cities (Zavertnik, 1925, 300, 359–360; Drnovšek, 1998b, 93–111, 215–218). There was also no shortage of craftsmen, loggers, and other workers, who were driven by initiative and eagerness to try and search for opportunities for a better life in foreign environments. Individuals moved abroad after the revolutionary year 1848 for political reasons. One of these *Achtundvierzigers* (forty-eighters) was Anton Fister (Füster), a priest, professor, and one of the March Revolution leaders in Vienna who set off for the United States via England and stayed there for 24 years. He worked as a public speaker and tutor in German immigrant communities in Boston and New York (Klinec, 1978; Stanonik, 1979). Missionaries were particularly visible in this pioneering migration period, with Friderik Baraga, Andrej Bernard Smolnikar, Franc Pirc, Lovrenc Lavtižar, Jožef Frančišek Buh, and other preachers becoming quite popular. Baraga, who from 1830 onwards worked among the Odawa and Ojibwe nations around the Great Lakes and, in 1853, was consecrated as the bishop of Cincinnati, Ohio, left a mark on US history with his ethnological and linguistic work (Baraga, 1850; Marković, 2017). In newspaper reports, the missionaries created an image of the "New World" and its possibilities for the Slovenian public. They considered the continent's colonization a part of their religious mission but also had improving the material circumstances of Slovenian farming families in mind, so the first Slovenian settlements in the United States are tied to their efforts. Two such examples in Minnesota were the villages of St. Stephen in Brockway Township and St. Anthony in Krain Township, which the immigrants established in the mid-1860s (cf. Klemenčič, 1984, 27; Kodrič, 1987, 1993; Drnovšek, 1998b, 94–95).

In the next phase, migration to the United States changed character and structural traits. Migrants were attracted in masses by the labor market, which proliferated because of the expansion of the US economy (cf. Fassmann, 1994, 296). Opportunities for immigration and employment increased with industrialization and urbanization in the states in the eastern US and around the Great Lakes—Pennsylvania, Ohio, West Virginia, Illinois, Michigan, Wisconsin, and Minnesota. Migration routes also went to other places in the United States, all the way to Colorado and California. The mass migration from the Slovenian territory was a part of the so-called "new immigration" from the eastern and southeastern parts of the European continent. Since the 1890s, it represented a reserve workforce for big industries in the United States and its renovated production system based on unskilled labor (cf. Irwin, 2003). Like other migrants from the European rural peripheries, Slovenians were driven by the desire to only work in the United States for a couple of years and then return home with their earnings. Their numbers fluctuated following the general patterns of the "new immigration" and contributed to its peak from the beginning of the new century until World War I (Willcox and Ferenzci, 1929, 168).

The Extent and Phases of "The Great Migration" to the United States

There are no reliable statistics about the dynamics and extent of migration from the Slovenian ethnic territory to the United States. The Austrian population censuses acquired data about migrants from their relatives' statements, so the numbers are much too low. Direct emigration statistics, which the Austrian authorities compiled using data from the European ports of migration, refer to all Austrian emigrants and do not break them down by crown land or other identifiable categories. More precise migration statistics from the Slovenian territory are only available from the 1890s onward and for Carniola alone. Political district authorities compiled it based on issued passports (Drnovšek, 1999, 413). The statistic of immigration to the United States based on ethnicity, recorded since 1899, is only partially useful. It does include Slovenian immigrants from all lands, including those who previously migrated to other countries, but only if they declared themselves as Slovenian, and additionally records them together with Croatians (cf. Klemenčič, 1978, 1987, 40; Valenčič, 1990b, 55). We must also consider that the US statistics do not include the Gottscheer/Kočevarji and other German-speaking people who made up a considerable proportion of the migrants from the Slovenian regions.

Table 1.4: Transatlantic (almost exclusively US-bound) migration from Carniola. Slovenian and Croatian US immigration and return migration, 1892–1915. Sources: Drnovšek, 1999 and author's estimates; Willcox and Ferenzci, 1929.

Year	Migrants from Carniola	Immigrants to the United States (Slovenian and Croatian)	Returnees from the United States (Slovenian and Croatian)
1892	2,107		
1893	2,955		
1894	1,996		
1895	1,664		
1896	2,898		
1897	1,118		
1898	1,619		
1899	3,328	8,632	
1900	2,857	17,184	
1901	4,504	17,928	
1902	5,119	30,233	
1903	6,512	32,907	
1904	2,883	21,242	
1905	6,334	35,104	
1906	5,837	44,272	
1907	6,528	47,826	

Year	Migrants from Carniola	Immigrants to the United States (Slove-nian and Croatian)	Returnees from the United States (Slove-nian and Croatian)
1908	1,814	20,472	28,589
1909	4,190	20,181	8,981
1910	4,404	39,562	7,133
1911	2,208	18,982	13,735
1912	4,326	24,366	13,963
1913	8,700	42,499	10,209
1914	2,000	37,284	14,440
1915		1,942	2,381
Total	85,901	460,616	99,431

The migration dynamics from Carniola (see Table 1.4 above), which we can take as an indicator of the migration process in the entire Slovenian ethnic territory, overlap—bar some minor deviations—with transatlantic migration from Austria-Hungary. This migration is divided into two development phases: in the 1890s, the phenomenon increased slowly and accelerated at the turn of the century, with an average of 5,000 departures per year. It peaked in 1902 and 1903, between 1905 and 1907, and again just before the war, in 1913—a record 8,700 people departed that year. Passport-based statistics reveal that between 1892 and 1914, around 86,000 persons migrated from Carniola to the United States, a quarter in the 1890s and three quarters in the new century. The US statistics record a large contingent of Slovenian and Croatian immigrants in 1914, but this number refers to the fiscal year and thus includes a large number of the arrivals from the second half of 1913. In reality, in 1914, immigration from Austria slowed down significantly before it almost completely stopped in July when the war started (Chmelar, 1974, 50–54). Considering the emigrants who went through Trieste, we can assume that in 1914 around 2,000 persons left Carniola for the United States.

Based on reports from district authorities, the emigration statistics from Carniola do not reflect transatlantic migration in its entirety. Many traveled abroad illegally to avoid military service or other circumstances. Others first moved within Imperial Austria, to Germany, or elsewhere in Europe, and only then to the United States. The unsuitable criteria for the census and the negligence and difficulties in keeping records contributed to these incomplete statistics (cf. Mulaček, 1913, 258; Valenčič, 1990b, 56; Drnovšek, 1999, 417–418). The entire Imperial Austrian emigration statistics suffered from this trait; the Imperial Austrian central statistical authorities failed to establish a more efficient system to monitor expatriations directly (cf. Englisch, 1913; Steidl, 2021, 110). Ivan Mulaček (1913, 257) estimated that the true extent of Carniolan emigration was 25 to 35 percent higher than the one officially shown. Jurij Trunk (1912, 380) believed that among the Slovenian and Croatian migrants, at least a third, around 150,000 people, were Slovenian. Later historians believe that the share of Slovenians was even higher. They estimate that approximately half of the 460,000 migrants of Slovenian and Croatian ethnicity shown in the US immigration statistics between 1898 and 1915 were Slovenian (Klemenčič, 1987, 39).

The migration dynamics depended greatly on the economic oscillations and the circumstances in the US labor market. During crises and negative economic predictions, the influx of migrants dropped and return migration intensified. An exceptionally high number of returns was recorded during the US crisis in the second half of 1907 and in 1908. In 1908, migration to the United States shrunk significantly. The number of returnees between 1908 and 1913 amounted to 40 percent of all Austro-Hungarian emigrants (Fassmann, 1994, 299). According to US statistics, Slovenian and Croatian returnees reached 35 percent of the immigration between 1910 and 1914. If we add the time from 1908 when, due to the crisis, the number of returnees was three times the average number, returnees from the United States to Europe reached almost 48 percent of the number of emigrants to the United States (Willcox and Ferenczi, 1929, 497).

We can only roughly assess how many migrants found their permanent (or long-term) home in the United States. Since 1910, the US population censuses have classified inhabitants—Slovenians included—according to their mother tongue. This classification includes immigrants and those born in the United States who declared Slovenian linguistic or ethnic belonging. The 1910 US population census thus found 184,000 Slovenian residents, of whom 124,000 were born in Imperial Austria, and almost 60,000 were their US-born descendants (Klemenčič, 1984, 29). Many more arrived—as we have seen—by World War I, but many also returned to their place of origin. Some estimates state that between 1900 and 1910, about 70,000 migrants stayed in the United States. At the end of the previous century, 50,000 were already there (Valenčič, 1990b, 58). The 1920 US census recorded 209,000 inhabitants with Slovenian as their mother tongue. This growth of "Slovenian America" is attributed to the 106,000-person-strong

US-born generation, which in this phase overtook the numbers of those who immigrated from the country of origin. By 1940, the number of people using the Slovenian language dropped to 184,000, of whom 76,000 were born in Slovenia, while 103,000 and 6,000, respectively, were members of the US-born second and third generations (cf. Klemenčič, 1984, 29). Many authors believe that due to methodological and other difficulties in establishing the belonging of immigrants, the actual extent of the Slovenian presence was greater than the official US statistics show. Some estimated that between the two wars, the number of Slovenians of all generations in the United States reached 300,000 (Govorchin, 1961, 63–65).

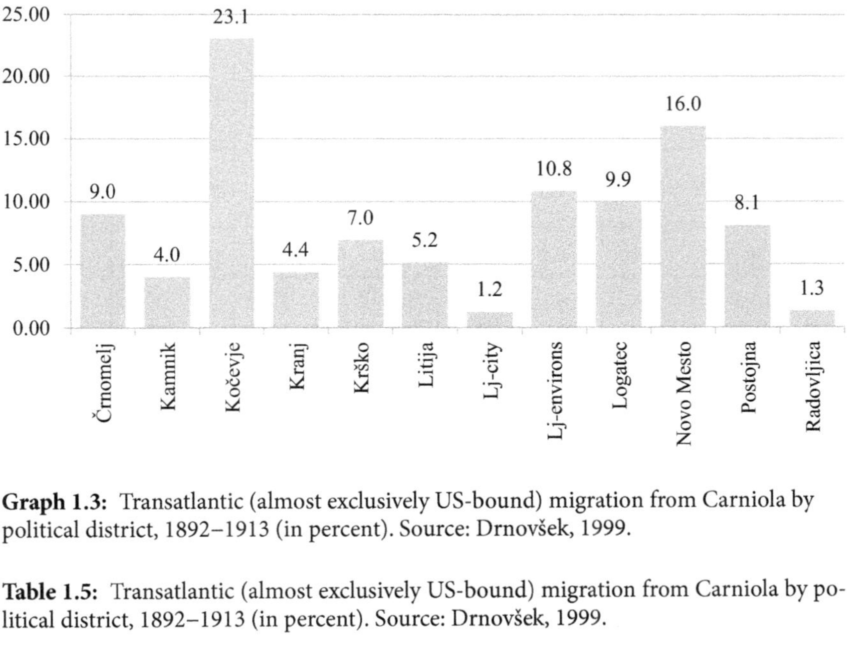

Graph 1.3: Transatlantic (almost exclusively US-bound) migration from Carniola by political district, 1892–1913 (in percent). Source: Drnovšek, 1999.

Table 1.5: Transatlantic (almost exclusively US-bound) migration from Carniola by political district, 1892–1913 (in percent). Source: Drnovšek, 1999.

Districts	**1892–1895**	**1896–1900**	**1901–1905**	**1906–1910**	**1911–1913**	**Total 1892–1913**	**% of 1900 population**
Črnomelj	11.6	13.8	9.7	7.4	4.2	9.0	15.8
Kamnik	0.9	1.2	4.0	4.8	7.0	4.0	5.3

Districts	1892–1895	1896–1900	1901–1905	1906–1910	1911–1913	Total 1892–1913	% of 1900 population
Kočevje	26.8	27.4	21.4	22.4	21.5	23.1	24.9
Kranj	1.9	2.8	4.2	5.1	6.4	4.4	4.2
Krško	10.9	4.3	7.2	6.5	8.2	7.0	6.1
Litija	5.2	4.6	5.3	4.9	6.0	5.2	6.8
Lj-city	0.6	0.4	1.7	1.5	0.9	1.2	2.1
Lj-environs	9.3	10.1	11.8	11.0	9.8	10.8	9.2
Logatec	12.3	9.8	10.6	10.3	6.6	9.9	12.5
Novo Mesto	19.8	15.7	15.4	15.7	16.9	16.0	15.2
Postojna	0.4	9.8	7.8	8.3	10.3	8.1	9.3
Radovljica	0.3	0.1	1.0	2.1	2.2	1.3	2.5
Total	100.0	100.0	100.0	100.0	100.0	100.0	9.7

Geographical Origins of Migration to the United States of America

The core of Slovenian migration to the United States was Lower Carniola, the region that provided the highest share of Slovenian migrants in general (see Graph 1.3 and Table 1.5). Almost half of all the Carniolan migrants hailed from here, particularly the parts known as Bela Krajina and Suha Krajina. According to the numbers, at the top—by far—was the Kočevje district, which contributed 23 percent of all Carniolan migrants. It was followed by Novo Mesto with 16 percent and Črnomelj with 9 percent of migrants. In Lower Carniola, the phenomenon spread earlier than elsewhere and was, throughout the period 1892–1913, the most intense. The number of migrants, as registered in the 1900 census, amounted to 25 percent of inhabitants of the Kočevje district, 15 percent of Novo Mesto, and 16 percent of Črnomelj. Inner Carniola/Notranjska, with its districts of Postojna and particularly Logatec, was the second most migration-intense region. Migration was already underway there in the 1890s when departures to

Brazil also contributed to the activity. In the Ljubljana-environs—although not the city itself—the phenomenon also became quite pronounced. In Upper Carniola/Gorenjska, it was the lowest because of industrial development.

Other Slovenian regions also contributed substantial shares to the migration to the United States, but numerical data are fragmentary. The transatlantic migration from Gradisca, the western (Friulian) part of Gorizia and Gradisca, was primarily directed toward South American countries. In contrast, the migrants from Goriška, the eastern (Slovenian) part, were mainly bound for the United States. The available statistics, limited to the Slovenian surroundings of Gorizia, show 1,300 transatlantic migrants between 1907 and 1913, which amounts to 42 percent of all migrants abroad (Srebernič, 1914). Some transatlantic migration was also present around Tolmin, in Istria, and in Trieste. Trieste was often a temporary migration destination, and after a short (or longer) stay, newcomers moved elsewhere, including to transatlantic countries.

Migration to the United States from Prekmurje in the eastern Slovenian ethnic territory was also quite intensive. Solid demographic growth in these predominantly agrarian and infrastructurally isolated regions within the Kingdom of Hungary strengthened migrations in the second half of the 19th century. Initially, these were primarily seasonal and directed toward other agrarian territories in Austria-Hungary. The earliest departures to the Americas were recorded in the 1870s, but mass transatlantic migration only started at the turn of the century. The migration was especially intense in the hilly areas of the region, where the 1910 census showed that about 20 percent of the population resided abroad. The 1910 US census found around 5,500 persons of Slovenian language and 1,500 of "Wendish" language living in the United States who were born in Hungary, so around 7,000 migrants from Prekmurje altogether (Černy, 1953, 132; Klemenčič, 1987, 52; Kuzmič, 2001, 30–31). Like migrants from other Slovenian territories, those from Prekmurje in the United States settled in industrial areas. The most famous Prekmurje immigrant settlement was, since 1893, the steel and coal town of Bethlehem, Pennsylvania. Another center of transatlantic migration was Benechia, an area on the extreme western margin of the Slovenian ethnic territory in the Kingdom of Italy. Likewise, Styria contributed to the migration to the United States.

Other Transatlantic Destinations

Migration from the Slovenian ethnic territory toward other great destinations of the European transatlantic migration was rare. Some migrants to North America were attracted to Canada—an emigration route that started in the 1880s and increased slightly at the turn of the century—but because of the country's immigration politics, which prioritized the agricultural workforce and favored those of English origin, Canada remained uninteresting to most. Immigrants from Imperial Austria were predominantly rural inhabitants from Galicia and German-speaking lands (Englisch, 1913, 105). Canadian statistics list 3,000 migrant Croatians and Slovenians from 1885 until World War I. The target destinations were the mining regions of British Columbia and Alberta, as well as Saskatchewan; there is evidence of several immigrants settling on farms (Genorio, 1989, 75–76; Kalc, 2018, 112). It was also typical for people to move between Canada and the United States in search of work opportunities. The US statistics for 1905–1918 list 6,500 Slovenian and Croatian immigrants from Canada and 3,000 who returned there (Willcox and Ferenczi, 1929, 492, 499).

Before the end of the century, migrants to South America[6] were primarily groups of small landowners who set off on a journey expecting to improve their material and social position, particularly in the 1870s when the general, specifically agrarian, crisis additionally endangered the existence of rural families. This emigration was enabled by the active roles of Argentina and Brazil, which based their economic policies on the ambitious colonization of uncharted territories and programs of immigration from Europe (Molek, 2019, 42). The countries used widespread propaganda to recruit migrants, as well as offering particular contracts for settlement and employment to the agrarian workforce or access to land property. In 1876, Argentina passed a law inspired by the principle "to govern is to populate" and implemented a systemic policy inviting the European immigrant population to set up colonies. In 1878, it entered into an agreement with Austria-Hungary and sent scouts to find rural families who would want to settle in Argentina. Several dozen families from Goriška and Postojna left for Argentina in 1878 and 1879 as part of these programs and settled in the Chaco, Formosa, and Entre Ríos provinces (Mislej, 1995a, 16; Kalc, 1995, 37–41; Brecelj, 1997, 217).

The propaganda for Argentina did not resonate in Slovenia, so no more than a few hundred people migrated there before the beginning of the 20th century (Brecelj, 1997, 218). In Entre Ríos, the Slovenian colony numbered about 350 people, but it is impossible to provide more accurate numbers of migrants (Bizai, 2006, 43). Slovenian rural emigration to Argentina and elsewhere in South

6 The first migrants to Argentina and other South American countries were, from the 17th century onward, Slovenian missionaries, before 1860, the Jesuits, and afterwards, mostly Capuchins and Salesians (Brecelj, 1997, 2016).

America was an extension of the emigration that had massively affected Gradisca and, particularly, Italian Friuli and Veneto. There, the situation of smallholders and farming tenants, as well as the agrarian proletariat on the territory of the *coloni*, deteriorated considerably because of the agrarian crisis and the dependency on the landowners. So, people responded massively to the invitations of migration agents and left, craving for their own plots of land (Kalc, 1995, 42–45). Emigration to South America from the hilly parts of Friuli also attracted numerous Slovenian families from Benechia (cf. Kalc and Kodrič, 1992). In the years before World War I, the groups of rural migrants were followed by spontaneous and individual migrations of people with different professional profiles that left for Argentina either as a part of their job or with the intent to join the Argentinean labor market. Among them were engineers, technicians, and other educated people. Many found employment in the public sector; some set up successful companies. It is estimated that around 1,000 persons of Slovenian nationality lived in Argentina before World War I (Molek, 2019, 59).

Slovenian migration to Brazil surpassed the one to Argentina. The first Slovenian migrants from Carniola and Styria migrated to Brazil along with other Austrians as early as 1825, at the invitation of the wife of Brazilian Emperor Pedro I, the Austrian princess Maria Leopoldina, who encouraged the rural population from Austrian territories to settle there. The Brazilian government provided different reliefs and allowed them to obtain land free of charge (Anić, 1988, 400). Migration increased from the mid-1870s onward, when Brazil, as a part of the reorganization of its agrarian system and development of agrarian potentials, and in the process of the abolition of slavery, began a planned import of rural settlers and agrarian workers from Europe. These campaigns echoed strongly in the Slovenian ethnic territory and were also met with interest, particularly in its western part (cf. Umek, 1968, 1979; Kalc, 1995, 39, 2001). While in the 1870s smaller groups migrated at their own cost (Brecelj, 1997, 216), in the 1880s the phenomenon was encouraged by the Brazilian government, which subsidized transatlantic journeys and startup funds for setting up farms. In 1888 and 1889, a large group of families and individuals from the Karst and the surroundings of Trieste moved to Brazil as a part of these programs (Kalc, 1995, 39). A new, far more extensive, emigration occurred in the next decade, particularly in 1896 and 1897, when around 3,000 persons left Lower Styria and Carniola; many of these migrants came from the Carniolan districts of Logatec and Postojna (Drnovšek, 1991, 75; Kalc, 1995, 39). Like Argentina, Brazil and other South American countries, particularly Uruguay and Chile, saw qualified workers, merchants, and intellectuals join agricultural migrants before World War I. A slightly larger colony of Slovenian migrants was created in São Paolo, where they connected with other Slavic immigrants from Imperial Austria. In 1908, they also established their first organization, the Jugoslovanski sokol (Yugoslav Sokol) (Štrucelj, 1909; Marušič, 1995, 63–65; Mislej, 2001, 244). Also worth mentioning is the emigration to Mexico in 1864 and 1865, when several hundred Slovenian men joined the political-imperialist expedition of Franz Josef I's younger brother, Maximilian, self-styled Maximilian I of Mexico (cf. Drnovšek, 1991; Ovsenar, 1993).

The German Empire

The German Empire was the second most typical destination of Slovenian migration abroad. After the unification, the country underwent a quick transition from "an agrarian state with a strong industry to an industrial state with a strong agriculture" (Bade, 1992, 311).[7] It also became the second-largest importer of workforce. To implement intensive and specialized methods to mass-produce industrial crops, its agrarian sector required high numbers of seasonal workers. Austria-Hungary, particularly its Slavic territories, became at this time—together with Russia, northern Italy, and the Eastern German (Polish) provinces—the main source of the seasonal workforce for the German agrarian economy (Caro, 1909, 39; Englisch, 1913, 130–140; Steidl, 2021, 73–79).

However, Slovenian migrants to the German Empire primarily migrated as a part of the growing workforce for heavy industry. German industry and mining attracted even more workers than agriculture, and workers employed in these sectors made up two-thirds of the immigrant workforce from the Austro-Hungarian Monarchy (Steidl, 2021, 79). The contingent of Slovenian migrant workers was made up mostly of miners going to the mining and industrial regions of Westphalia and Ruhr (Wörsdörfer, 2017, 49). A small-scale migration to the German states had existed even before: workers migrated to work in the textile industry and the separate phenomenon of the Gottscheern (Gottscheer German peddlers) went back centuries. However, mass labor migration to Germany started in the 1890s when the first large group of miners went there (Drnovšek, 1991, 68). While people migrated to Germany from the entire Slovenian ethnic territory, most of them originated from mining regions, particularly in the most important mining area in Slovenia in the Zasavje region, that is, the Trbovlje-Zagorje-Hrastnik "black" (coal) mining district on the border between

7 All translations of citations from non-English sources are the translator's unless otherwise noted.

Carniola and Styria. This migration was born, much like so many of that time, from the crisis in the agrarian economy and other traditional economic sources in the rural areas, which encouraged the shift to non-agrarian activities and new forms of work and social structure in the Slovenian countryside. More specifically, it was caused by the expansion of mining and heavy industry and the international employment situation in the mining sector. Zasavje, a rather important center of intensive economic development, was a region of demographic growth caused by immigration. Trbovlje was among the fastest growing towns in Slovenia: its number of inhabitants increased sixfold in 40 years and reached 9,000 before World War I (Studen, 2005b, 105–106). In the second half of the 1890s, the demand for Slovenian coal from its primary buyer, the Southern Railway, and clients in the Austro-Hungarian and foreign industries started dropping rapidly due to the competition from Czech coal, which was of better quality and cheaper. In the 1880s, there was union unrest and dismissals in the Slovenian mining centers, while the demand for workforce increased in Germany. The first departures for Rhineland and Westphalia coincided with that, and new departures soon followed (Wörsdörfer, 2017, 54–65).

Migration to Germany is also difficult to quantify due to insufficient data. The data from Imperial Austrian censuses are not helpful because the citizens in the German Empire are not divided according to the region of origin. The German population census of 1910 found 9,200 persons who reported Slovenian as their mother tongue (6,500 men and 2,700 women), mainly from Carniola and Styria. This number is too low, while the figure of 70,000 Yugoslavs, mostly Slovenians, reported in Yugoslav sources and studies after the war, is overestimated. The extent of the Slovenian presence gradually changed because the mining population, like the one in the United States, was relatively mobile. A large Slovenian migrant community evolved around different mining and industrial towns in Westphalia. Thanks to the writings of priest and politician Janez Evangelist Krek, the community was given the name "Westphalian Slovenia" (cf. Drnovšek, 2007a). In 1910, around 800 Slovenians, all miners, were in the regions of Alsace and Lorraine.

Egypt

From the 1850s, a prominent destination of the Slovenian migration was Egypt. The migration to Egypt mainly occurred in the Austrian Littoral, and Slovenian migration centered around Gorizia. When Egypt chose to modernize and integrate into the global trade system, it also opened up to European economic interests. It became attractive to European workers, particularly in the Mediterranean countries, as an alternative to transatlantic and continental moves. From the beginning of its process of implementing structural reforms, and during a period of rapid economic growth, Egypt recruited European experts to this end. A significant shift occurred in the 1860s with the construction of the Suez Canal, which solidified Egypt's importance in maritime traffic. At the same time, the economy grew because of the expansion of cotton production, which benefitted from the decline of US production during the US Civil War (cf. Panjek, 2011, 59; Biancani, 2015). Economic growth and favorable living circumstances caused many Europeans to migrate there, and this migration was structurally different from the population directed toward the "New World" and other mass destinations. It was a migration of the colonial type, which included all classes, from the entrepreneurial and petty bourgeoisie to the lower classes, members of an entire spectrum of professional profiles, including public administration. Between 1848 and 1907, the number of Europeans in Egypt increased from 10,000 to 150,000 (Panjek, 2011, 58–59). The community of Austro-Hungarian citizens was usually around 8,000 members strong from the 1880s onward (around 5 percent of all foreigners in 1907). They mostly lived in Alexandria and Cairo, with a smaller number in Port Said (Lever, 2012, 25).

A more significant Slovenian migration to Egypt dates back to the construction of the Suez Canal, when many workers from the Austrian Littoral and Carniola found employment there (Valenčič, 1990b, 53; Drnovšek, 1991, 72–73). Most of these temporary migrants returned home after their contracts ended. The migrations from the Austrian Littoral to the Middle East changed in the 1870s. People from all walks of life—from merchants and tradesmen to workers and clerks, individuals and entire families—moved there from Istria and Trieste. For example, many were directly attracted by the British authorities to work in the police or public administration (Grossutti, 2013, 27). An exceptional case was Antonio Lasciac, whose career trajectory was linked to the development and opportunities in Egypt. An architect of Slovenian origin from Goriška, he migrated to Egypt where he made his name as the planner of important private and public buildings and was later awarded an aristocratic title as a person of merit (cf. Mamić, 2008). We could also list other persons who created solid economic and social positions for themselves. Many worked in trade and transportation and were closely linked to Trieste. Before World War I, about 400 persons from Trieste lived in Egypt. The community also helped promote Slovenian wood industry products in the Egyptian market, from semi-finished wood from the forests of Inner Carniola and elsewhere, sold as intermediate goods, to fashionable furniture designed and produced in carpentry workshops in Solkan (cf. Kalc, 2015a; Žagar, 2015).

Migration from Goriška and, on a smaller scale, from other Slovenian regions comprised mostly (young) women. They were employed as domestic workers—as chambermaids, cooks, and maids, as well as wet nurses, nannies, and governesses. As the wealthy middle class in Egypt expanded, and regular travel lines and greater trade and economic exchange between the territory of the Littoral and Egypt appeared, so did the employment possibilities for women. This employment coincided with the general decline of opportunities for domestic work in European cities (Hahn, 2012, 115) and with the specific crisis of the Triestine economy in the 1870s when the numbers of domestic workers—a line of work in which women from Goriška had always been prevalent—also dropped significantly (La popolazione, 1878, LXVI). Individual domestic workers from Goriška appeared in Egypt as early as the mid-19th century, but the phenomenon became a mass one in the 1860s and 1870s. Although women from other Slovenian territories left for Egypt, it was particular to Goriška (Markitan, 1913, 24).

Estimates about the Slovenian presence in Egypt vary. For the mid-1870s, the claims are 2,000 people, and at the end of the 19th century, between 5,000 and 7,000 people (Slovani v Egiptu, 1897; Drnovšek, 1991, 73). The official Austrian statistic from 1900 shows 4,500 citizens from Imperial Austria living in Egypt, of which half were from the Austrian Littoral. Almost two-thirds of them were women. Ten years later, the total number had not changed; the Austrian colony in Alexandria decreased, but the one in Cairo increased (Österreichische Statistik, 1905, XXX, 1919, 38*). Evidence from the Austro-Hungarian consular service on Austrian citizens reveals that between the mid-19th century and the beginning of World War I, around 9,000 Slovenians, primarily women, came to Egypt to work for a shorter or longer period (Humar, 2019). The center of *aleksandrinstvo*, the female work migration to Egypt, was the immediate Slovenian surroundings of Gorica, which overlapped with the administrative district Gorica-environs. Between 1907 and 1913, 1,400 women migrated abroad from there. Of them, 72 percent went to Africa. The actual center of this phenomenon was the village of Prvačina in the Lower Vipava Valley, from where 120 women migrated, which according to the 1910 census, represented 11 percent of the population. Other villages heavily affected were Renče (111 or 4.3 percent), Bilje (96 or 7.1 percent), Kojsko (87 or 1.9 percent), Dornberk (72 or 3.3 percent), and Vrtojba (70 or 2.7 percent) (Srebernič, 1914; Kalc, 2015a, 62).

Aleksandrinstvo, although small when compared to the transatlantic migrations, was, in its time, a very particular phenomenon in Goriška because of its geographic orientation and gendered structure (Markitan, 1913, 24). It was part of the broader female labor migration to Egypt, which was also known in Italy—in certain parts of Friuli (cf. Grossutti and Boz, 2009) and more massively in

Calabria and Marche (Garroni and Vezzosi, 2009; Gobbi, 2011)—and in Greece. Unlike the Slovenian migration, the Italian and Greek migration of domestic workers was part of much more extensive migration processes from these two countries to Egypt, where the migrant colonies themselves generated the need for household staff. According to the registrations at the Austro-Hungarian consulate in Alexandria, aleksandrinstvo in Slovenia was the strongest in the last 20 years of the 19th century. Its decrease is probably linked to the economic development and expansion of urbanization in Goriška and Trieste and the improved situation in agriculture.

From Switzerland to Bosnia and Herzegovina and Beyond

The paths of Slovenian migration also led to other countries. However, their low numbers cannot be compared to those going to the Americas, Germany, and other territories of Austria-Hungary. The 1910 census records a small group of Slovenian migrants in Switzerland (Österreichische Statistik, 1919, 52). Switzerland primarily employed construction workers, for example, for major infrastructure works, such as the construction of the railway network, electrification, and residential construction. Construction workers traditionally left for Switzerland from the Karst villages in Goriška and Benechia, and their migration was mainly seasonal (Sedmak, 1994, 84–87). Lacemakers from Idrija and Žiri, employed by lacemaking enterprises in the canton of St. Gallen/Kanton Sankt Gallen, represented a particularly peculiar case of migration to Switzerland (cf. Strle, 2015a, 113).

Before World War I, small groups of Slovenian migrants lived in France and the Netherlands. In 1911, 16 percent of the Austro-Hungarian citizens in France were from south Slavic nations, most from the Slovenian ethnic territory (Čačić and Kumpers, 1989, 289). These routes were also a part of the international employment network for mine workers. Wages and working and living conditions in France were less attractive than those in Germany. Mining in Limburg in the Netherlands was in its early stages before World War I and only expanded later (Drnovšek, 1991, 72; Mulders, 2010, 19).

A more important destination was Bosnia and Herzegovina, where, after the military occupation (1879), the Imperial Austrian authorities took up planned rural colonization. These programs met with a relatively large response in Galicia, from where around 10,000 Poles and Rusyns migrated to the Banja Luka area, more precisely to the Prnjavor district (Caro, 1909, 144). A small Slovenian colony was created there in the 1890s. Migration of those in non-agrarian professions also increased, among them were foresters who had been working in Bosnia even before the occupation. In the first years after the occupation, around 1,000 persons from Slovenia's Kobarid and surrounding villages found temporary employment in Bosnia (Kalc, 1995, 36; Marušič, 1995, 61–62). Among the migrants were miners, foundry workers, blacksmiths, mining experts, and persons of other qualifications, from artisans and small merchants to public servants, military officers, educators, and other persons for whom mobility was part of their work or career trajectory. In 1880, there were almost 1,000 migrants from Slovenian regions in Bosnia and Herzegovina; the census of 1895 found more than 5,700 of them, mostly from Carniola and Styria. In 1910, 3,108 inhabitants declared Slovenian as their language of communication (*Umgangssprache*) (Strauss, 1897; Österreichische Statistik, 1919, 23). Most lived in Sarajevo, Banja Luka, Zenica, and Tuzla and their surroundings (cf. Valenčič, 1990a, 67; Drnovšek, 1991, 61–62, 2010a, 66–67; Koblar, 2008, 9–16; Sarić and Štimac, 2010, 397–406). The migration routes also led to Serbia, where, like in Bosnia and Herzegovina, Slovenian migrants were mostly attracted by non-agrarian professions. There was a lively migration to the Serbian mining centers, for example, the Timok Valley, where documents show the presence of Slovenian workers as early as 1840. At the beginning of the 20th century, around 1,500 lived there (Piljušić, 2013, 5–9). Another example of a more numerous Slovenian presence was a miners' colony in the Vrdnik coal mine in Vojvodina (Lukić, 2017, 39).

Migration destinations in the Ottoman Empire (other than Bosnia and Egypt) attracted mainly migrants from the Austrian Littoral. The majority of those whom the 1857 census registered as absent in Goriška and Istria lived in the Ottoman Empire, with the Istrian workers mostly working in construction or in Austrian companies, particularly glass production, which Triestine Jewish entrepreneurs set up near Istanbul (Grossutti, 2013, 32–33). Some Slovenian migrants went to Russia, Bulgaria, and Romania, mostly foresters but also other professionals, including intellectuals (Drnovšek, 1991, 63–64, 2009, 478, 479). The small-scale Slovenian presence spread far: a group of Slovenians from Benechia settled near Vladikavkaz in southern Ossetia. These peddlers, who sold their merchandise throughout Eastern Europe in the 19th century, obtained a vast plot of land from the imperial authorities and set up a colony named after their country of origin, Italijanski hutor/Italian Khutor (hamlet). This community was known in the Slovenian and the wider Russian surroundings as a specific case of colonization that merged agrarian and trade-industrial activities. It disintegrated during the Russian Revolution and the civil war, and its members either returned home or scattered throughout the Soviet Union (Ruttar, 2002). We can also find smaller groups of Slovenian migrants in many other countries and places worldwide, including sub-Saharan Africa and faraway Australia.

Migration Patterns: Continuities and Ruptures

Migrations in the Slovenian territory from the second half of the 19th century up to World War I included multiple forms of mobilities. Many migration patterns continued from the past and represented tested behaviors rooted in social and economic strategies. Others were new, with different geographic directions, distances, timings, purposes, and impacts. Some traditional movements and routes disappeared or were transformed and acquired new geographic, structural, and dynamic connotations. The migration variants reflected the functional choices made according to the structural specifics and circumstances in the places of origin and the work and life inclusion opportunities in the receiving contexts. The continuities and the changes in these movements' triggers and tendencies reflect macrostructural shifts and general societal events that contributed to the expansion of the migration phenomenon and, within it, the objective socio-organizational aspects and subjective choices. The different weights and roles of migrations in particular social contexts must also be considered. The intensity of spatial mobility and the geographical scope of migrations in this period are striking, permeating all social, economic, and even cultural dynamics. Nineteenth-century mobility nevertheless continued to coexist with stationary life. Although mass migrations meant the Slovenian ethnic territory witnessed altering demographic shifts and "losses," most of the population continued to live within its borders. While migrations and different forms of spatial movements within these territorial limits were lively, the statistics about the domicile and other data show that large segments of the population did not significantly become more mobile or less stationary in their narrow traditional living environment than before. In not-so-rare cases, structurally similar communities responded differently to the challenges despite the same socio-economic situation and circumstances: some employed a certain kind of "migration strategy," and others made choices where migrations were not at the forefront. Although some were mobile and some forms of migration represented established social behaviors, in other cases, migration was episodic, both socially and individually.

Old and New Circular Migrations

Like elsewhere, Slovenian migrations in the mid-19th century increased primarily through circular mobility, which was essentially cyclical and consisted of short-term departures and returns (cf. Valenčič, 1990a; Moch, 1992, 17). These migrations, characteristic of the pre-industrial society, represented a traditional way of overcoming imbalances between the numbers and needs of the rural population and the insufficient economic sources in their places of residence. Despite some common general features, they differed in specific traits, for example, professional specializations that communities developed to carve out market niches or occupational opportunities. Such traditional circular migrations were often linked to commerce. While modern development dealt a fatal blow to traditional transportation activities, for example, transportation of goods by pack animals, the peddling of rural craft products and bought merchandise persevered and even increased in many parts of the Slovenian territory. Based on capillary mobility "from door to door" (clearly seen in the dialect names for it, *hausiranje* or *pohiševanje*), it occurred most often in the winter months, when the agricultural cycle required less workforce, but also in other seasons of the year, or over longer periods.

The centers of such itinerant commerce were the Lower Carniola districts of Kočevje and Ribnica and the district of Črnomelj in Bela Krajina, where these activities had existed since the 15th century. In 1859 the Trade Ordinance promoted the expansion of peddling, and the development of rail transportation sped up traveling and allowed entry into new markets (Mrkun, 1940; Šega, 1998, 153–154). Gottscheers—peddlers from Kočevje whose trade in German was named *Gottscheerei*, after Gottschee, the German name for their town and region—obtained the right to practice their trade from Emperor Friedrich III of Habsburg in 1492. They focused on the northern part of the Austrian Crown Lands and the German Empire, where they sold leather goods, textiles, small household utensils, and other trinkets (Drnovšek, 2007c, 86). Trades similar to *Gottscheerei* were *nemškarstvo* ("Germaning")—the departure for Bavaria and Switzerland with a wicker basket on their shoulders—in Bela Krajina and *suhorobarstvo*—the selling of *suha roba*, or "dry ware," essentially woodenware—around Ribnica. The *suha roba* peddling, at least from the 17th century onward, went hand in hand with rural wood handicraft production. At the same time, the Gottscheer and peddlers from other regions switched to commercial trade only. At the end of the 19th century, among the 4,400 inhabitants of the Ribnica Valley, 272 were peddlers who traded goods from Ribnica through Imperial Austria, Croatia, and Vojvodina. As late as the interwar years, 360 were peddlers (cf. Lokar, 1911, 4–5; Šega, 1998, 153–154; Drnovšek, 1999a, 33). Another example of itinerant commerce was the selling of lacemaking products. This was a female specialization in the Idrija area, where lacemaking developed into a substantial female home activity and source of income for the miners' families (Terpin, 2008).

Picture 1.1: *Suha roba* (dry ware) peddlers from Ribnica on their route. Author: unknown. Courtesy of: Slovenski etnografski muzej (Slovene Ethnographic Museum), Ljubljana.

The peddlers from Bela Krajina traveled as far as Sweden, Norway, Denmark, England, and France (Drnovšek, 1999a, 33). The routes of *žokarji*, or sock (*žoka*) and glove sellers, went to Trieste, Graz, and Vienna, along which they sold their goods made by rural families in the mountains above Idrija (Zbornik Šentviške planote, 1991, 146–148). Benechia, at the western edge of the Slovenian ethnic territory, was known since the 18th century for its *guzirovci* (or *hauzirovci*, from the German word *hausieren*, hawking goods from door to door) who made long journeys selling holy images and decorative paper from the famous papermaker and printer Remondini in Bassano del Grappa near Venice all the way to Bohemia, Romania, and Russia (Zanini, 2004). The peddler tradition was even older in the Resia Valley/Rezija because the agricultural soil was scarce and poor. Back in the late Middle Ages, the merchants from Resia worked mostly in Bohemia and Moravia; in the 17th century, they set up a proper colony specialized in ironware commerce in Prague (Perusini, 1948). In the 19th century, Resian artisans joined peddlers as seasonal migrants; among them, knife grinders soon gained fame—they were, in fact, so famous that, in the areas where they offered their services, *Rezijan* became a synonym for a whetter (Kalc and Kodrič, 1992, 202–204). *Hausieren* was also common in the mountain districts of Goriška, where also "opticians" lived, that is, sellers of spectacles, binoculars, and thermometers (Sedmak, 1994, 84–87; Dular, 1996, 166–167). Another example is selling roasted chestnuts, a trade that migrants from the Velike Lašče area in Lower Carniola took over in the mid-19th century and developed into an identifiably Slovenian activity in Vienna and beyond (cf. Šega, 1997).

These activities and specializations could either be for mere subsistence or, with entrepreneurial techniques and investment, a means to rise to a higher economic level. One such successful example is the case of the Resian merchants Lettich and Di Lenardo, who evolved from peddlers to importers of citrus fruits and set up wholesale and retail companies and shops in the biggest towns and cities of Imperial Austria (Kalc and Kodrič, 1992, 204–205). Some peddlers from Kočevje also became wealthy merchants in Vienna, Budapest, and elsewhere (Šmitek, 1981, 42). Earnings from peddling were invested in solidifying the "home"—village—part of this combined economy or in expanding business abroad: roasted-chestnut selling thus developed as a supplementary activity along with peddling. As a result, migration occasionally lost its seasonal character. Chestnut-sellers who did not return home in spring found other sources of income that were congruent with the alternating conjuncture of work seasons, be it selling fruits, refreshing drinks, or woodenware (cf. Mrkun, 1940; Šega, 1998; Kante, 2000).

Itinerant artisans and merchants in the mentioned areas were characteristically mobile, and the money they earned with their activities was, at times, the principal source of family income. Population census data draw attention to the importance of this phenomenon. The 1910 population census registered 1,400 Kočevje district-born citizens in Vienna alone (Österreichische Statistik, 1903, 49)—and as the census was conducted in December, most of them were temporary migrants working in the capital over the winter season. The numbers of more prosperous merchants varied, too, as most of them returned home several times a year. Usually, their families stayed back home, while the merchants often shared apartments in the cities to optimize the overhead costs of their commercial activities. They employed people from their home areas as assistants and apprentices who worked for them until they paid off their apprenticeship or started their own enterprises. Based on these networks and the hierarchical structure, the system reproduced itself. The newcomers and those without start-up capital had access to the profession, and consequently their families had a source of "outside" income (cf. Šmitek, 1981, 42–43). These activities and trades were passed down through generations.

With the expansion of the economic sectors in which the work followed seasonal cycles for natural, climate, or organizational reasons, seasonal labor migration increased. Lumberjacks from Lower and Inner Carniola and around Tolmin in Goriška went to work from late fall to spring, first to Croatia and Slavonia (and thus nicknamed *Hrvatarji* at home to mark their connection to Croatia) and then onward to Hungary, Galicia, Transylvania, Romania, Russia, and Bosnia (Bilec, 1891; Mrkun, 1940; Drnovšek, 1999b, 421). These migrants also chopped and worked wood in forests and sawmills at home. Mostly rural laborers and smallholders who had no draft animals departed from the Bistrica area (Postojna district) for seasonal work. The wealthier stayed home and made money transporting wood from forests and delivering it to customers. However, the crops from small farms were only sufficient for half a year, so the farms were often not self-sustainable for families. Therefore, many smallholders worked in forestry more than agriculture. From the 1860s onward, more than a thousand forestry workers from the villages in the Logatec, Postojna, and Tolmin districts migrated to work seasonally (Goršič, 1914, 320).

The origin of this labor migration is linked to the loss of income from feudal times when the village community bought the right from the gentry to cut and sell wood (Valenčič, 1980, 85, 1989, 14). The drop in timber prices and the shrinkage of forests owned by rural landowners due to overcutting (Ribnikar, 1889) further decreased earnings. The migration of forest workers was organized collectively and in agreement with timber wholesalers. It was intended to provide a basic annual income and take advantage of the higher income opportunities offered by the international labor market. There was no shortage of work for foresters at home: the area around Bistrica had to fill their ranks with foresters from Italy and Tolmin over summers due to the shortage of local workforce (Valenčič, 1989, 4–5; Postojnsko okrajno glavarstvo, 1889, 37, 79, 93). The forest workers departed in organized companies and accepted work based on a contract with the forest's owner or user. Such companies were led by middlemen called *kasatorji*, and this piecework usually lasted from October to April when the workers returned. The forest migration and activity were based on negotiated agreements and legal stipulations that defined the rights and duties of all the parties in the system (cf. Goršič, 1914). The departure and return of workers were accompanied by "ritual" celebrations in which entire villages participated. Other places in the same districts were known for different migrant occupational specializations. For example, in the Postojna district, mason work in Trieste and Rijeka became a standard component of farming families' income and a form of combining the agrarian economy with non-agrarian income accessible through seasonal migration. These and other construction jobs, such as stonemasonry in the Karst and around Tolmin, were passed down through generations and shaped entire working careers through seasonal and other forms of circular migration (Škrlj Počkaj, 2011).

The agrarian sector contributed significantly to the increase of seasonal migration, as it traditionally needed a temporary workforce. Itinerant agrarian laborers—men and women—departed from many Slovenian districts for occasional or regular seasonal work. They were employed on mid-sized or large estates, for example, as mowers, harvesters, and other agricultural workers around Celje and in Upper Styria, Upper Austria, Croatia, and Slavonia. Generally, the season lasted from March to October/November. Labor bureaus organized their departures, and foremen or middlemen (*Winkelagenten*) managed their work (Wörsdörfer, 2017, 37; Wanderarbeit in Steiermark, 1913, 124). The demand for wage labor also grew in the Slovenian territory, for example, on larger estates around Idrija, where smallholders and tenants preferred to find work in the Idrija mercury mine, which caused a shortage of agricultural laborers (Ribnikar, 1889). Another example is Lower Styria, where hop growers needed an additional workforce to pick hops from August until the end of September (Wanderarbeit in Steiermark, 1913). Such extensive migration to other parts of the monarchy and abroad created a massive shortage of seasonal agricultural laborers in the regions with the highest emigration rates. It was solved by hiring workers from Croatia and raising wages. The mobility of agricultural laborers was an established practice even in previous centuries. However, modern agrarian seasonal work and forest work acquired new dimensions by the mid-19th century—spurred by the expansion of intensive monoculture farming and the development of transportation, which enabled moving across longer distances.

Seasonal migration of agricultural laborers became a true pillar of the regional economy in Prekmurje and Porabje (together known as Vendvidék in Hungarian). The specific circumstances that caused it came from the property structure of the region and the socio-economic system of the wider surroundings, where there were no cities and important industrial centers that would partly absorb the workforce surpluses or offer possibilities for non-agrarian income. After the emancipation of the peasants and the concomitant agrarian reforms, the land in Prekmurje remained partly the property of large landowners and partly divided between smallholders. They worked in agriculture and viticulture, and increasingly with animal husbandry, which contributed to the relative improvement in their living standards. However, only about a third of those who owned their land could secure their existence from these sources. The rest complemented their income with seasonal wage work. Moreover, many people owned no land at all and were completely dependent on wage work. With demographic growth, fragmentation of small agricultural estates, the growing proletariat, and the decrease of revenue from farms, the need for seasonal work, particularly in Goričko, the hilly northern part of the Prekmurje region, was constantly growing. Compared to the rest of the Slovenian ethnic territory, the seasonal work in Prekmurje was rooted in the work obligation of 100 days per year that, until 1848, the Prekmurje peasants owed their lords instead of material tributes in return for possessing the land (Olas, 1957, 179).

Following the 1884 agrarian crisis caused by the import of Russian and American wheat, large and mid-sized farms shifted to intensively growing sugar beets, potatoes, corn, and fodder crops. With the reorganization of work and the introduction of mechanization, the properties' needs for permanent staff and workers shrank, and the demand for seasonal work increased. By the 1880s, seasonal work in Prekmurje appeared in the form of moves from estate to estate, where the workers cut grass and harvested and threshed wheat. Later, they were increasingly needed—and for more extended periods—to hoe, thin, pull out, and other labor-intensive tasks necessary in sugar beet production. This seasonal work, which included other tasks, was poignantly called "going for the beet" (Maučec, 1933, 114; Olas, 1957, 179–180; cf. Horvath, 1989; Munda Hirnök, 1992).

This labor migration was aligned with the seasonal cycles of the intensive agrarian economy and distributed according to the needs of large estates. Its characteristic was also that the workers were partly paid in produce, with which they met their own annual food needs or sold for cash. This migration was mainly directed toward the Kingdom of Hungary and, in the northern areas of the region, to German-speaking Styria because wages were significantly higher there. Thus, the price of agrarian labor also increased in the northern part of Prekmurje and the entire county of Vas megye/Železna županija/Komitat Eisenberg. The wage differences were also reflected in the qualitative diversification of the seasonal workforce and its destinations. The most capable among them aimed for the areas where they could earn better, but many others had to settle for a lower payment and worse working conditions (Munda Hirnök, 1992). The seasonal agrarian migration functioned as a strategic component of the Prekmurje and Porabje economic and social systems. It ensured a regular influx of outside income, but this income contributed to the "immobility" of the land economy and the slow arrival of innovations. With population growth causing the fragmentation of land property, the unfavorable structural relationship between demography and the region's economy became even worse. Employers hired workforce from less developed areas that suffered economic depression because they agreed to lower wages and very basic—often makeshift—living conditions, thus lowering production costs. The economic success of the market and capitalist-led agriculture was essentially based on this differentiation.

On the western edge of the Slovenian ethnic territory, Benechia and Resia (under Italian rule) are illustrative cases of quantitative and qualitative changes in seasonal migrations. Like elsewhere, these changes happened in the second half of the 19th century when the local socio-economic factors pushed people to look for different work options, and the international labor market environment appeared as an attractive alternative. Until the 1870s, the migrants were primarily itinerant vendors and tradesmen who continued the tradition of pre-modern circular migrations. Afterward, large infrastructural worksites, construction, and the agrarian sector throughout Imperial Austria and wider Central Europe demanded a workforce, thus generating the mass phenomenon of this "new seasonal migration" (Kalc and Kodrič, 1992, 198–200). The professional profile of migrants changed. Some specialized in masonry and worked in construction as skilled and unskilled workers. Another group comprised ordinary laborers in brick factories, diggers, and other manual laborers in the construction and maintenance of road and railroad networks, tunnels, bridges, and other structures. The third group was agrarian laborers. The economic importance that migrations acquired in this period is clear from the fact that more than 40 percent of the male population in the Resia Valley left their homes at the turn of the century. While the older style of seasonal work was aligned with the ecological cycle of the agrarian economy in their place of origin, taking advantage of the "dead" months of the agrarian-mountain economy, the new one subjected itself to the time dynamics of the labor market. It lasted from spring to fall, requiring the reorganization of work and roles in the traditional local economy. Although many migrants returned home to cut grass, the work and management of rural households fell squarely on the shoulders of those who stayed home: the women, the young, and the elderly (cf. Pahor, 1990, 18–25).

The patterns of the "economy of absence," as defined by scholars of upland communities, and the gender division of economic roles, aimed at integrating the local and "external" activities, thus structured new migration systems which linked Resia and the valleys of Benechia with the regions that employed migrants, reiterating the annual encounter of workforce offer and demand. The monetary influx from seasonal work eased paying taxes and debt, increased money circulation and spending, further solidified combining traditional and external sources, and strengthened the systemic role of seasonal migration in the economic logic of families and the community. The effects of property fragmentation were visible here, too, and not only on land property. Houses and individual rooms within them were divided among the descendants and thus belonged to numerous different owners. In Resia, demographic growth surpassed the economic sustainability of the valley, in which its morphologic closeness and difficult passages prevented the development of transport or other non-agrarian local sources of income. Migration was also reflected in specific social and cultural patterns, for example, in the professional specialization of migrants in individual Resian villages, in entering economic alliances through marriage strategies, and in the gradual decline of the traditional rural economy. Migrants who spent more time on their work routes or settled more permanently where they were economically active retained close ties and represented permanent extensions of the social fabric and economic system of their places of origin. Among the characteristics of the Resian social strategies was emphasized endogamy, which contributed to migrations and the increasing scope of those who spent extended periods outside the Resia Valley (Morassi and Panjek, 1984).

From Farmland to Farmland

Significant changes in agriculture and pressures faced by rural estates prompted smallholders and agricultural laborers to move to secure more beneficial foundations for their existence, to eschew class diversification and proletarianization, or to find themselves work and an opportunity to rise above the agrarian proletariat. This was the case for many among the Slovenian migrants to Croatia, where, after serving as farmhands, they often married and settled down as smallholders. In 1880, one-quarter of the Slovenian migrants to Croatia worked in agriculture (Drnovšek, 1999a, 34). Transfers from farmland to farmland were common elsewhere in the Slovenian ethnic territory. From the 1870s onward, for example, villagers from the hilly Baška Grapa in Goriška moved to Carinthia (Lesjak, 1996), while in Lower Styria, peasants from Prekmurje took over the land from smallholders who emigrated elsewhere or switched to non-agrarian activities. (Olas, 1990, 592–593). The surroundings of larger urbanized centers became attractive, among them Koper, where agriculture, influenced by the food market in the neighboring Trieste, developed significantly and attracted not only the seasonal workforce but also permanent settlers from inland Istria. At the end of the 19th century, these newcomers purchased the property of the *paolan*—the city-dwelling farmers of northern Istria. At the time, this traditional agrarian segment of the urban population was in a crisis, and landowners shifted to shipping and other non-agrarian activities (Titl, 1965, 111, 117).

This period was when moves from farmland to farmland offered new possibilities and geographical destinations. In the 1860s, when the tax burdens, lack of money, debt, and other effects of the transition from the feudal corporative society to a liberal socioeconomic order worsened the current situation and future prospects for farming, interest increased in searching for more fertile land and better farming conditions in distant countries. A particular case of such wishes was *meksikajnarji*: the 433 Slovenian volunteers who set off in 1864 with the Habsburg heir apparent Maximilian on his transatlantic adventure. They joined him for patriotic reasons and for the land they were promised in Mexico as a reward for the six-year paid stint in the military. Most of them had already completed their regular military service, but some who volunteered for the expedition were exempt from compulsory military service at home (Janša Zorn, 1962; Ovsenar, 1993, 14–17).

Land acquisition was possible in eastern Slavonia (Serbia where, in the years 1858 and 1865, they passed the Colonization and Settlement Act), in the Caucasus of Russia, where the authorities likewise supported rural settlements, and in North America, from where Slovenian missionaries reported about the fertile American plains and benefits for settling farmers. Despite being advertised as a better alternative to transatlantic migration due to geographic proximity and linguistic and cultural similarities, people in Slovenian regions did not overwhelmingly respond to the colonization programs in Serbia and elsewhere in Southern and Eastern Europe. If, on the one hand, Serbia offered fertile land, good settlement conditions, tax breaks, exemption from military service, and other benefits, on the other hand, settlers were subject to various conditions related to the nature of their settlement (Drnovšek, 2009, 476–477; Lukić, 2017, 37). News about unsuccessful colonization attempts and reservations toward the Orthodox environment raised skepticism on whether such relocations made sense. Another problem was the lack of a "mediator" to help people organize such a radical step. Before World War I, the Slovenian presence in Serbia thus included only a couple hundred people. Agrarian settlers were also rare in Bosnia and Herzegovina, where the number of immigrants from the Slovenian Lands was larger (Josipovič, 2014, 74; Drnovšek and Kalc, 2014, 94; Kržišnik Bukić, 2007, 20).

The establishment of Slovenian rural communities in the United States reflected the early active role of the missionary Franc Pirc, who encouraged the first groups of settlers. In Upper Carniola, Pirc was known, among other things, as the promulgator of fruit growing and sensible farming. Once he left for the missions, he also attracted numerous German farming families to Minnesota as early as the 1850s. In 1865, he was appointed the federal immigration commissioner in the United States and managed the colonization of the Upper Mississippi River area. The Slovenians who settled in the different parts of Stearns County were convinced by the good economic prospects. They also trusted the overall Catholic context of colonization, specifically Pirc and other well-respected missionaries. Already in 1864, Pirc had used Slovenian newspapers to invite rural people to come to the United States once the American Civil War ended. He planned to create "a new Slovenia with Slovenian parishes and Slovenian priests." This promise came true in 1871 when the first Slovenian parish was established and the church built in St. Stephen of Stearns County under the supervision of missionary Buh. This settlement, and the nearby St. Anton, were home to more than 50 Slovenian farm owners at that point (Zitka, 1969; Drnovšek, 2003b, 2010a, 40–41, 63; Kodrič, 2003, 53–62; Klemenčič and Šeruga, 2019, 130–131). Migrants covered the cost of migration by selling, renting out, or pawning their property. Once in Minnesota, some initially took employment in the local lumber industry to get the funds necessary to purchase land or homesteads. Most settlers soon created a sound economic foundation for themselves. In the next couple of years, writers who visited the Slovenian ethnic settlements reported about families and individuals who were successfully managing their Minnesota properties and the many unmarried ones among them who turned to their homeland to find life companions and start families (Drnovšek, 2009, 470, 480).

The participation of Slovenian emigrants in the rural colonization of the United States was undoubtedly a marginal phenomenon. The US agrarian colonization model—based on isolated and market-oriented farms—aligned with neither Slovenian farmers' financial capacity nor their knowledge and experience. Due to the work conditions and often harsh climate, living off such farms was also difficult (cf. Velikonja, 1996, 384). Despite the availability of land and the establishment of ethnic communities such as St. Stephen and St. Anton, Minnesota saw no significant influx of Slovenian farmers. Minnesota had neither subsistence nor small-market farming with which the Slovenian farmer was familiar. After the American Civil War, the Midwestern United States was dedicated to the wheat farming economy. The tree, the scythe, and the rake were replaced by modern technology and family workers by hired farmhands. Intensive monoculture also demanded occasional diversification and substantial capital investment (Gilman, 2000, 3–4).

Slovenian migrants mainly joined the ranks of farmers during the second phase of their American experience, after they had first worked in mining and other industries. Like other Slavic migrants, they invested their savings into buying land and returned to their "rural roots" (Balch, 1910, 337–339). In his book, *Amerika in Amerikanci* (America and the Americans), priest Jurij Trunk offered a broad insight into the Slovenian immigrant reality in the United States and the American way of life. He advised the migrants to try and become farmers because—due to the fertile soil, mechanization, and farm management—an American farmer was better off than one in Slovenia. Like a "fortress," the farm was supposed to guarantee survival under any circumstances (Trunk, 1912, 350, 409–410). Migrants who found employment in mining and manufacturing were primarily interested in a good income. Only gradually, once settled in the United States, did they choose farming, which meant a form of economic emancipation and a step away from the often precarious dependent work. In short, rather than via colonization, the relocation from the Slovenian to the US farmland usually went via proletarianization and then by acquiring existing farms.

The accounts from the time between the two world wars tell about numerous successful Slovenian farmers near mining and industrial centers. Farming was positioned as an alternative to wage-work and urban life or in combination with it. Many farm owners continued to live and work in cities yet farmed for themselves or as a supplementary source of income (Trunk, 1912, 386; Lipoglavšek-Rakovec, 1950, 15–18). Before World War I, the largest Slovenian farming community developed in Willard, Clark County, Wisconsin. Since 1908, dozens of Slovenian families from different manufacturing, mining, and other towns settled there on homesteads between 40 and 120 acres in size. In this case, the settlement's founding was linked to the workings of a single Slovenian land agent who used the Slovenian migrant press and other channels to encourage many to invest in the initially demanding but, eventually, lucrative farming. At the end of World War II, there were around 200 Slovenian farms in Willard, with another hundred scattered across different places in Wisconsin (Prisland, 1950, 272–273; *Spominska zgodovina*, 1992). Slovenian migrants also started farming in other states. Around Cleveland, Ohio, for example, the descendants of Slovenian immigrants still produce wine. Unlike in the "old country," farming in the United States did not always have a constant character. Many only farmed for a limited time and then moved on to other economic sectors. Some were attracted by the utopian farming colonies, such as the Eden Valley Ranch in California in the 1890s. This Catholic settlement was founded in 1895 by the Slovenian priest Josip Jeram to pull Slovenian migrants from physical and spiritual destruction in cities and industrial centers. The plan was met with an enthusiastic response, but merely years later, financial problems, disputes, and embezzlement caused the community to disintegrate. Similar farming colonies based on common property and some socialist basis also did not work out for long (Cesar Nezbala, 1990).

The immigration and colonization programs of the South American countries proved relatively more attractive and efficient, as they opened the path to the land even for poorer rural groups. A well-oiled organizational machine behind them actively promoted and managed the migration process from the starting points to the end destinations. Slovenian migrants to South America embarked on these migration routes in different historical moments and circumstances. Some at their own cost, others with their voyage pre-paid, they all left with a desire to set up new estates with the money they got selling their old homes before the agricultural economic crisis at home pushed them into debt and the proletariat (Benvenuti, 1982).

The migration to Argentina from the Austrian Littoral in the 1870s occurred in the most fertile territories where smallholders and tenants (*coloni*) lived better than elsewhere and had the means to finance the transatlantic move. On the other hand, among the migrants to Brazil at the end of the 1880s and 1890s, many destitute farmers, cottagers, and farmhands were in great existential distress and had no financial reserves. To them, the subsidized voyage, which they then paid off as coloni or agricultural workers on *latifundia*, provided hope for an exit from an unenviable social situation. They were driven by the desire to persevere and establish themselves in the agricultural economy and hopefully get their own farms. This motivation is clear from the fact that the first people to respond to the campaign for Brazil came from the immediate surroundings of Trieste, where there was no shortage of possibilities for non-agrarian income. The Morpurgos—the Triestine entrepreneurial family whose members served as Brazilian honorary consuls and organized this migration—estimated that around 5,000 persons from the Austrian Littoral and Carniola expressed interest in Brazil. However, far fewer actually migrated because the authorities prevented further recruitment due to misleading propaganda (Umek, 1968, 1979; Kalc, 1995, 45–50).

In Brazil, immigrants often encountered completely different circumstances from what the agents had painted, and many promises and contracts signed on departure turned out to be deceiving. Living conditions were harsh, particularly in territories where the arable land still had to be reclaimed from the wilderness. Some soon returned disappointed (Kalc, 2001). However, most of the migrants to Argentina and Brazil stayed there, some because they could not afford to return. For a long time, some of them could not get out of the system that subjugated the tenants to the interests of the owners of latifundia and kept them in a subordinate position that often resembled semi-slavery. However, some obtained property and achieved a decent standard of living or even wealth (S. P. B., 1959). Many turned toward non-agrarian professions. Despite their significant numbers, the type of settlement meant these migrants were scattered and did not have the opportunity to congregate in Slovenian ethnic colonies. They mostly lost ties with their homeland and soon assimilated into German or Italian migrant communities (Lipoglavšek-Rakovec, 1950, 38–39; Marušič, 1995, 65; Kalc, 2001). Some moved from agrarian areas into cities and economic centers, and many left for other countries, particularly the United States and its industrial centers.

Migrations in the Era of the Transnational Industrial Labor Market

The migration that took so many Slovenians into the world belonged to the shift from the agrarian to the industrial society that tended to move people from the countryside to cities or other urbanized areas. It occurred as a part of migration systems that connected the Slovenian ethnic territory with the regions of accelerated industrialization, economic progress, and urbanization in Europe and the Americas, and synchronized the supply and demand for human resources. During this time, the labor market gained transnational and transcontinental dimensions. Individuals entered it in various ways: they moved shorter or longer distances, undertook sometimes discrete—permanent or temporary—and sometimes iterative migrations that found new routes, changed destinations within the host countries and, in many cases, transformed into permanent residence. Nevertheless, migrations often retained a temporary character. Similar to the erstwhile circular migration, they also functioned as economic strategies to satisfy the needs and improve the status of migrants in their places of origin. This process was anything but linear and one-way. It ran back and forth between the "workforce pools" and centers of industrialization and urbanization and, notably, between the centers themselves. It knew many variants, stages, and shapes that could change along the routes or morph into each other (Steidl, 2021, 5; Moch, 1992).

The Case of the United States of America

Because of its scope and importance, migration to the United States is illustrative of how this economic and social modernization process appeared in Slovenia. The foremost Slovenian geographer Anton Melik claimed in the 1960s that mass migration to transatlantic and other "foreign" (non-Slovenian) destinations was "a special transition from the countryside and peasant to the urbanized and industrialized socioeconomic environment" (Melik, 1963, 337). Here, Melik was referring to the explanatory concept of "rural drain" or "escape to the city," which has since been surpassed as it failed to provide a full picture and a deeper understanding of the migration processes of the time (cf. Rosental, 1999; Moch, 1992; Hochstadt, 1999; Steidl, 2021), which were far more complex, as already revealed, and also entailed intra-rural moves. However, what Melik primarily wanted to emphasize was that, for Slovenians, this "special transition" only partly occurred in the Slovenian ethnic territory, which rarely provided occasions for such moves. It mainly occurred through the population's departure from this territory. In a particular phase, the clearest destination became the United States of America (Melik, 1963, 337).

In the first phase of the transatlantic migration to the United States, artisans and non-agricultural laborers moved there alongside rural migrants. This social composition was also seen in some other immigrant groups that started to join the overseas labor market as an alternative to the European one. In the 1870s and 1880s, there were smaller Slovenian colonies in some US cities, towns, and mining centers: New York; Chicago, Omaha, and Joliet in Illinois; and Calumet in Michigan (Zavertnik, 1925, 300; Colakovic, 1973, 38). Part of the "Copper Country," Calumet is the home of the oldest continuous Slovenian community (founded in the 1870s) and of the first Slovenian association established in the United States (Grill, 1979, 136; Herle, 2020). Before the 1870s, the US industry mainly depended on small and mid-sized industrial plants that operated in a regime of free competition and used a production model that employed a semi-skilled and skilled workforce. Mass migrations around the turn of the century coincided with the crucial changes in the US industrial production system, which enabled the expansion of the US economy. Mass production, led by monopolist corporations and concentrated capital, played a leading role.

The "new immigration" from Eastern, Southeastern, and Southern Europe became a key factor of the new methods of production and an endless source of the workforce which the employers used to push out qualified American and immigrant workers and annihilate their union achievements. As Dirk Hoerder states, compared to the migrants from European countries who arrived in the United States in previous decades and centuries as members of the conquering ethnicities, the multitude of people of different nationalities from European peripheries (including Slovenian migrants) arrived in the "New World" as a socially and culturally subordinate proletariat (2002, 339). The employers found it easier to subjugate them to their interests and treat them as disposable. Therefore, depending on the oscillation of their needs and, unlike the qualified and unionized workforce, employers could hire and fire them easily, move them from one production line to another, and, above all, pay them less than the American workers. An important part of the strategy was also providing a reserve workforce, be it for the concrete needs of the production or to create competition between workers for wage-wars or strike-breaking (cf. Bock et al., 1976; Bock, 1987).

Slovenian migration to the United States in this phase accurately reflected the structural traits of the new US labor market. According to the US immigration statistics for 1898–1914, three-quarters (73.1 percent) of the Slovenian and Croatian immigrants were categorized as "laborers," that is, ordinary workers, of whom 29 percent were farm laborers and servants, and the rest were industrial and other laborers. Independent farmers who also came to the United States to work for industrial wages comprised 1.6 percent, 13 percent were people without a profession, most of them wives and children of migrants, and just under 8 percent were professionals and intellectuals (Klemenčič and Maver, 2017, 172). The United States Congressional Joint Immigration Commission, known more widely as the Dillingham Commission, reported that only 5 percent of the Slovenian and Croatian migrants who arrived in the United States between 1899–1910 were in skilled occupations, almost 33 percent were farm laborers, just under 54 percent were laborers, 8 percent were workers in other occupations or had no occupation, and only a handful had higher education, among them clerics (Colakovic, 1973, 44). The professional structure of migrants reflected the structure of the Slovenian society at that time, which despite urbanization and the expansion of the non-agrarian economy, retained a predominantly rural character. The sample of male Slovenian immigrants reviewed by the Dillingham Commission reveals that, before migrating, 24.4 percent were paid workers, 15.2 percent were farmers, and 58.4 percent worked without payment, presumably family members who worked on family farms. Among female Slovenian immigrants, 41 percent were without occupation, 13 percent were paid workers, mostly farm laborers and domestic servants, and 44.7 percent had done unpaid work on their family farms (Reports of the Immigration Commission, 1911b, 89–98). The social background and financial situation are also quite clear from the funds that Slovenian migrants brought to the United States to cover their immediate needs. In 1905 and 1906, the average Slovenian or Croatian migrant came to the United States with 15.50 dollars, while the Czechs had 28 dollars and the Germans had 43 (Caro, 1909, 23).

Migration to the United States followed the chain model. The first migrants paved the way to employment for those who followed by providing them access to personal networks and information. In this way, colonies grew into larger or smaller communities as more immigrants settled. Their social organization functioned as a point of reference for future arrivals. The globalization of the labor market at this time changed the nature of transatlantic migration. Although in the previous phase, migrants had set off to the United States wishing to create a new life, returning was not wholly alien to them, not even for those who wanted to remain farmers and settle in agrarian areas. However, from the 1880s and 1890s onward, Slovenians, like other migrants, immigrated en masse to take the opportunity to make money or, as Klaus Bade wrote, for economic-speculative reasons (2001, 158). Most of them joined the transatlantic migration system intending to stay in the United States for a limited period and return home with the money they had saved. As the migration increased, so did the number of those who stayed. Still, the decision to settle in the United States was formed later, when migrants adjusted and integrated into their new environment, created stronger economic foundations for themselves, and married. It was based on objective factors and subjective choices. Settling in immigrant environments was a long process that planted many migrants firmly in the United States, but it did not exclude the possibility of return.

Picture 1.2: Slovenian foresters in Pennsylvania, 1907. Author: unknown. Courtesy of: Slovene Ethnographic Museum, Ljubljana.

Since the 1890s, this aspect of transatlantic migration was reflected in the high level of return migration, which affected Slovenians and Croatians more than any other nation of the Austro-Hungarian Monarchy. Between 1908 and 1910, one of the peak periods of this two-way migration process, the returning Slovenians and Croatians amounted to 56 percent of those leaving, with the number of returnees in 1908, the year of an industrial economic crisis, surpassing the number of emigrants. The longer period between 1908 and 1923 shows the rate of return was 51 percent (Čizmić, 1990, 156). Lawyer Juro Adlešič wrote in 1909: "Our farmer goes to America, not to become a worker, but to earn some money and return as quickly as possible to his inherited homestead [...]" (1909, 173). That same year, "on the occasion of the departure of 50 young men and women, aged between 18 and 30," *Narodni list* (National Gazette) from Celje wrote: "These days, people from the Upper Savinja Valley travel to the United States and back like going to the nearest market" (Amerikanci, 1909). Many left to make money in the United States several times, and this practice increased during the period of the highest migration: in 1900, 9.8 percent of Slovenian and Croatian immigrants had been to the United States once before; in 1906, this share increased to 12.6 percent (Balch, 1910, 464). Migrants returning to Benechia worked in the United States for four years on average. Very few returned after less than two years; if they did, they were most often forced to by an economic crisis. Three-quarters of them stayed in the United States for between three and six years the first time they went there (calculation is based on the passenger lists).

Return migration studies show that a five-year stay was the average cut-off mark, after which the probability that the stay in the United States would turn into a permanent settlement grew exponentially (Steidl, 2007; Gould, 1980). The Dillingham Commission statistics confirm this tendency: 81 percent of the returnees between 1908 and 1910 left the United States after less than five years of living there. The information, of course, only shows the prevalent (and oscillating) tendency. The same is true for the finding that the arrival of the family and family reunification was a step toward a permanent settlement in the "new homeland." Microanalyses reveal that the reality was infinitely more diverse. If a longer stay, marriage, and the birth of children were the elements that worked in favor of permanently settling in the "new homeland," there was no shortage of returnees who returned home, alone or with families they had either called for or started in the United States, after 10, 15, or even more years (Drnovšek, 1989, 206). There were also families whose members lived apart for decades, families that got separated when some members returned home, and families that were never united and their members lived separately on two (or more) continents. In short, the patterns of migration aligned with the ever-changing pragmatic plans and choices, as well as deviations from them and coincidences. In the Slovenian case, statistical sources do not allow for a more detailed observation of return migration over a longer period or for the scope and dynamics of the repeat transatlantic migration. Both were dependent on US labor market fluctuations and other events on the one hand and, on the other, on migrants' plans, situations, and decisions. Studies examining long-distance migration show that these transatlantic moves spread at the turn of the 20th century, once the economic and logistic circumstances allowed them. In 1912, Trunk estimated that, since the beginning of the migration, about one-quarter of Slovenian migrants returned home and three-quarters remained in the United States (381).

Considering the purpose of leaving for the United States and the characteristics of the labor market, migration was mostly individual. Mulaček's statistics for Carniola show that two-thirds of the migrants were single and a third married. Of those married, only 7 percent either took their wives and families with them or were joined by them later (Mulaček, 1913, 260, 265). The US statistics for 1899–1914 show that 91 percent of the immigrants were aged between 14 and 65 (Chmelar, 1974, 113; Klemenčič and Maver, 2017, 172). The calculations for Slovenians from Benechia, based on more detailed data and ship passenger lists, show that 65 percent of the newcomers were aged between 15 and 30 and 25 percent between 30 and 45. Single migrants, of whom there were more than 60 percent, most often set off between the ages of 20 and 22. Among those over 30, many were coming for the second time, and 70 percent were married. Married migrants were more likely than single ones to enter the US labor market more than once in order to satisfy their subsistence and other needs in their places of origin (Kalc, 2018a, 111–113). The gender structure of this migration was overwhelmingly male.

The US-bound migration from the Slovenian territory—as from other Eastern and Southeastern European countries—found its impetus in higher wages. Although wages in Slovenia increased on average by 2.6 percent a year in the decades leading up to World War I, and a decrease in workers' wages accompanied the rationalization of production processes and restructuring of the US labor market, migrants in the United States could secure a better income for themselves, earning sums unimaginable compared to those on the home labor market (cf. Kresal, 2005). Trunk writes in his book that the real cause of emigration "was to be found in the difference in earnings in the old country and America" (1912, 393). The same was also reported in the 1911 Dillingham Commission report on immigration, which singled out "the smallness of the amount paid for labor" compared to the United States as the most apparent characteristic of the (Imperial) Austrian economy (Reports of the Immigration Commission, 1911c, 362–363).

In Bela Krajina, where the workforce shortage raised the wages of agrarian workers higher than elsewhere in Slovenia, a mower earned 2 kronen per day at the beginning of the 20th century (Šuklje, 1905; Dražumerič, 1987, 187). During the mining strike in Zagorje, workers fought to increase the wages of diggers to

3 kronen a day, trammers to 2.60 kronen, and those of women and children to 1.60 kronen (Rozman, 1979, 215). In Slovenia, in 1905, wages were highest in railway, metallurgy, and mining: between 22 and 24 kronen per week. The textile industry predominantly employed women and paid between 12 and 15 kronen per week. In Styria, the average wages were 6 percent higher than in Carniola or the Littoral. In Vienna, wages were 20 percent higher than in the Slovenian Lands (Kresal, 2005).

Between the early 1890s and 1910, the working-class wages in Trieste increased by 50 percent. Nevertheless, ordinary laborers, for example, assistant masons and diggers, working on construction sites and public works, still only earned 4 kronen a day at most. Ordinary laborers in industrial plants earned up to 5.40 kronen, and skilled workers 7 kronen (Alberti, 1911, 86–90). In the United States, the average daily wages of Slovenian migrants in the manufacturing and mining sectors, which the Dillingham Commission took into account, were 2.13 dollars per day for men and 1.31 dollars a day for women (Reports of the Immigration Commission, 1911b, 104, 106). The conversion rate at the time was 5 kronen to a dollar.

The earnings differed in the United States, depending on the economic sector and the type of work, economic cycles, temporal oscillations in the production processes, and other variables. In Pennsylvania, in 1905, the wages of ordinary factory workers without knowledge of English were between 1.35 and 1.63 dollars for a 10-hour workday. In the underground mines, an ordinary worker earned 2.36 dollars in 8 hours, and the worker who did piecework earned 3 dollars (Balch, 1910, 285). In Steelton, Pennsylvania, in 1908, during the economic crisis, ordinary workers earned 1.44 dollars for a 12-hour workday and semi-skilled and skilled workers earned up to twice or thrice as much (Bodnar, 1977, 22, 181–182). Tradesmen, for example, carpenters, wheelwrights, joiners, blacksmiths, and similar, earned between 12 and 18 dollars a week in 1908 in the United States, while foresters and sawmill workers made between 4 and 7 dollars in addition to food and lodging (J. P., 1903). The comparison between wages alone cannot precisely explain the differential ratio between work and earning at home and in the United States. We have to consider the cost of living, the level of consumption, the steadiness or fragmentation of employment, and other things. Workers in industrial towns lived together in "lodging" or "boarding houses." According to the Dillingham Commission estimates, they paid between 14 and 18 dollars per month for lodging, food, heating, and personal care (Reports of the Immigration Commission, 1911a, 83). Other sources report that the monthly cost of living was around 30 dollars (Brunnbauer, 2016, 100).

The possibilities to earn and accumulate means were high enough to justify the investment in a transatlantic voyage and the cost of living (Trunk, 1912, 393–394). A worker could save up to 300 dollars in a year with favorable working conditions. Even with lower wages, under normal circumstances, the immigrants in the United States could provide a higher standard of living for themselves, particularly those who did not have enough economic sources at home to survive and no opportunities to supplement them with additional work or with opportunities that were, like for agrarian workers, limited to short periods. The descriptions of the United States and life there usually stated that meat was an everyday food, while in Slovenia it was served only on Sundays and holidays if at all (Prisland, 1968, 53). Clothes were also more accessible and of better quality. A pair of shoes could be bought with the earnings of one or two workdays (Reports of the Immigration Commission, 1911a, 81). The economic advantage of leaving for the United States was clearly shown in a report that a municipal clerk from Knežak presented to the Postojna district authority in 1900. He stated that the local migrants earned well in the United States and sent a lot of money home. "Often, everyone earns or saves up to 2000 kronen every year. If the work continues, our municipality will have a lot of money. If the work stops, everyone has enough saved to be able to come home" ("A Report on the Migration of Persons Living in Knežak to the United States between January 1 and December 31, 1900").

At the time of the most significant migration at the beginning of the 20th century, traveling also became more accessible. On average, it paid off in four to six months of work. With technological advancements, the all-around expansion of traffic on land and at sea, and spatial mobility as a characteristic of modern times, traveling also lost its adventurous ring. The perception of the transatlantic distance changed. The voyage across the ocean became an almost routine practice and, from the economic aspect, a less taxing item when calculating the costs and benefits of a short- or long-term migration plan (Colakovic, 1973, 43; Keeling, 2012, 7). In 1899, historian Ivan Vrhovec pointedly noted: "Our compatriots departing to America […] say farewell to their people with such ease as their fathers felt when they went to a neighboring parish for a church blessing […] as long as the traveler has left Ljubljana behind […] America is not far anymore" (Drnovšek, 1997, 198).

Regardless of professional skill and work experience, a large majority of Slovenian immigrants from the 1890s onward started their working life in the United States on the unskilled labor market; only a number of them progressed to better jobs. As they came from a predominantly rural environment, with poor skills or without a specific profession, and from a country or region with low living standards and requirements, in the United States, they could aim for the most difficult, most dangerous, and least paid jobs and less than average living standard (Govorchin, 1961, 81). They were often employed in heavy industry and mining, where few labor profiles required professional skills beyond physical strength.

Migrants faced many obstacles in their career trajectory, for example, not knowing the language, which meant that even experienced and skilled workers first had to "swing the pickaxe" (Dražumerič, 1987, 186). Organized skilled workers, who protected their position from the influx of cheap competition, represented another obstacle (Martellone, 1980, 52–56). In his work about Slovenians in New York, Janez Arnež finds that few stayed there after arrival. To find work in this city, one needed some sort of professional skill and knowledge of at least German if not English. For this reason, most took their sheer physical strength and went to sell it elsewhere at a low price (Arnež, 1966, 8). Knowing the language was the key to being promoted to positions requiring more communication—particularly to positions of foremen, who had to manage larger groups of workers. As the Dillingham Commission found, around 1910, 56.4 percent of male Slovenian immigrants worked in manufacturing and mechanical pursuits, 34.4 in mining, and 6.4 were general laborers (Reports of the Immigration Commission, 1911b, 100). At this time, many already worked in factories, workshops, and other better positions than they had 10 or 15 years earlier.

More research on the entry into the labor market and social mobility as specific aspects of the history of Slovenian migration to the United States is required, but these aspects do not generally differ from other ethnic groups. The US labor market was clearly segmented within individual industrial branches. The place in the hierarchy of positions and pay grades depended not only on generally belonging to the "new immigration," whose members were allocated jobs or working conditions that American workers or older immigrant groups no longer accepted but also on other factors, such as ethnic belonging. A typical example of this is Minnesota's Iron Range, one of the richest iron ore deposits discovered in the 1880s, where, before World War I, around 65,000 inhabitants lived. More than half of them were immigrants and comprised 85 percent of the workers in the mines. Slovenians appeared there in the early 1890s and, of the 43 ethnic groups, were the second-largest ethnic group after the Finns.

With the increase in open-pit mining sites at the beginning of the 20th century, the ethnic structure of the region changed radically. Open-pit mining employed a predominantly unskilled workforce, while underground mines, where the work was technically more demanding, required skilled or semi-skilled and specialized miners. With this shift, the work pool for the Western European immigrants shrank, but opportunities for the "new immigrants" from Eastern Europe increased. The pay regime also changed: if experienced miners earned 4 dollars a day underground, ordinary open-pit workers had to accept 2 dollars (Lubotina, 2015). In a workers' hierarchy, in which higher positions were reserved for American workers or North European immigrants, Slovenians occupied the top rung of the social ladder among the workers from Eastern and Southern Europe. Based on their physical ability and work ethic, they were categorized as more appropriate for underground mining, more reliable, and more predisposed to work. According to these stereotypical definitions, perpetrated by economists of the Progressive Era, and the implementation of the rationalistic, scientifically supported management of industrial processes, different ethnic groups were allocated for different types of work within the system of production in mining (LaVigne, 2017; Sirjamaki, 1946; Leonard, 2005).

Also noteworthy is how the workforce was recruited. "Boarding bosses" (known in Slovenian as *salunarji*, after "saloon keepers") also appeared among Slovenians. *Salunarji* were often former mine workers who operated as intermediaries between workers and employers and offered various other services, from bureaucratic to financial. These relationships were less prescriptive than the *padrone* system, in which the ties—and obligations attached to them—spread among the places in the old homeland and the United States. However, Slovenian immigrants did have to sign binding contracts that subjugated them to their employers, forbade their employment elsewhere, and curtailed their personal freedoms in other ways too (cf. Klemenčič and Šeruga, 2019, 26). The employers also used religious differences and antagonisms among different migrant groups for their own benefit and control. In Chisholm, Minnesota, for example, Slovenian and Croatian migrants lived separately from the Serbian migrants, and the Catholic and the Orthodox Church discouraged their socializing. When employers in 1907 used Serbians and Montenegrins to break the strike of the Italian workers (whom the Slovenians and Croatians supported), the relationships deteriorated further. The reconciliation only came at the establishment of the common Yugoslav country after World War I (Lubotina, 2015, 35).

The arrival time or sequence of a contingent of an ethnic migrant group's entry into the labor market also influenced the work hierarchy because employers used the inclusion of new migrant groups into the labor relations to reduce wages and thus increase their profits. The circumstances and mechanisms that influenced the social stratification along ethnic lines and work positions can also be seen in the case of the ironworks center of Steelton, Pennsylvania. Slovenian and Croatian migrants moved here en masse ten years before Serbian, Macedonian, and

Italian workers came at the beginning of the 20th century. In this case, Slovenian and Croatian workers were, wage-wise, the first among the Slavic people but still behind the African-American workers. The latter entered the labor market before them and were also more frequent among the semi-skilled workers. Slovenian and other "new immigration" members were more numerous among the staff with low specializations, working non-blue-collar jobs (assistants and similar). Some "new migrants" also entered the ranks of specialized workers, while such posts remained inaccessible to African Americans (Sori, 1979, 378–381).

The case of Steelton also shows how the steel company and the dominant social classes stratified immigration through employment policy, favored ethnic segregation of immigrants in workplaces and housing estates, and limited the social mobility of the members of the new migration. In this way, they also maintained the social hierarchy in town and preserved the dominant positions for natural-born Americans and older immigrants of Anglo-Saxon and Germanic origin (Bodnar, 1977, 35–37). The significance of work promotion is clear from the differences in pay: while, for example, an ordinary open-pit miner in Hibbing and Chisholm in Minnesota in the early 20th century earned 48 dollars a month, a foreman earned between 80 and 120 dollars, plus production bonuses (Lubotina, 2015, 35–36).

Most immigrants' working and living conditions were hard, and the work in steel mills, mining, construction, and other worksites was physically exhausting and often dangerous. It was dictated by demanding timetables and production deadlines and, particularly when it came to piecework, had a faster pace than people had known at home (La Vigne, 2017). The migrant correspondence, personal testimonies, and pages of the Slovenian press emphasize how hard and how much people worked in the United States. There were many injuries, illnesses, and work accidents. In the open-pit and underground mines of the Iron Range in Northeast Minnesota, five workers per one thousand were killed on the job every year between 1905 and 1910.

Mining companies did not pay much heed to that and did very little to improve workplace safety (Gilman, 2000, 6). Since the 1890s, two seemingly antagonistic traits coexisted in the mining districts. On the one hand, there was progress in mechanization, which is why the mining districts, steel mills, and industrial centers became a synonym for the technological development and work organization that enabled the increase of production and the rise of the United States among the world's industrial superpowers. On the other hand, the extensive abuse of the cheap unskilled immigrant labor force, recruited using primitive contractual methods and included in the production process with scientific precision, reduced production costs (cf. Peck, 2000). For example, the trigger for the great strike in Michigan's Copper Country in 1913 was the introduction of new drills, which permitted greater productivity while employing fewer skilled miners. Such productivity, however, meant an increased risk for workers (Michigan Copper District Strike, 1914). In Minnesota's Iron Range, the companies only started investing in safety systems more systematically in 1912. During World War I, care for safety intensified alongside the working classes' increased negotiating ground, and the government introduced incentives (Klemenčič and Šeruga, 2019, 65).

The living conditions in workers' shacks or boarding houses, which in many places were a long-standing widespread form of accommodation, particularly for single workers and those who came to work for limited periods, were difficult and unhealthy. The living standard of Slovenian workers was reflected in the below-average rents they were paying for accommodation and the above-average number of persons per apartment or room. Although its sample's representativeness is questionable, the housing statistics of the Dillingham Commission show that Slovenian immigrants, along with Croatian, Macedonian, Serbian, Mexican, Bulgarian, and Slovakian, were those with the lowest standards of living. One-third of Slovenian households included boarders, nevertheless, it is significantly fewer than in Croatian and Serbian households, which hosted boarders in 59.5 percent and 92.8 percent of cases, respectively. With 3.75 boarders per boarding apartment and 1.43 boarders per room, they belonged to the communities with high—but not the highest—living density (Reports of the Immigration Commission, 1911e, 420–432). This image is confirmed by the specific case of Steelton, Pennsylvania. In 1900, nine out of ten Slovenian and Croatian immigrants were boarders, with Croatian and Serbian families in the forefront of the percentage of those hosting tenants (Bodnar, 1977, 33–34). All this was a part of the fundamental strategies of the work migration so that it would be as economically viable as possible. Like renouncing privacy and being satisfied with lower wages and worse living quarters, other specific behaviors are linked to the same logic: disinterest in workers' rights, scabbing, insusceptibility to labor organizing, and similar. Of course, there was the other side of the coin: class consciousness solidified among Slovenian workers and radical ideas spread. The socialist movement gained importance and was the only Slovenian political or ideological camp that organized as a political party (Yugoslav Socialist Alliance) and became part of the Socialist Party of America. Slovenian workers, also influenced by other ethnic groups, engaged in workers' rights movements and union organizations (Klemenčič, 1985).

The income was somewhat unstable due to production oscillations and lowered wages, and immigrants often faced unemployment or precarious employment. According to the data from the Dillingham Commission, only 40 percent of male

Slovenian migrants had year-long employment, 35 percent worked between nine and twelve months, and the rest worked less than nine months. In coal mines, only 17 percent of the workers were employed throughout the year, while in the mines with lower quality coal, only 40 percent worked more than nine months per year (Reports of the Immigration Commission, 1911b, 134, 235). In such cases, it was difficult to save and even more difficult to send remittances home. Savings erosion and debt were common. Additional sources of income were also important, such as having some land and seeking alternative employment, with migration always being counted as a part of the deal. Among the "new immigrants," the living standard of those from Slovenian regions was certainly not among the worst (Reports of the Immigration Commission, 1911b, 142–143, 155). Studies and some personal testimonies emphasize that the statistics recorded the lowest levels of pauperism among Slovenians and other Slavic immigrants from the Austro-Hungarian Monarchy. Rare were those who ended up depending on public social services, which should be attributed to the high level of self-organization in different benefit societies and other types of organizations (see the Chapter 2). They were also considered good payers and reliable when paying off debt (Bicha, 1982, 32).

Pronounced internal mobility caused by the US labor market was encouraged by the economic cycles, production oscillations, changes in production systems, depletion of mines and closing down of industrial plants, union fights, redundancies, unemployment, or creation of new industrial centers and employment opportunities. The Iron Range in Minnesota thus first attracted Slovenian workers from the mining areas of Michigan and Pennsylvania and only later directly from Europe (cf. Sirjamaki, 1946, 207, 210). Work was seasonal in many mines, particularly the open pits in unfavorable climates. In fall and winter, when the work was stopped or slowed down due to low temperatures, thousands of workers moved elsewhere (cf. Klemenčič and Šeruga, 2019, 24–25, 62–63). Industrial plants, too, often only offered seasonal or even shorter employment to the unskilled workforce. These workers constantly moved throughout the year and shifted from mining to agrarian, industrial, or other sectors. Single workers, or those without families, were particularly subjected to that, but entire families moved too, although they strove to be stationary (Bodnar, 1977).

The number of inhabitants in many Slovenian mining communities, and even communities created in industrial towns, fluctuated significantly due to economic cycles and the uneven demand for work. Until the mining production systemized at the end of the 19th century, the population of the Iron Range in Minnesota—including Slovenian immigrants—was very fluid (Nemanic, 2007, 41). It was similar in the steel town of Pueblo in southern Colorado, where around 3,500 Slovenian migrants lived in 1907. When many moved to other industrial cities or returned to the homeland during a significant economic downturn, the smelting plants closed down, and the Slovenian community shrunk by a third in the next three years (Zaitz, 1934, 136; Klemenčič and Šeruga, 2019, 77). In pursuit of higher wages and better work opportunities, motivated by the desire for better living circumstances and social advancement, many were constantly on the road, and some moved from mining towns to industrial centers[8] at the beginning of the 20th century (Zavertnik, 1925, 360; Govorchin, 1961, 77). The numbers, dimensions, and geography of Slovenian settlements in the United States reflect the described objective influences, subjective choices, and development phases of migration events.

Picture 1.3: Greetings from the Iron Range's Corsica Iron Mine near McKinley, Minnesota, around 1910. Author: unknown. Courtesy of: Znanstvenoraziskovalni center Slovenske akademije znanosti in umetnosti, Inštitut za slovensko izseljenstvo in migracije (Research Centre of the Slovenian Academy of Sciences and Arts, Slovenian Migration Institute).

8 One of these cities was Detroit, where only a handful of Slovenian and Croatian migrants lived and worked in the automobile industry. In the years before World War I, people started arriving from northern Michigan because of higher wages and strikes in the mining districts (Govorichin, 1961, 77).

In this sense, the small case of Benechia provides a clear picture of the extent and effects of geographic mobility, dynamics of density, and scattering. In the first five years of the 20th century, the first groups of migrants from this area lived in 10 states, increasing in the following decades to 28 states and the Canadian province of British Columbia. However, 42 percent of all Benechia-born migrants were concentrated in Illinois, particularly Chicago, and 37 percent in Indiana, Ohio, and Pennsylvania (Kalc, 2018a, 117–118).

The communities in larger cities—though not only in larger cities—with a more diversified industry and extensive labor market were strengthened and created a broader and more stable settlement core. One of them was Cleveland, which from the 1890s onward—particularly at the turn of the century—attracted an increasing number of Slovenian immigrants, becoming the Slovenian American metropolis. The Slovenian communities in the eastern part of the city and nearby neighborhoods of Collinwood and, later, Euclid were created around the industrial plants and other worksites where migrants were employed. Most of them were concentrated along St. Clair Avenue where some sort of regional Slovenian neighborhoods emerged, from Inner Carniolan and Lower Carniolan to Littoral. Slovenians also settled in other parts of Cleveland and, after World War I, some moved to healthier and calmer districts, away from factories and workshops. The 1910 census recorded 20,000 Slovenians of both generations in Cleveland: in 1920, almost 25,000. Some 18,500 lived around the four Slovenian churches and the rest in other parts of town. This area also had the most Slovenian institutions, from religious centers to national homes and cultural organizations (see the Chapter 2). Although the Slovenian community was not among the rich and influential ones, like, for example, the German or the Czech, it was nevertheless recognizable and respected (cf. Klemenčič, 1995).

If mass migrations to the United States retained its predominantly proletarian character throughout time, the increased migration also caused professional or economic differentiation and social stratification among immigrants. Slovenian workers advanced to positions of foremen, shift supervisors, and similar roles in the mining and industrial centers. These advances also improved their material situation, although promotion in the workplace hierarchy and social climbing, as John Bodnar's study about Steelton points out, was made relatively difficult for the new immigrants (Bodnar, 1977, 37; Sirjamaki, 1946, 208). In time, their income grew. In the first five years, as evident from the Dillingham Commission study sample, only 25 percent of Slovenian workers earned a salary higher than the average. However, after a ten-year stay, this share was 45 percent (Reports of the Immigration Commission, 1911a, 24). In about half of the cases, wives and children contributed to the family income. Some women were employed, but in most cases, they provided service for other workers-tenants at home. Marriage or the arrival of the spouse and children from the old place were not only intended to establish or reunite a family. They also had an economic significance as households often transformed into boarding houses. The growth of a family—once its able young members found employment—increased its economic potential. The family statistical pattern used by the Dillingham Commission also shows that in the mining communities and the areas of heavy industry, around 20 percent of the family income was provided by children (Reports of the Immigration Commission, 1911a, 293–297).

Boarding, a widespread economic strategy in working-class communities, was present in 30 percent of Slovenian households, which was less than in other immigrant communities: the Dillingham Commission found that Slovenian households had on average 1.23 boarders per household, while Croatian households had 3.80 and Serbian 6.78 (Reports of the Immigration Commission, 1911e, 425; Govorchin, 1961, 182). Taking in boarders partly depended on the differences in gender configuration, patterns of migration, and migrants' activities.

Some migrants rose to become more skilled workers and helped themselves with the knowledge they brought from the old place, the work experience and education acquired in the United States, and self-initiative. This situation was most often linked to moving from the proletarian ranks to independent activities or employment in other sectors.

Migrant settlements themselves created circumstances for different activities. One characteristic activity of the first generation of Slovenian migrants was opening inns (or saloons), shops, food establishments, and other trade businesses. In the mining towns of the Minnesota Iron Range, more and more Slovenian migrants and their descendants took up commerce, trades, hospitality, and other activities. In Ely, in 1910, there were several dozen Slovenian enterprises, from general stores and clothes stores, taverns, tailors' and cobblers' shops to a jeweler's store, a public services company, and an insurance company. At the end of the 19th century, a Slovenian–Croatian investment company was operating here, and, at the beginning of the 20th century, some were working in real estate. In 1912, Slovenian investors were the majority shareholders in a local bank, while in 1916, they initiated the city chamber of commerce (Klemenčič and Šeruga, 2019, 28, 93, 98, 104, 149). Ely is a specific case: because of the size of the Slovenian community, its members had a particular place there, which is reflected in its economic and social structure and politics. Slovenian representatives held influential positions in the city administration and local politics from the beginning of the 20th century.

In industrial centers and big cities such as Chicago and Cleveland, specific development and social improvement were also noticeable before World War I. As the Slovenian community grew and diversified—socially and economically—the opportunities for employment and integration of newcomers also increased. "American Slovenia" was what they started to call it, referring to the national belonging and solidarity organized on ethnic grounds. It soon developed not only as a cluster of communities but as a structured and connected social fabric, in which a space opened for professional and social differentiation. The communities themselves became generators of immigration and, with their economic and organizational structures, attracted migrants from the old place, including those with academic certificates and professional skills. The second generation progressed significantly in society. In the second generation, what accelerated most were social mobility and a shift to more skilled professions and activities.

In connection to this, we must underline that most Slovenian migrants nevertheless approached improving their material and social situation primarily through work rather than schooling. The majority even opposed educating their children for a long time (Grill, 1979, 35). Their motto was: "We came to America to take the opportunity to work and earn money." There is no statistic available for Slovenians, but the indicative data says that, before World War I, only 2 percent of Slavic children continued their education to the secondary stage (Bicha, 1982, 28). Dropping out of primary education to work[9] was typical, as was the second generation following in the professional footsteps of their parents, finding employment in the same or similar industrial sectors and companies. This practice was linked to the economic structure and employment opportunities in their places of residence.

In school, the young generations came in contact with the United States in a way that differed from their parents' contact and thus created new values. In addition to the acquired linguistic knowledge, which adult immigrants lacked and felt socially stigmatized for, as well as the basics of American history and national spirit, the young were increasingly interested in education as a vehicle for

9 By the second grade, a quarter of the schoolchildren dropped out, by the fourth grade a half, and only a fifth of the children made it to eighth grade. The data from the Dillingham Commission show (although we must take into consideration that they are based on a very small sample and refer to two towns in Pennsylvania) that, before World War I, the attendance of Slovenian children beyond the fifth grade dropped below the average for immigrant children. The same was true for secondary education: only 0.7 percent of Slovenian children attended, compared to the 5 percent average (Reports of the Immigration Commission, 1911d, 702).

social advancement. This interest is reflected in the interwar image of "American Slovenia," where a growing scope of professions and a share of people with secondary and post-secondary education can be seen. The Dillingham Commission data for 1907 and the population census of 1910 show that the vast majority of people in Slovenian settlements in Cleveland were workers and that few among them were skilled; some were merchants and others self-employed. Twenty years later, this landscape had changed significantly: around 67 percent now worked industrial jobs, with more than 32 percent working (many self-employed) in the tertiary sector, more specifically, in commerce and different white-collar jobs. Before World War I, and particularly after it, a class of successful businesspeople established itself. By organizing and connecting, "career" opportunities opened to many who rose from the working class to provide services in ethnic organizations, not only economic ones but also those enriching social, political, and cultural life. The 1930 census shows a structure that was even more clearly articulated. In his memoirs, Vatroslav Grill states that, in 1931, "hundreds" of second-generation Cleveland Slovenians completed secondary and tertiary education (Grill, 1979, 626). This image reflects the situation in the biggest Slovenian community, which offered more opportunities for non-blue-collar and service activities. However, it is not representative of the whole of the Slovenian immigrant reality, which, as studies warn, in smaller and more industrially profiled environments, continued to remain socially immobile but very geographically mobile (Bodnar, 1977). During the interwar period, "American Slovenia" reached its peak in numbers and in terms of progress. The numbers and the progress, in turn, brought along the change in the social physiognomy of the community.

The Case of the German Empire

Migration to Germany was similar to the one to the United States. However, it had certain specific traits linked to the professional structure, the European migration regimes, and employment rules. Migration to Germany was a part of the international labor market and the mining workforce's mobility system. Most Slovenians who worked in the US mines took up mining once they moved across the Atlantic. Miners working in Germany, however, arrived directly from the Slovenian mining districts, and were thus a population with an established mining tradition (Drnovšek, 2007a, 164). Among them were skilled and experienced miners—between 25 and 30 percent of all Slovenian miners in Germany—and ordinary workers, who formed the auxiliary workforce (Werner, 1985, 75). The market for mining work included other mining regions, for example, Serbia, Bosnia and Herzegovina, France, the Netherlands, and especially those in German-speaking Styria and Lower Austria. The Styrian mining centers, such as Leoben, also attracted workers because wages were higher than in Slovenia (Wörsdörfer, 2017, 54–65; cf. Rozman, 1979).

The migration of mining workers was closely linked to the need to acquire a workforce for the strategic, productive sectors of national economies and was subject to harsh international competition. German mining companies were very aggressive in this fight. They attracted workers with their wages—which were the highest in Germany, followed by those in France and Belgium (Strikwerda, 1993, 1120)—social transfers, and housing policies. These approaches created a union conflict among the workers in the Zasavje region and elsewhere in Imperial Austria. The first migrants from Zasavje in the 1880s were also a consequence of strikes for higher wages, an eight-hour workday, and general improvement of workers' circumstances and rights. The movement for higher wages was repressed by force by the government, and the strikes ended with massive layoffs and departures of workers (Drnovšek, 2007a, 164). From the 1890s onward, wage differences increasingly impacted workers' moves toward the German mining regions and the regions with heavy industry, which made the Imperial Austrian mining sector suffer.

The recruitment system had an essential role in the strategy for attracting workers. Unlike the US companies, German mining companies were allowed to advertise in the press and use agents to actively scout skilled mining workers in Slovenian and Imperial Austrian mining regions. They organized transportation, sorted out the red tape and all the necessary things for their settlement in Germany, and often cared for their family members. The mobilization also took place in secret, and the measures that the Austrian government used to curtail the operations of the German scouts and stop the departures were unsuccessful. In Zasavje and in the Styrian mining centers, the workforce shortage became a problem that needed to be solved by bringing in workers from Serbia, Italy, and Poland (cf. Rozman, 1979, 211; Drnovšek, 2007a, 164; Wörsdörfer, 2017, 76–85). Rolf Wörsdörfer believes that these recruitment strategies and immigration policies contributed most to maintaining mining-related links to Germany, even when Slovenian emigration was massively geared toward the United States. The latter was Germany's main competitor in the battle for the workforce. Like the United States, Germany, France, Belgium, and Luxemburg included an international workforce in their mining and heavy industry. Many mining and industrial multinational companies preferred to employ foreigners of different nationalities over locals to have more power over them; local workers preferred to seek employment in manufacturing and other sectors. On the other hand, the demand for a workforce strengthened workers' negotiating power. Employers preferred to concede to improving working and living conditions for workers over widening the union rift (Strikwerda, 1993, 1120).

Migration to Germany, like the one to the United States, appeared in different patterns. It partly had a permanent character and contributed to the growth of immigrant colonies. Partly, migrants left these colonies and returned several times to work for a shorter or longer period. Similar was true for those working in the German-speaking parts of Styria. Because the primary goal of migration was to enter the labor market, a high rate of returnees was a constant among the migrants to the United States and Germany. Temporary migration was most widespread among those who were tied to agrarian estates and thus divided their efforts between subsistence farming and paid work. Professional miners and others dependent mostly on wage work left for extended periods and were inclined to settle in Germany. Mine workers were also very mobile in Germany, mainly those who were single and found it easier to move in search of better wages and working conditions. This migration was, due to its working-class nature, overwhelmingly male.

The mining centers in Westphalia developed a broad economy of services catering to the mining population, which created additional possibilities for migration and provided opportunities for women. In large numbers, the wives and daughters of Slovenian miners worked as domestic servants for management and administrative employees of the mining and other companies. A particular economic segment that was the domain of women was, like in the United States, renting out rooms, cooking, laundry, and caring for workers who lived in hostels or as subtenants. Employment opportunities for multiple family members and the accumulation of resources of family income reinforced the circumstances and reasons for permanent residence, with mixed marriages with immigrants from other nationalities being a crucial factor (Wörsdörfer, 2017, 97). Like in the United States, the Slovenian migrants in Germany also dabbled in other activities. They moved into industry and combined mining with employment in other sectors, including agriculture. They also worked in trades, but more than 90 percent of active immigrants from the first generation remained employed in mining, which gave the migration to Germany a specific character (Wörsdörfer, 2017, 103–111).

Picture 1.4: The Saint Barbara Society of Slovenian miners in Westphalia. Author: unknown. Courtesy of: Research Centre of the Slovenian Academy of Sciences and Arts, Slovenian Migration Institute.

Migration Transportation Industry

The revolution in transportation was one of the cornerstones of modern mass mobility. Expanding railway and maritime networks and routes made overcoming distance and transporting goods and people significantly simpler. Transporting migrants was a crucial element of the transatlantic migration system that, with all its bureaucratic and organizational logistics, became a true "migration industry" in the 1880s. It mobilized and shipped the European proletariat to the US labor market and was thus a key element in the Atlantic economy. At the beginning of the 20th century, it became the greatest business in the North Atlantic and Mediterranean shipping industry (Keeling, 2012; Feys, 2013; Light, 2013). When mass migrations started from the Slovenian ethnic territory, transatlantic crossings were already in this development phase.

Once ocean liners became more powerful and technically advanced, and more companies maintained regular transatlantic lines that specialized in passenger traffic, the voyage could be made in a week: it became cheaper, safer, and—with time—more comfortable. It was also lucrative: high risks and occasional losses due to the fluctuating numbers of passengers that were a result of labor market crises notwithstanding, scheduled lines for passengers, 90 percent of whom were migrants, provided half of the income to the shipping companies (Keeling, 2012, 44). The transportation industry was a link in the chain of economic and national interests. It supported investments in the technological development of shipping, port, and other transportation infrastructure and, consequently, national economies and the international economic influence of countries.

Like elsewhere in Austria-Hungary, the shipping companies kept their agencies in the main Slovenian cities, offering information, selling tickets, and assisting migrants on their transatlantic voyages. Mass transatlantic migration shifted toward the east and southeast of the European continent, the Slovenian territory became also a transition migration area. Carniola's capital, Ljubljana, was a node of international railway traffic, making it a hub of activities connected to the "emigration business" and a junction of migrations from the wider international space. Together with Vienna, Chernivtsi in Bukovina, Buchs in Switzerland, and many other stops along the way between the migrants' places of origin and the ports, Ljubljana was one of the links in the organizational apparatus on which the "migration industry" depended. With more migrants, in addition to the regularly scheduled trains, extra train lines from Ljubljana to different destinations were introduced.

The offices of shipping companies were located on the aptly named Kolodvorska ulica, the street leading to the Ljubljana central train station (*kolodvor*). Hostels, travel inns, and hotels near the train station made good money off migrants in transit. The Ljubljana offices were subagencies of big European ones. They represented Europe's largest shipping corporations, from Bremen's Norddeutscher Lloyd, which was the first one present on the Slovenian market (since 1878), to Hamburg's Hamburg–America Line, the French Compagnie Générale Transatlantique, Belgian Red Star Line, Dutch Holland–America Line, British White Star Line, and the domestic Austro-Americana of Imperial Austria (Drnovšek, 2002b, 367).

Most Slovenian migrants sailed from northern European ports, particularly those in the German Empire and France. In the early days of the great transatlantic migration, migrants from Austria-Hungary had no opportunity to set off from their national ports. For a long time, neither Trieste nor Rijeka had regular shipping lines for migrants and were not a part of the big business of the migration industry. They lacked entrepreneurial initiative, financial resources, and political support. In Imperial Austria, the *Auswanderungsfrage*, or the "emigration question," divided politicians and representatives of the state's different nationalities as well as the state apparatus and its interests. Because of these antagonisms and contradicting opinions about the advantages and disadvantages of migration, Imperial Austria, until its end, had no law systematically regulating the migration question and favoring the development of a domestic migration port (Chmelar, 1974, 140–143; Bader-Zaar, 2003, 149). Private Austrian shipping companies were too weak to enter the business independently and without broader support. The Austro-Americana, for example, which to solidify its freight lines to North America, also considered passenger traffic, sought a financial partner in vain. The problems were also geographical: while the northern Adriatic was well connected via rail with the regions in the east and southeast of Europe that were the source of transatlantic migrations, the sea route to North America was significantly longer from there and the Eastern Mediterranean than it was from the ports in the north of Europe. This distance prevented investment into transatlantic passenger lines for migration traffic.

The northern European shipping companies presented the second serious obstacle. Organized in a cartel, they monopolized traffic from Eastern and Southeastern Europe and secured most of the enterprise for themselves and their home ports. North European ports and shipping corporations were well organized, equipped with advanced infrastructure, and supported by their countries. The organization was perfected in Germany, which in the 1890s developed a police and health security control system that registered passengers in transit to German ports already on the borders with Russia and Austria (Sielemann, 1994). Germany protected its shipping companies by limiting the operations of foreign companies on its territory and harassing those migrants who were in transit toward ports outside Germany. Italy also tried to secure the migration business in its territory to Italian shippers, for example, by banning foreign ships from operating out of Italian ports for not complying with construction requirements. In Imperial Austria, the organization was worse and the shipping companies were mostly left to fend for themselves. For this reason, tens of thousands of migrants from Eastern Europe—and Austria itself—rode trains past the national port of Trieste to Italian emigration ports (cf. Kalc, 1992, 2013c, 2014b).

In 1903, Trieste and Rijeka finally entered the game—as rivals in the competition between the British company Cunard Line and the North Atlantic shipping cartel. Because of the supremacy of the German partners and too low transport quotas, the British company stopped collaborating with the cartel and became its competitor. In agreement with the Hungarian government and in collaboration with the Adria shipping company, it set up its own agency for migrants as soon as the Hungarian half of the monarchy passed its emigration law in 1903.

Picture 1.5: Waiting to depart to the United States in front of the Pension Austro-Americana in Trieste/Trst. Author: unknown. Courtesy of: Narodna in Študijska Knjižnica – Biblioteca Nazionale Slovena e degli Studi Trieste (Slovene National and Study Library in Trieste).

In response, the cartel, presided over by German giants from Bremen and Hamburg, opened a competitive line in Trieste, operated by the Triestine shipping company Austro-Americana. The company was recapitalized by German partners and, with the explicit purpose of acting as a competitor to the British in the Mediterranean, was accepted as the cartel's outside partner (Kalc, 1992, 481–482, 2013c, 134–135; Chmelar, 1974, 77–84).

The emigration from the Slovenian ethnic territory (and the monarchy as a whole) continued mostly via the established routes and north European ports.[10] Along with the cartel agreements and longer routes, the reasons were the lack of legal instruments and state incentives to redirect the routes toward Trieste and the shortage of lodging and other infrastructure that could accommodate higher numbers of

---

10 Less than 7 percent of migrants from Austrian Lands emigrated via Trieste between 1903 and 1914, while 25 percent of the migrants from the Hungarian half of the monarchy emigrated via Rijeka (Kalc, 1992, 482).

migrants.[11] The acceleration of migration traffic through a national port—a fixture of parliamentary debates on migration and maritime-economic policies and its activity to formulate migration legislation—was slowed down in Trieste by the very contracts between the emigration shipping cartel and the Triestine Austro-Americana. Despite the price reductions that the railways gave to Austrian citizens on their way to Trieste, the company was not interested in attracting a larger number of Austrian migrants and fighting for a greater share of traffic from Eastern Europe for its home port. It filled its ships with migrants in other Mediterranean ports, where it did not depend on the quotas from the cartel agreement. Austro-Americana was the only company allowed to operate in every municipality in Imperial Austria, while other companies could only have bureaus in the crown lands' capitals. Its agencies, however, preferred to work for the benefit of the company's German cartel partners. In addition, Austria's economic and political relationships with Germany limited the country's policies of "nationalizing" transatlantic emigration, rerouting migrants through its national ports, and retaining the profits and indirect economic impacts of the "migration industry" at home. The traffic organization, particularly in Imperial Austria, was thus left to the system of agencies and local sub-agencies, which directed the migrants to the established channels. This system made it easier to process the migrants who left without the required documents and permissions and would otherwise be intercepted by the controls in Trieste. The cartel (particularly its German members), whose aim was to prevent the opening of ports for migrants in their countries of origin, eventually used a strategic capital investment to open the Habsburg "national emigration port" in Trieste. However, they only conceded a small share of migration traffic to Austro-Americana and thus continued to secure the bulk of the operation for their own ports. The position of Trieste as an emigration port was strengthened from 1908 forward thanks to regular routes to South America, which Austro-Americana set up outside the cartel system. The traffic increased further in 1913 when the routes to Canada were established.

11 Only Austro-Americana had an appropriately outfitted building to receive passengers, while the passengers on the Cunard Line and the Canadian Pacific Railway Company waited for their departure scattered in guesthouses and private rooms (Kalc, 2015b). The Austro-Americana, too, had to find additional rooms because its facilities had a limited capacity of 900 guests. They often accepted a lot more, which created all sorts of health and other problems. Only in 1913 did they modernize the building and increase its capacity to 3,000 guests (Bertuzzi and Fait, 2010, 81; Kalc, 2013c, 132). In Rijeka, the organization and circumstances for the migrants were better because, in 1908, the state built the Hotel Emigranti for this purpose which could host 2,500 guests and had high standards (Rotim Malvić, 2015).

Slovenian emigration via Trieste increased, but less than one-third of all Slovenian migrants left through the national port in the final years before World War I. Between 1903 and July 1914, nearly 22,000 migrants from the Crown Lands with a Slovenian population left the country via Trieste: 10,700 from the Austrian Littoral, 9,300 from Carniola, and 1,600 from Styria.[12] In total, there were 220,000 migrants, Austrian and international combined.[13] Seventy-three percent left for the United States, 22 percent for South America, and 5 percent for Canada. The return traffic of many returnees was slightly higher because the returnees from the United States preferred to use the national carrier to return to the ports closest to their end destinations. The return traffic share was higher because the Cunard Line also disembarked its return passengers in Trieste (cf. Kalc, 1992, 2015b; Dubrović, 2012).

Some Social, Economic, and Cultural Aspects

The Coexistence of Migration Patterns

From the 1890s, but particularly in the 20th century, the United States was a real magnet. The country attracted a growing number of migrants and represented an alternative to other previously established migration routes. Its attractiveness also impacted the regions of Prekmurje and Porabje, which, in the mid-19th century, jumped from the bottom of the list of regions affected by emigration to being the primary regions of US-bound migration. The reasons for these changes and deviation from the established regional migration scheme, based on seasonal agricultural labor, lie in the structural break that ended the previous demographic and socioeconomic development cycle. Despite the absence of noticeable shifts toward modernization in that cycle, the combination of rural economy and income from agrarian seasonal migration provided the small farm owners, rural tenants, and agrarian proletariat with a modest but relatively

12 While most migrants from Carniola and Styria were US-bound, one-third of those from the Austrian Littoral left for South America. The vast majority of them were Friulians from Gradisca.

13 Eighty-eight thousand were from the Austrian half of the monarchy (most of them from Galicia and Dalmatia), 36,000 from the Hungarian half (of them, 19,000 from Croatia), 8,000 from Bosnia and Herzegovina. Among migrants from other countries (89,000), the largest share were Russians (48,000), the rest coming mostly from the Balkan states and Italy. Among the migrants from Galicia and Russia, in particular, many were Jewish, so a committee to help Jewish migrants operated in the city from 1908 onward (Kalc, 2013c, 137–139).

sustainable existence. By maintaining a low standard of living and low levels of consumerism, the region had avoided the population drain and even experienced demographic growth until, at the end of the 19th century, the relationships between this model's factors changed.

The increased demographic pressure and the unrestrainable fragmentation of rural estates contributed to the shift. In the first decade of the 20th century, the number of landholdings with more than 5 hectares of land dropped from 5,600 to 2,600. In contrast, the number of small landowners increased from 1,700 to 4,900, which also meant a drastic drop in agricultural production per family or individual inhabitant. The number of those who made a living from farming decreased by 5,000. As more people began to depend on seasonal work as a source of income, this arrangement could no longer guarantee the system's sustainability. The solution was to move to the rural areas of neighboring regions, depart for industrial centers in Styria, or migrate to the Americas. Like elsewhere, many crossed the ocean temporarily to fix or improve their agricultural situation at home with the earnings. However, an increasing number of migrants remained on the other side of the Atlantic and created a new life there. One of the consequences of this permanency was an increased share of women migrants, who eventually comprised about 40 percent of those from Prekmurje. Small landowners and people of "substance" left for the United States as they possessed the necessary funds. However, cottagers and proletarians continued to do seasonal work, with a number of them moving permanently to other destinations within the empire (cf. Horvath, 1989, 182; Olas, 1990, 592–594; Kuzmič, 2001, 26–27). Two large migration systems thus coexisted in Prekmurje before World War I: one reproduced the old socioeconomic model, and the other set up a transatlantic extension of the economic and social Prekmurje economic structure and social fabric.

Benechia—the small part of the Slovenian ethnic territory in the Kingdom of Italy—saw a similar shift: at the beginning of the century, when the seasonal migration typical for the Friuli region was at its historic peak, migration to the United States took off but affected a very limited area, while in the majority of the region, it was very modest. The discrepancy between locations was so pronounced that it immediately attracted attention from labor experts and academics. In Benechia, too, the reason was the transition from the traditional to the new pattern of temporary migration, caused by, on the one hand, the widened gap between the needs and the usual sources of income, and on the other, the allure of the prosperous US labor market. Many overseas migrants from Benechia also settled in the United States and established permanent communities in different states. The transition to transatlantic migration changed the attitude toward Benechia's traditional rural socioeconomic system. Long-term migration and permanent stay in the United States allowed people to step out of it en masse, while transatlantic migration transformed into an alternative economic and social strategy for individuals and families. The United States did not simply become a new economic source; transatlantic migration caused a deviation from reproducing the traditional ecologically based socioeconomic model (Kalc, 2018a, 122–123). Šentviška planota above Idrija also witnessed a similar shift in the directions and patterns of migration. Although in the second half of the 19th century the local population was used to leaving for seasonal work, they started to leave for the Americas at the beginning of the 20th century (Horvat, 2014). Studies also find that in other areas characterized by short-term circular migrations, transatlantic migration appeared later as a variant of the established migration system and as a break with the established social and economic tradition (Steidl, 2007, 86).

Other regions also experienced the shift toward transatlantic migrations for structural and circumstantial reasons. A part of the transatlantic current that appeared as *stage migration*, or phase migration, originated from the regions where the migrants had first settled. Between 1898 and 1914, almost 1,600 Slovenian or Croatian nationals who moved to the United States were born in Germany, which shows that many Slovenian migrants and their families continued their journey to the United States or their children migrated there (Wörsdörfer, 2017, 48; Klemenčič and Maver, 2017, 171). Likewise, migrants to German-speaking Styria and other Austrian Crown Lands turned toward the United States. In Inner Carniola, where the population worked in forestry and in sawmills, the oscillating demand for timber, as well as its prices, did not guarantee a regular income. For example, one year, a poor harvest of citrus fruit in southern Italy caused a decrease in orders for wooden crates in Inner Carniola from southern Italy and an increase in the share of workers who substituted the loss of income at the sawmills by leaving for seasonal work (Valenčič, 1989, 23). Seasonal migration did not guarantee a stable or good enough income, so many set off for the United States.

Occupational and Professional Opportunities

At the peak of the transatlantic migration, America's attractiveness also reached the centers of urbanization and economic development, for example, Ljubljana. Ljubljana, of course, cannot represent the Slovenian ethnic territory as a whole because its urban character and the configuration of people migrating from it set it apart. With not even 1,000 documented emigrations to the United States between 1892 and 1913, the Ljubljana-city district contributed the lowest share of emigrants by far. However, these numbers reveal how migration changed in the years before World War I when the possibility of integrating into the American society for migrants also expanded because of the growth of "American Slovenia." Migration from Ljubljana was particular in that 20 percent of emigrants were workers, while 20 percent, primarily children and spouses, had no profession. The remaining 60 percent included a wide range of professions, from wheelwrights, blacksmiths, builders, painters, and carpenters, to cobblers, tailors, locksmiths, typesetters, bakers, and shop assistants. There was also no shortage of technicians, clerks, engine drivers, and teachers (Zgodovinski arhiv Ljubljana, Reg II/21 (V.); Reg 1/1004, fol. 308–312). While previously, the predominant group of migrants had been unskilled laborers, this group of skilled laborers and professionals grew steadily after 1903. This emigration was not only a response to the demands of the US labor market but also the consequence of "American Slovenia," which migrants no longer saw only as an ethnic entity but also as an attractive, structured socioeconomic and cultural reality with expanding overseas opportunities.

Promising opportunities, which triggered specific migration chains, also opened for certain specialized professions. Among them was the migration of straw-hat makers from the Upper Carniolan town of Domžale (Kamnik district), particularly women. At the end of the 1860s, Domžale became one of the centers for straw-hat making in Imperial Austria. Tyrolean merchants contributed to this by opening mechanized factories in Domžale and Mengeš, thus changing the centuries-old rural handicraft into an industrial activity. Thanks to this industry and immigration from places near and far,[14] the number of inhabitants in Domžale doubled. This increase happened alongside massive emigration from the underdeveloped Kamnik district (of which Domžale was part), primarily to other parts of Imperial Austria. The area's backwardness was one of the push factors. The modernized straw-hat-making industry was another, leading to a decline in home braiding of intermediate straw products that did not meet the factories' higher quality standards. Factories replaced the lesser quality local straw with imported straw; this switch caused a drop in income for local hatmakers who had previously profited from using the cheaper, lesser quality local straw (Valenčič, 1959, 177–179). In the 1880s, some skilled female straw-hat makers went to the United States with an entrepreneur from Domžale who

14 In 1910, Domžale and the surrounding villages counted 276 persons with the right of domicile in Tyrol among its inhabitants (Valenčič, 1958, 64–65).

opened a factory in New York and employed them as forewomen. At the same time, many spent the winter months engaged in seasonal work around Central Europe where they were employed as skilled workers, production managers, and employee trainers in hatmaking factories and workshops in large cities from Vienna to Prague and Budapest (cf. Roškar, 2014).

Picture 1.6: Straw-hat makers from Domžale in New York. Author: unknown. Courtesy of: Slamnikarski muzej, Domžale (Domžale Straw Hat Museum).

Nevertheless, the real "American epidemic" broke out in 1905 when scores of workers started leaving for the United States. Emigration was boosted by a strike in Domžale in 1907. Established personal connections functioned as a job agency for the US straw-hat-making labor market and supported this specialized professional migration. In 1908, 200 people from Domžale and the neighboring villages lived in the United States and, in 1923, there were 384, almost all of whom had immigrated in the years before World War I. They mostly worked for New York hat manufacturers but also in Cleveland and Chicago (Bernik, 1923, 246–252; Valenčič, 1958, 63–64, 1959). The second, lesser-known example of specialized professional migration to the United States were glassblowers from Zagorje ob Savi/Sagor, themselves descendants of the glassblowers who migrated to the Slovenian territory in the 1820s from Alsace to work in the glass factories in Zagorje. In the United States, the glassmakers settled in Virginia where they started a new, successful economic activity with their specialized professional knowledge.

Regional Realities and Particularities

Toward the end of the 19th century, the United States became an almost mythical destination for those Slovenians turning to emigration to escape economic distress or to provide for their future. Lower Carniola, where transatlantic migration was the most intense, best shows the emigration's characteristics, impact, and effects. There, poor transportation connections restricted the selling of produce to other Slovenian territories and even to bigger local towns, thus limiting the agrarian economy and stumping the region's natural potential. Roads crossing the Kolpa River into Croatia were also poor. The railway line to Bela Krajina, which this subregion had demanded since the 1860s, was built only in 1914. In the 1890s, a phylloxera outbreak caused a drop in income and new costs when vineyards had to be replanted. In 1882, the iron foundry in Gradac near Metlika stopped operating. In 1884, a newly opened brickyard provided some jobs but other possibilities for non-agrarian employment were scarce (Balkovec, 1994, 45–47). Large groups of young cottagers and smallholders alike set off on the transatlantic voyage because their prospects for mere economic survival, let alone starting a family, were slim in the rural environment. Even ostensibly better-off people sometimes decided to leave. In many cases, the designated heirs to property needed to earn American money to pay out their siblings and take over the farms; some needed the money to get out of debt or renovate and strengthen their estates. The interest rates on monetary loans in Lower Carniola surpassed the Slovenian average due to the stagnating economy and a small farm could not pay them off, so work migration was the most promising option to acquire the means for any of the purposes mentioned.

Migration manifested as a chain and pulled in several members from the same or extended family. Family relationships, which at home excluded family members or limited their ambitions in life because of the inheritance system and the need to rationalize the rural households, became a foundation of mutual help and solidarity once in the United States (cf. Ravnik, 1981). It was not just concrete economic causes and motivations that led people to America. The transatlantic journey simply entered the mindset; it became an everyday option of social planning for individuals and families and was supported by kin and social networks. In his work about rural households in the time of capitalism in Bela Krajina, Janko Marentič wrote: "The father had no qualms about what to do with the boys once they grow up: to America!" (Marentič, 1957, 109 cited in

Dražumerič, 1987, 173). A transatlantic ticket was a payout to a son for his share of the house. In *Carniola, Izvestja Muzejskega društva za Kranjsko* (Carniola, the Newsletter of the Carniolan Museum Society), Dr. Janko Lokar believed that in the years before World War I, people from Bela Krajina got so used to foreign climates that "some consider it their homeland even before they set off." Moreover, in the United States, they got unused to field work and used to good salaries and nutritious food, which their home soil could not provide. "Foreign glory" enticed them, and only the "bitter experience abroad" managed to "bring back some sons to their homeland" (Lokar, 1911, 3–4).

In the years before World War I, Bela Krajina (like the Kočevje region) had three-quarters of its strongest rural workforce and homestead owners across the ocean. Interviews conducted during fieldwork revealed that families saw migration to the United States as an option for those children deemed less fit to work on farms, hoping they could secure suitable work and living conditions (Dražumerič, 1987, 173, 180). Emigration demanded a reorganization of work and a shift in roles on farms and in households. While the men were overseas, the women managed farms, sometimes with the help of relatives, often also with the help of paid workers. Migration to the United States thus created opportunities for employment in local agricultural activities and encouraged immigration from elsewhere. The wives of emigrants took on roles and positions within the family and in social dynamics that used to be the domain of men. Rare were the migrants who sold their estates; most rented them out. All sorts of arrangements were made, all in line with family structures and their migration strategies. The funds their "human capital" in the United States sent back to the old place were crucial for Bela Krajina to survive. They helped maintain and modernize agriculture, overcome financial distress, pay off debt, and improve the standard of living. At least when it came to basic consumption, the families with members in the United States generally had a slightly higher standard of living than others (Dražumerič, 1987, 188). The effects of migration were noticeable in the renovated and newly built houses that the "Americans" constructed instead of wooden ones and the slightly increased spending on basic goods (Lokar, 1911, 6). The consequences of mass migrations were also seen in the many abandoned houses and estates because many never returned.

Migration patterns throughout Slovenia were complex and diverse. In some areas, a single type of migration and one geographical specialization dominated. In others, migrations of different lengths, distances, and destinations coexisted. Goriška was a typical example of the diversity of migrations as a response to the needs and opportunities in nearby and faraway labor markets. In his assessment of the migration patterns in Goriška, Jože (Josef) Srebernič wrote that the inhabitants of mountainous places far from towns and other industrial centers were used to leaving their homes at a very young age. They then "travel abroad like migratory birds every year, until old age prevents them from doing so" (1914, 6). He was describing the agrarian workers who worked in Carinthia and Upper Styria, the foresters who went to work in the eastern and Balkan states, and the masons from the Karst and lower Vipava Valley who earned their living in Austria, Germany, and Switzerland (Sedmak, 1994, 84).

Around the same time, the miners from Tolmin and Bovec left to work in Austrian and German pits and, from the 1890s onward, to Cave di Rabil/Rabelj/Raibl, changing their migration into weekly work mobility (Srebernič, 1914, 5; Sedmak, 1994, 94). Every summer, women from Goriška Brda migrated to Austrian cities and holiday resorts—from Merano/Meran to Bolzano/Bozen/Bulsan, Innsbruck and Salzburg, Semmering, Graz and the towns around Lake Wörth/Wörthersee/Vrbsko jezero—to sell fruit from Goriška. This migration was made possible by faster travel connections, expansion of the city markets, and the birth of mass tourism, all of which expanded the market range for fresh produce from Goriška. Fresh produce and dried fruit were delivered by rail (the Bohinj line), so a post and telegraph office opened in Goriška Brda (Kozanski, 2002, 9–10; Gomiršek, 2010, 350). From the 1860s onward, Goriška Brda, the Gorizia-environs, and the lower Vipava Valley were the center of aleksandrinstvo. People also constantly migrated—back-and-forth or permanently—to regional economic centers, particularly Trieste and Gorizia. After the departures to Argentina in the 1870s (Brecelj, 1997, 216–217), temporary and permanent migration to the United States started in the 1880s with a group of miners from Bovec who went to Montana (Sedmak, 1994, 92).

In the Course of Urbanization

The contextualization of migration patterns and the reshuffling of the population in the period researched cannot neglect the migration processes within the Slovenian ethnic territory. If we limit ourselves to such processes between the countryside and the developing urban and industrial centers, we must highlight the attractive role of regional centers like Ljubljana or Maribor. The population of Ljubljana increased from 32,000 to 42,000 between 1869 and 1910 by migration alone, as natural growth was negative. Similar is true for Maribor, where, over the same period, the population increased from 13,000 to 28,000. The most impressive case was Trieste, on the western edge of the Slovenian ethnic territory. In the 19th century, Trieste became one of the centers of urbanization in the Austrian and Eastern Mediterranean space. Like other cities and centers of rapid development, this main port of Imperial Austria, with its economic growth and the expansion of urban tissue, attracted and retained growing numbers of inhabitants from the immediate surroundings and wider international space. In 1875, 125,000 persons lived in the Trieste municipality. Thirty-five years later, in 1910, there were 230,000. In the last decade of the 19th century and the first decade of the 20th, the population increased by 50,000, and, in the respective decades, 79 and 72 percent of those were due to immigration. In 1890, the Triestine population born elsewhere stood at 40 percent; in 1910, it was 43 percent (Bufon, 2021, 181).

Trieste was the most frequent end destination for migrations from the Karst, the northern part of Istria, the Gorizia-environs rural district, and Postojna and Logatec. From the mid-19th century onward, Trieste absorbed almost three-quarters of the annual natural population growth of the Sežana district—which shared a border with the Trieste municipality. Excluding the administrative, commercial, and transportation hubs of Sežana and Divača, the population of this Karst Plateau district stagnated or barely increased due to migration to Trieste (Kalc, 2013a, 2018b, 174–176). Research into the socioeconomic transformations in the Koper countryside and the effects that the development of Trieste had on them shows migration from villages at higher altitudes and more remote rural locations to villages closer to the coastline where the agrarian economy for city consumers was more developed, as well as further migration to Trieste itself from the entire area (Titl, 1965, 111). In Trieste, newcomers from Istria were the highest-growing group in the decade before World War I. Studying family histories showed how the inhabitants of Istria often approached the big city gradually, almost in stages, which could last years, even generations. Depending on the typology and size of individual families, the property size, type of inheritance, marriage strategies, and other factors, some members remained in their home villages; others left for Trieste either directly or "via detours" (Ravnik, 1996, 244–246). The number of migrants from other parts of the monarchy also increased. Still, it was significantly lower than those coming from the immediate surroundings or nearby places. There were, however, many migrants from abroad: the 1910 census recorded 38,000 inhabitants born in other countries. The majority came from the neighboring Kingdom of Italy, mainly from Friuli and Veneto, but also from other regions of the Italian peninsula, particularly those along the Adriatic coast (Cecotti, 2010).

Migration to domestic cities differed from migration to the Americas, Germany, and other more remote destinations. Shorter distances allowed for a wider spectrum of relations between the places of origin and the city. In the case of Trieste, too, the migrants swelled the ranks of the diverse Triestine proletariat.

The district of San Giacomo/Sv. Jakob was the working-class quarter in the city, where the migrants from the surrounding Slovenian hinterland were most numerous. A similar working-class quarter developed in the industrial area below Sèrvola/Škedenj where, among other factories, the Krainische Industrie Gesellschaft (Kranjska industrijska družba, or the Carniolan Industrial Corporation) ironworks operated since the end of the century; the population grew from 2,600 to almost 5,400 between 1890 and 1910 due to immigration. While San Giacomo developed as part of the new urban district near the shipyard and other plants along the coast, the population in Sèrvola concentrated in the old village where the local population converted farmhouses and barns into workers' dwellings. In all Triestine working-class neighborhoods and suburbs, where urbanization clawed into the suburban villages, the immigrants from the Slovenian ethnic territory represented between 60 and 75 percent of the newcomers in the first decade of the 20th century (Bufon, 2021, 182).

We can also find Slovenian migrants in Trieste's elite central quarters. In some, the number of Slovenian-speaking inhabitants grew from 12 to 23 percent between 1890 and 1910 (Cattaruzza, 1979, 43). Some were domestic workers for local families, but most were the growing Slovenian middle class. The social and professional scope of migration to Trieste was relatively strongly profiled. The migration to the city was linked to the process of social mobility and social stratification of the Slovenian city dwellers. Trieste was traditionally a place of vocational training and training for some trades. People from all over the Slovenian ethnic territory arrived to become apprentices and assistants in commerce, trade, and other activities. Many immigrants and their descendants set up new enterprises in these fields; many relocated their enterprises from their places of origin and enriched them economically in the city. Some activities were Slovenian specialties, for example, bakeries. Large commercial and other enterprises moved here, too, as did numerous workers employed in the railway industry, postal clerks and clerks for private companies, public servants, intellectuals and educators, and others whose careers brought them to Trieste (Verginella, 2001). The private Ciril-Metodova šola (School of Saints Cyril and Methodius), which Slovenians opened in 1888, became one of the largest in the municipality and attracted teachers from the surrounding Slovenian territories. Slovenian teaching staff also worked in Trieste's Italian and German secondary schools (Rupel, 2018).

The rural parts of the Trieste municipality, where the Slovenian population was prevalent, attracted new settlers and agricultural laborers employed in gardens and vegetable farms that supplied the city's farmers' market. This economy was structurally dependent on the seasonal workforce from the closer Slovenian hinterland. On the one hand, the city was the destination for constant migration and integration into the urban socioeconomic reality. On the other, through temporary migration—from daily or weekly mobility to seasonal or back-and-forth moves—its immediate or slightly more distant surroundings used it as a source of income. Settling in the city did not necessarily mean the ultimate exit from the economic and social system, or the family circle, in the place of origin. In many cases, migrants lived somewhere "halfway." The masons from Bistrica came to build year after year, and girls from the Vipava Valley served for a couple of years to save up for marriage or a couple of months to contribute to the family budget. Men from the Karst, employed as porters or port drivers, moved weekly between the city and their villages. The agrarian workers from Istria came three times a year to do seasonal work on vegetable farms and vineyards surrounding the city.

Likewise, migrants who settled in the city permanently and worked in the port, industry, construction, and other sectors, often returned to their places of origin to help with farm work and other tasks (Titl, 1965, 107; Kalc, 2018b, 182–184). Quite a few members of middle and upper-middle classes also lived between Trieste and their towns of origin. For example, the Žagar family, who employed around 300 workers and clerks in their industrial and commercial wood company in Inner Carniola and Trieste, lived between their apartment and offices in Trieste and their estate with a large sawmill in their hometown of Rakek (Žagar, 2015). Artisans, merchants, publicans, workers, and other migrants employed apprentices and other workers or provided accommodation for their families or acquaintances from their places of origin. They formed stations along people's routes, from which they could enter the city's economic and social reality, thus anchoring them in the new environment. They entered the city labor market through some form of temporary migration and created social mobility in migrants' places of origin. In the villages around Trieste that were at a distance that allowed daily or other short-term labor migration, changes occurred in family structures, life habits, and relationships in the traditional agrarian social hierarchy (Verginella, 1990).

Immigration also caused radical changes in the traditional social order in the city itself. The immigrants from the Slovenian hinterland contributed crucially to the growth of the Slovenian population in Trieste, which in the 19th century and until World War I was nicknamed the biggest Slovenian city. At the last Austrian population census in 1910, 56,000 people, or almost 25 percent of the municipality of Trieste, listed Slovenian as their *Umgangssprache* (the language of everyday communication). By comparison, Ljubljana was largely Slovenian but only had 42,000 inhabitants that same year, of whom around 6,000 were

German speakers. The number of inhabitants in Trieste is just a partial indicator of the changes that the city's Slovenian population experienced in this period. Migration contributed greatly to the breadth of the Slovenian social structure in Trieste, which grew significantly not only among the working class but also among the middle and even upper-middle classes. Mass migrations and integration into the city's social and economic structure coincided with the Slovenian nation-building process from the mid-19th century onward. In these decades, the Slovenian presence in Trieste acquired the character of a stratified, multifaceted, and organized national community.

At the beginning of the 20th century, this community had over a hundred cultural, economic, social, and political associations and organizations. Among them were savings and lending societies, insurance companies, and an investment bank with shareholders from the wider Slavic territory of the monarchy. The Slovenians of Trieste had political representatives in the city council, the regional council, and the parliament in Vienna. This structure was mainly the domain of the liberal national political society. The working class had parallel benefit and educational organizations and found its place in the Jugoslovanska socialnodemokratska stranka (Yugoslav Social-Democratic Party, or JSDS). The rise of the Slovenian community was best symbolized by the Narodni dom (National Home). Built in 1904, the architecturally prominent palatial building in the new part of Trieste housed the Slovenian community's most elite enterprises and activities, including a theatre, a bank, a hotel, a gymnasium, a café and restaurant, and other cultural, economic, and political organizations. It was a testament to the Slovenian and other Slavic communities' contribution to the development of modern Trieste. All this questioned the Italian character of Trieste, which was the line that the dominant Italian liberal-national politics supported, referring to historical tradition. The newcomers from the Slovenian hinterland were no longer Italianized through their contact with the city's majority Italian culture as they had been in the past. Instead, they filled and diversified the Slovenian ranks, took part in organized life, and strengthened the self-confidence of a national community that demanded an equal position on the grounds of historical presence. The model of the dichotomous division of the municipal space into the Italian city and the Slovenian surroundings, including the identification of the city dweller with all things Italian and the Slovenian with peasantry, was interrupted (Verginella, 2008). This counter position swelled into a national confrontation, which intersected with class and ideological conflicts. In the confrontation with all things Italian, Trieste was one of the centers of the Slovenian nation-building process and, simultaneously, a field of Slovenian social stratification. Ivan Tavčar, a Slovenian writer and politician, wrote: "Ljubljana is the heart of Slovenia, and Trieste is its lungs!" (Cankar, 1976, 121).

Gender Asymmetries

When it came to gender, migration patterns differed significantly. Migration routes from the Slovenian territory directed abroad and into other parts of the monarchy were predominantly male. The significant surplus of men in international, particularly transatlantic, migration was not particular only to Slovenians. It was widespread and often even more emphasized in the studied period throughout Eastern and Southeastern Europe (Donato and Gabaccia, 2015). Inequalities in the relations between genders depended on the various positions and roles of men and women in the family and society, on the purpose and forms of migration, and on the labor market that offered men and women different possibilities for employment, favoring some over others, or directed them toward different destinations and migration patterns. The notion and status of male and female mobility were also differentiated, allowing less freedom to women and subjecting them to various systems of protection and control.

Emigration from the Slovenian ethnic territory to the United States was particular in its clear imbalance between the number of men and women. While in the total migration numbers from Imperial Austria between 1876 and 1910 the share of women was 35 percent, there was only 18 percent among the Slovenian and Croatian immigrants from the end of the 19th century until World War I (Colakovic, 1973, 42; Drnovšek, 1999b, 424). These percentages were aligned with the industrial sectors that employed Slovenian and most other Slavic immigrants and their "migration projects" or migration patterns. The imbalance resulted from the large numbers of temporary immigrants or the generally temporary nature of migrations. Female migrants to the United States seldom sought independent and direct paths to the US labor market; most of them accompanied the men or joined the previous cohorts in more or less settled communities. Thus, wives accompanied husbands or often followed them, fiancées reunited with their partners, and daughters joined fathers, brothers, relatives, or acquaintances. Of course, some set off to America by themselves, often at the invitation of the employers, such as Slovenian publicans or, as they were called, *salunarji* (Mädchenschutz. Mädchenhandel von Krain nach Nordamerika, 1911).

Female emigrants could enter the United States only if accompanied or with clear proof that they were joining a trustworthy person. If they arrived alone, they were subject to special checks, which, following the law from 1891, checked their "honesty" and ability to sustain themselves (Steidl, 2017; Kulihova, 2019).

Female migration also depended on the fact that many mining districts and centers of heavy industry offered few possibilities for women's employment or demanded adjusting to harsh living and working conditions (Steidl, 2009, 216). The ratio between the genders changed according to the period and the form of migration. The low numbers of women were, in the cases of individual labor migration, typical for the time when migrants were not yet economically strong, lived in precarious work conditions, and were not yet clear on whether or not they would stay in the United States (Colakovic, 1973, 41–42). Female migration was curtailed by men's work and social position: employment in the lower segments of the labor market, low salaries, and slow social promotion and stabilization.

The circumstances for female migration improved with the establishment of larger and socially better-structured communities. The research on transatlantic migrants from the Dual Monarchy shows that the Czech- and German-speaking communities—which experienced migration earlier than other nations in the monarchy and had a more articulated social and professional structure—included over 40 percent of women in 1910, while the number of women among Hungarian migrants in the years before World War I even surpassed the number of men (Puskás, 1982, 40; Steidl, 2009, 215). Slovenian migration was in the phase where the number of women had only started to grow: by 1907, they represented 14 percent of the Slovenian and Croatian immigrants, and in the following years, until World War I, 25 percent (Willcox and Ferenczi, 1929, 432–443). The data about migration through the port of Trieste confirm this trend as the share of women in the years before the war increased to 30 percent. The share of women was higher among the Gottscheer population; as the male emigration to the United States began earlier in that region than elsewhere, so did the permanent settling in the receiving country. By World War I, one-third of this German-speaking population from the Kočevje district had already moved to the United States. Among them were over a thousand young women who worked for families in New York, Cleveland, Chicago, and other big cities (Die Auswanderung aus Gottschee nach Amerika, 1913). Migrants from Prekmurje also saw an above-average number of women (Olas, 1957, 183). And, as we might expect, in the straw-hat-making community from Carniola's Domžale, women surpassed men among the area's migrants to the United States (Drnovšek, 1999b, 424).

The gender differences in the destinations and scope of migration expose the social networks that operated in the background and channeled migrants according to professional profiles or activities. Male networks reproduced the movements mostly following the male labor market, while female networks maintained the channels of access to employment opportunities for women (cf.

Ramella, 2009). This method also retained access to the labor market to which Slovenian migrants, male and female, entered predominantly, if not exclusively, through the door of unskilled work or prevalent activities and "specializations."

Among those migrants who stayed in the United States, the gender ratio was less imbalanced because female migrants did not return in such numbers as men and their arrival was often linked to settling. The data from Benechia can shed some light on these dynamics and the role of women as a "stabilizing" factor in the migration plan of families and the wider process of migration. According to the ships' passenger lists, 90 percent of the male migrants from the valleys of Benechia arrived on US soil between the turn of the century and the beginning of World War I. Only 40 percent of all registered female migrants crossed the Atlantic in the same time span, and 60 percent arrived in the early 1920s. More than half of adult women arrived with children to reunite with their husbands, while single ones followed their fiancés and relatives. In the latter group, most arrived after the war and almost all declared to the US immigration officers that they were arriving to stay (Kalc, 2018a, 114–115). Similar dynamics and patterns of transatlantic moves were found in the studies on migration from Bela Krajina (Dražumerič, 1987, 180). The migration plans of Slovenian female migrants who went to the United States differed from their male counterparts. According to the prevalent tendencies of the time, a transatlantic migration meant a permanent change of domicile for the majority (Steidl et al., 2017, 220).

Female migration to the German Empire—and 30 percent of the Slovenian migrants to Germany were women—was also the consequence of male migration. Few of them went there independently, without joining their families or acquaintances working in the mining districts (Wörsdörfer, 2017, 97–99). Men also prevailed in the industrial centers of Upper Styria, while there were more women in Middle Styria, Lower Austria, and the German-speaking parts of Carinthia (Valenčič, 1990a, 52, 60). The ratio in South America was more balanced because the migration route included more families. Seasonal migration followed the division of economic roles and activities that the migrants conducted and, again, was almost exclusively male in certain places. Almost all peddlers were male. The only exceptions were women who sold fruit door to door and during times when men were called to other duties, such as war. In Benechia and Resia, the economic system was based on the division of roles by gender, with women traditionally caring for the farms and men bringing in "outside" sources of household income through seasonal migration (cf. Cosattini, 1903).

The same division of roles by gender was widespread among the Gottscheer peddling families in Kočevje (Gruden, 1992, 254). As studies of family strategies tell us, this distribution was not a result of a migration "passivity" among women as young women often went to work in cities before getting married. Female "sedentariness" and managing the home sources were a prerequisite for male mobility and the family's complementary "outside" income. The seasonal agrarian labor market kept opening for women, for example, in Prekmurje. Employers profited from the competition between the female and male workforce because women were better at performing specific tasks and received less pay than men for the same work. The same thing occurred in some non-agrarian activities, such as brickmaking. Some women, and even children,[15] occasionally appeared among the seasonal workers from Benechia who worked in brickyards; out of necessity, they conformed to extremely harsh and exploitative working conditions (Cosattini, 1903, 58–59; Picotti, 1909).

Female presence in the paid labor market increased in the decade before World War I. In 1910, women in Slovenia represented 30 percent of those employed in industry and trade (Fischer, 1984, 12). Such employment was also connected to migration. The first textile factories, established in the first half of the 19th century, employed a predominantly female local and immigrant workforce (Kresal, 1976, 30). The tobacco factory Tobačna Ljubljana, the city's largest industrial plant, employed over 1,000 women in 1900, of whom only one-third were locals. The rest came from nearby places, some from further afar, some daily migrants, and some moved to the city and married there. Working in a factory and moving out allowed many to experience personal economic emancipation, which was also visible in other personal and social forms of emancipation. Other women combined factory work with the agrarian economy at home (Kremenšek, 1976, 96–97). Female straw-hat makers were different in that their profession took them abroad. They also gained a reputation as the breadwinners of their families and representatives of skilled migrants, all of which was also—as a local priest from Domžale emphasized—a point of pride for their hometown (Bernik, 1923, 232). They were publicly acknowledged because they were in high demand and did not work for low wages or as unskilled workers (Milharčič Hladnik, 2015c, 47–48). A similar case, albeit less numerous and publicly resonant, was that of the lacemakers who worked in the canton of St. Gallen before and during World War II or took employment as lacemaking teachers (Svoljšak, 2004, 157; Terpin, 2008).

Like in the decades before, women who migrated individually mostly migrated to work in domestic service. This market most directly helped co-create

15 Children, aged 12 and over, commonly worked among the seasonal agrarian workers from Prekmurje (Olas, 1957, 181).

the geographical configuration and scope of female labor migrations. The labor market for domestic servants employed them in regional and other big centers. Therefore, the share of women among Slovenian migrants was higher in Vienna and other Austrian cities, and at times, as in the case of migrants from Carinthia, it was prevalent. Slovenian maids, cooks, chambermaids, and other domestic servants were common in Zagreb, Italian cities, and Belgrade. Still, the main labor market for them was Trieste and Rijeka. Most women migrated to Trieste with the intent of finding employment. They were prevalent among the migrants from Goriška, particularly those from Carniola and Carinthia.[16]

Working for urban families was left to rural women, who made up 90 percent of those working as domestic servants in Trieste. Local women and migrants who lived independently avoided it because it presupposed living with the employer in a subordinate position that limited personal choices. They preferred to compete for other professions and employment opportunities. On the contrary, for country girls, a live-in employment situation—a prerequisite and pattern of cityward moves—provided access to the city labor market. It also satisfied the conventional rules regarding safety and morals, which considered particularly unmarried women without a chaperone as endangered and needing special protection from the perils of the new environments. Among the charities supporting migrants in Trieste, the Slovenian Zavod Sv. Nikolaja (St. Nicholas's Institute) was established in 1898 based on self-help and the support of respectable members of the Slovenian middle class. The association provided migrants shelter and employment in moral and nationally conscious Slovenian families. This institution was the first of its kind in the Slovenian territory. Its initiators belonged to the progressive circle of women who also published the first Slovenian women's magazine, *Slovenka* (Slovenian Woman), in Trieste (Verginella, 2011, 259, 2017).

Most female migrants were single and a few years in service were intended to improve the economy of their original family while they gained valuable work experience for entering married life: domestic service was considered a premarriage work phase that allowed young women to make some money and gain housekeeping experience. It could be followed by returning home, but it was frequently the initial step in transferring from the countryside to an urban environment, which was also a large marriage market. In many cases, marriage in an

16 In 1910, 53 percent of in-migrants from Goriška, 60 percent from Carniola, and as many as 66 percent from Carinthia were women. A pronounced surplus of women from these lands was clear from the 18th century onward. It was connected to the specialization of employment for domestic servants (Kalc, 2008, 2014a, 15).

urban environment usually meant settling down (Kalc, 2014a, 19–20). Daughters were usually the first ones to leave their original families and used migration to shift from rural and agrarian to city economic and social environments (Ravnik, 1996, 144–145).

For many young women from Goriška, taking an extra step and going beyond Gorizia and Trieste to work in Egypt was just another short or long phase on the path to starting one's own family. Nevertheless, Egypt-bound female migration had a broader function because it became a structural economic factor for women's families and entire villages. Many wives and mothers took part in it also, and their work in Egypt played a crucial role in providing for the existence and development of their families. Stemming from the traditional female migration to Trieste and Gorizia, it developed into a cross-Mediterranean migration system that operated until World War II and partly until the postwar decolonization. It was based, like the migration to the United States, on the differential value of female work among the elite Egyptians and the opportunities for domestic service. In Alexandria or Cairo, the maids, cooks, and chambermaids working for wealthy European entrepreneurial families earned incomparably more than in Gorizia and Trieste, where many young women worked for less prosperous middle-class households, and the favorable exchange rate made their earnings go even further. The most highly paid were wet nurses who, together with nannies, were a special category of this female migration. Research shows they were valued because they established a good rapport with their charges (Koprivec, 2013).

Aleksandrinstvo was a part of the broader commercial and labor network crossing the Mediterranean. Colonial Egypt—at the time only one border away—was relatively well interconnected with Trieste in economic and maritime terms. Regular transport lines, economic exchange, and the position of Trieste and the wider littoral space in the traffic and trade networks of the Eastern Mediterranean made Egypt a part of the economic logic for families from Goriška (Kalc, 2015a, 69). Regular ship lines of Austrian Lloyd provided a safe voyage supervised by the "home" crews in the familiar Adriatic or Mediterranean mixed social environment, and, from the beginning of the 20th century onward, the volunteers of the Österreichische Liga zur Bekämpfung des Mädchenhandels (Austrian League for the Suppression of the Traffic in Young Women). From 1908 onward, the School Sisters of St. Francis of Christ the King who worked in the Asylum of Franz Josef in Alexandria also supported them. Notable members of the Austro-Hungarian community established the asylum to provide shelter for the migrants, care for them, assist them in finding work and act as a meeting point and a center of social life. Migrants also had access to the Austrian Prince

Picture 1.7: Brochure for the Austrian Lloyd Postal Route Trieste–Alexandria. Author: unknown. Courtesy of: Muzej novejše in sodobne zgodovine Slovenije (Museum of Contemporary History of Slovenia).

Rudolf Hospital and to free treatments at a private doctor's office by the Slovenian physician Karol Pečnik (cf. Però and Vascotto, 2011).

This migration, too, entered the Slovenian migration map as an option intended to fulfill all kinds of family and individual needs, plans, and aspirations, from solving material distress to earning funds to fix houses and farm buildings, from purchasing land and improving the standard of living to educating children and attempting to advance socially. Aleksandrinstvo was a family economic strategy based on the value of paid female work and the position that the aleksandrinke acquired for themselves on the Egyptian labor market of care providers. It presupposed a specific reorganization of family life, not only its economic aspects but also emotional roles and relationships. Working in Egypt could be a singular—short or long—work episode or it could repeat and meet new goals each time. At times it extended into a long-term absence of wives and mothers from home and resulted in a "transnational" form of family life. Sometimes, the children and husbands of these migrants followed them to Egypt; some aleksandrinke started families in Egypt with other migrants or Egyptians and settled there, and some remained unmarried and remained in service into their old age (cf. Koprivec, 2013).

Aleksandrinstvo, a frequent practice that ran through generations, was a tremendous emotional burden, particularly for married women and mothers. However, it was established as a proper professional and migrational specialization and a form of female participation in a gainful economy, which at the same time also meant initiation into working life and the broadening of horizons beyond the village and regional borders into the cosmopolitan world. "I could barely wait to become an aleksandrinka," is how a migrant described her growing up and expressed the youthful expectations of many of her peers who took the Egyptian path (Škrlj, 2009). Many girls who had to settle for lesser jobs or were unemployed and dependent on the help from the Franz Joseph Asylum also attest to the seductive call of Alexandria and Egypt (cf. Però and Vascotto, 2011). Aleksandrinstvo deviated from the predominant female and male migration scheme in numbers and geography. Between 1907 and 1913, female migrants made up 70 percent of all migrants from the Gorizia-environs district abroad. While most male labor migration remained within the Austrian state borders, more than 72 percent of female migrants abroad were aleksandrinke (Srebernič, 1914, 12; Kalc, 2015a, 60–61).

The Manifestations of Gender Asymmetry

The differences in the extent of male and female migration are reflected in the uneven gender structure of the population of the places of origin. Carniola was the crown land with the highest surplus of women in Imperial Austria—1,102 women per 1,000 men among born residents. In certain parts, the balance was even worse: in 1900, the Kočevje district recorded 1,219 women per 1,000 men because of migration, and in Bela Krajina, 1,308 women per 1,000 men (Österreichische Statistik, 1903). Emigration also influenced the changing structure of employment, specifically women's employment. In the decades before World War I, Austrian professional statistics place Carniola among the lands with the highest percentage of women in gainful employment. This phenomenon was most pronounced in the districts with the highest emigration rates; in Črnomelj three-quarters of women were registered as gainfully employed while in Carniola the average was about half. With the growth of migration, the share of women officially registered as employed on farms also grew because they replaced the men who migrated (Žnidaršič, 2015, 122–124). Male migration opened the social field for women, who entered it to take over the male roles. Applications for passports to travel abroad testify to the emphasis on the importance of family and wider social alliances. Such applications often explicitly specified how the applicant's wife would be able to manage their estate without her migrant husband, only with the help of their children and relatives (Kalc, 2021b).

The imbalance in numbers and different directions that men and women took in their migrations also created a gender imbalance in the receiving societies; the gender asymmetry of the marriage market became a problem in the search for a partner. Slovenian migrants, like migrants in general, sought to bond within the community and set it up on an ethnic basis. When facing a foreign environment and other immigrant groups, language and ethnic belonging and the reinforced (or newly discovered) national consciousness were the glue of the process of unification, aided by the already mentioned policies in the places of immigration and the interest of employers to keep the workers separated along ethnic lines. In the same vein, perhaps even more vital were other identification factors, such as belonging to a particular locale or region, cultural and social affinities, religion, life habits, and a system of social networks, with the help of which migrants left for new countries and upon which they still relied when there.[17] The migration patterns, entry into the labor market, and everyday needs led them to establish serried neighborhoods, where it was easier to maintain solidarity and organize formal and informal social life.

Marriage and starting new families within the Slovenian ethnic context were integral to the described social dynamics. In both principal countries of Slovenian immigration, the United States and Germany, there were often not enough opportunities for Slovenian young men to choose a partner from Slovenia.[18] So

---

17 Certain parts of the Slovenian community along St. Clair Avenue in Cleveland were, for example, named after the Slovenian regions or narrower areas from which the migrants living in Cleveland originally arrived. Among Slovenian migrants, particularly before World War I, regional identity was strongly felt and often came before national identity.

18 In Ely, on the border of Minnesota and Canada, for example, the ratio between permanent female and male residents of Slovenian origin was 100 to 165 in 1895, and around 1910, it was 100 to 200. The male majority among Serbian migrants was even more

single men, and sometimes widowers, used letters or personal connections to invite young women to join them, or they returned home to search for a life companion. The transnational and transatlantic marriage market thus became a constitutive aspect of mass migration, with migration and the marriage prospects of young women often being elements of the same strategic plan. Research shows that before World War I, Austro-Hungarian migrants to the United States largely married within a few years before or after coming to America, with women often doing so within a year after their arrival (Steidl et al., 2017, 230–232).

For Slovenian and most other migrants from Austria-Hungary, the choice of partners was generally endogamous. Exogamous marriages favored partners (female but also male) from communities similar in language, values, and traditions, such as Croatian and other, particularly Slavic, nationalities. These were definitely unions between members of the "new immigration" because marriages to American-born citizens or other European immigrants were extremely rare. Likewise, Slovenian miners in Germany also mostly married brides from other migrant communities, although marriages to Germans were more common here (Wörsdörfer, 2017, 102–106). The choice mostly depended on the available "offer," while religious adherence rather than ethnicity defined exogamous unions (Melik, 1956, 254; Nemanic, 2007, 47–50, 189; Steidl et al., 2017, 212–222). Gender asymmetry, or the availability of marriage-ready partners, differed from settlement to settlement, and with it, the number of ethnically mixed unions. In larger communities, and in time, as the number of women grew (also because of the coming of age of the second generation, born in emigration and balanced in gender), the possibilities for reproduction within the ethnic communities were higher, not only because of the more beneficial numeric ratio but also because of the ties in the community, its social life, and more opportunities for socialization and stabilization.

Marriage practices and the history of family among Slovenian migrants are entirely unresearched. Thus, we do not have a deeper insight into the said dynamics and behaviors linked to the imbalances in gender structure and generational changes. These behaviors may have been quite different, given the different social and economic situations and expectations to create families in different settlement areas and times. Some international studies show that mixed marriages in the Slovenian community, like in other Slavic nations of the Austro-Hungarian

---

pronounced as the community had eight men per every woman (Mackaman, 1998, 12, cited in Klemenčič and Šeruga, 2019; Nemanic, 2007, 189). In 1930, the Cleveland Slovenian community still had 20 percent more men than women (Klemenčič, 1995, 105).

Monarchy, were rare or a minority in the first generation, even after World War I (Steidl et al., 2017, 224). They started increasing in the generation(s) of their descendants. With the Americanization and adoption of the values and lifestyle of the "new homeland," the prejudice against mixing relaxed, and obstacles to stepping outside ethnic borders were reduced. A study on marriages in Croatian parishes in Cleveland between 1903 and 1914 shows that 90 percent of Croatian and 85 percent of Slovenian immigrants married partners of the same nationality. In the first generation of their descendants, three out of four marriages were mixed (Čizmić, 2002, 46–47).

After World War I, the interrupted arrival of new migrants from the old place because the United States closed its doors caused a shift in the norm. The research further shows that women deviated more from the prevalent ethnic marriage model than men and were more likely to enter into a marriage with a partner of a different nationality. Studies based on a small sample of data from the 1910 population census show that there were 1.9 percent mixed marriages among Slovenian men who got married in the United States, and among women, 5.2 percent; in 1930, these shares increased to 6.6 percent for men and 11.6 percent for women (Steidl et al., 2017, 224). A wide range of factors influenced the choice of a partner, in addition to different affinities, from emotional to socioeconomic and accidental. The studies of ethnic and social endogamy and exogamy find that the differences between male and female behavior patterns are linked to the possibility of women using marriage for social advancement—as a means of improving their social position.[19] For men, this heavily endogamous behavior was, at least in the first generation, influenced by the intention to return home (Steidl et al., 2017, 225).

The gender imbalance and the shortage of women were not the only problems of migrants who wanted to start a family. The core problem was creating immigrant communities and their structural "normalization." The imbalanced demographic, with a clear majority of men, came hand in hand with several negative phenomena, even more pronounced in isolated mining and industrial places.

19 This tendency is observable also in migrants from Kočevje where the research finds even more pronounced levels of endogamy, which was also linked to the status of the German community in the Kočevje region and its social and economic strategies. The marriage registers from the All Saints' Church in Brooklyn, which registered 817 Gottscheer marriages between 1886 and 1908, show that the Gottschee men married exclusively brides from the same area, while women occasionally married other partners (Hutter, 1999, 92–93 quoted in Moric, 2016, 134–135).

High levels of alcoholism, prostitution, crime, violence, and other deviant behavior were only some of them (Nemanic, 2007, 49; Klemenčič and Šeruga, 2019, 36–43). Female presence balanced and recalibrated communities: it helped stabilize and develop migrant settlements into structured social communities. The traditional and new social roles women took on created the circumstances for the growth and consolidation of the migrant community and organized social relationships within it. Although female presence ensured demographic reproduction and promotion of family life, women's roles were far from restricted to just family and economy; out of need and under the influence of the American principles and cultural models, it also reached the spheres of public and organized life (see the section Organizing and Connecting Women Migrants in the Service of Two Homelands: Marie Prisland). Women were often at the forefront of the fight for workers' and social rights; they played a vital role in union movements and solidarity organizations (cf. Stek, 2015; Penn, 2018).[20] Furthermore, just like men, they recognized themselves in different worldviews and values and were divided: like in the "old country," "American Slovenia" was divided into the conservative Christian, liberal progressive, and socialist camps.

Similar processes also took place in Germany and other countries of mass migration, both internationally and within the borders of Imperial Austria. We must read them considering the specific socioeconomic circumstances, legal frameworks, and cultural and political contexts: the marriage market and marriage strategies would make these processes most "legible," but they remain inaccessible to us as they take place on the micro levels of family and community, and thus far, no studies have been done on this subject. The homogamous and endogamous, or mixed, marriage relationships between members of different social classes or different ethnic or geographical origins are indicators of the segregation tendencies of a particular ethnic group or its openness and susceptibility to intertwine with other communities. They are also interpreted as indicators of the ways and levels of participation and integration into the environment into which they migrated (cf. Lorenzetti, 2000; Lorenzetti and Schumacher, 2005; Kalc, 2013b.)

Marriage dynamics were even more important in the Slovenian ethnic territory, as they took center stage when considering the "national question" and in the political measuring of national forces for territorial rule. In Trieste—a

20 Let us mention Ana Klobučar Klemenc, better known as "Big Annie of Calumet," a well-known personality in the big mining strike in the copper range of Michigan in 1913–1914 (Stanley, 1996; Comstock, 2013).

meeting point of the Slovenian (and Slavic) and Italian (and Romance) culture—the Slovenian community experienced the fiercest national confrontations; marriage acquired a clear national-political connotation and played a key role in reinforcing national positions for the Slovenian population (Verginella, 2006, 73–93). A mixed marriage between a Slovenian and an Italian was an accelerator of Italianization because the everyday family language in such unions automatically became Italian. The descendants, also under the influence of the Italian schools and urban environment, in which the Slovenian language—compared to the dominant Italian—was marked as culturally inferior, left the Slovenian national ranks. The Slovenian community thus stigmatized the mixed Slovenian-Italian unions and branded them as renegades. The circumstances were, of course, different than in the United Stated or elsewhere abroad. The Slovenian Trieste was at the edge of the extensive Slovenian hinterland, which was at the same time the original pool of the large part of immigration. The question of gender asymmetry was not so pronounced in Trieste; it was quite the opposite due to the female slant in migration. Mixed marriages of Slovenian migrants were among the factors that contributed to the loosening of the national roots and withdrawal from the Slovenian environment. This process, which outside the national-centered perspective could be read as a strategy for integrating into the city's social structure, continued in the decades before World War I. However, assimilation reduced significantly due to the increasingly more articulate social and economic development, national and political organization, and strengthening of the national consciousness in Slovenian Trieste.

Between the Need and the Benefit

From the 1850s onward, migrations became an increasingly important factor in the microeconomics of both families and the entire regions and countries of emigration. Mass inclusion in the internal and international labor markets decreased demographic pressure on the economic structures and resources, eased social burdens, and, at the same time, allowed migrants' "near" and "extended" homelands to receive the inflow of money and other goods. The economic effects of migration and migrants' remittances are hard to measure on the micro- and macroeconomic levels and belong to the less researched areas of the migration processes (Esteves and Khoudour-Castéras, 2009). The economic "profit" of migration depended on the destination, the labor market, and the pattern of migration. The biggest profit came from temporary or permanent migration to the United States. Seasonal migration, particularly that of agricultural laborers, brought in less income and was seen as the "poor migration." However, its economic worth was not marginal since it had to be considered within the systemic combination of sources.

According to the studies of remittances from Austro-Hungarian migrants, the extent of the annual influx of money from the United States in the early 1890s was more than 67 million kronen (13.5 million dollars). The economic crisis in 1893 reduced this influx by two-thirds, but it started rising sharply at the end of the century. Along with the growing number of migrants, the increased share of temporary migrations contributed to this decrease. In 1901, almost 120 million kronen of migrants' remittances arrived in Austria-Hungary. Ten years later, it was almost 627 million kronen (125 million dollars) (Morys, 2010, 50). The yield of seasonal migration to Germany, Denmark, and Sweden was estimated to be more than 6 million kronen in 1902 (Wanderbewegung und Zahlungsbilanz, 1912), while from 1908 onwards about 35 million kronen yearly (Bartsch, 1911, 154–155). Of course, the travel and other expenses generated by migration must be subtracted from the income from remittances. The active net worth was undoubtedly precious money for the state and helped offset the foreign trade deficits (Chmelar, 1974, 133–134). Between 1910 and 1914, migrants' remittances amounted to 20 percent of all foreign payments and surpassed the value of the external monetary influx that the treasury had from transportation and tourism. In 1880, US money orders were thought to contribute 0.8 percent and, in the years before World War I, between 2 to 3 percent of the gross domestic product of the Habsburg Monarchy. The funds came from abroad via different channels: as postal orders, via big banks, and primarily via small banking institutions that migrants had set up in the United States as well as other agencies or intermediaries, which was riskier because of frequent fraud and appropriations (Steidl et al., 2017, 282–286). A large share of funds came to the old place simply in cash or as personal checks in returnees' pockets (Chmelar, 1974, 134). Because of the diverse and informal channels that eschewed administrative control, the statistics about the influx of remittances are insufficient and very approximate.

Calculations based on data published by the Austrian financial ministry show that overseas migrants sent or brought to Carniola an average of 11.4 million kronen yearly in the period 1906–1910 and 14.4 million kronen in the period 1911–1913. Thus, subtracting the travel and other expenses, the net amount was 9.6 and 12.6 million kronen, respectively. The remittances to the territory of today's Slovenia in the last years before WWI were estimated at a gross amount of 22 million kronen per year or a net of about 19 million kronen (Kalc, 2023, 151–154). The influx continued after the war, when it peaked. Authors generally find that Slovenian and other "new immigrants" in the United States were characteristically frugal (Govorichin, 1961, 101). Such behavior was, after all, part of the strategy to optimize the individual or family "migration plan." Accumulating and sensible spending of savings can also be assumed because of the many Slovenian savings and credit societies, cooperatives, and insurance companies (see also the section The Organization and Scope of Activities: The United States). Migrants carved out relatively solid material and social positions in the United States, bought houses, built churches, national homes, and schools, and established and maintained parishes and different organizations. In addition to their waged work or other economic activities, many supplemented their income by investing in real estate, which they then either re-sold at a higher price or rented out (Kristan, 1927b, 866; Adamic, 1940, 152; Klemenčič and Šeruga, 2019, 94). In the mining region of Allegheny County, Pennsylvania, a mid-1930s socioeconomic questionnaire found that most bank deposits were those of immigrants and that many "Carniolans" had funds for building a house, opening shops, and other economic activities available (Klein et al. after Govorchin, 1961, 101, 290).

Even before World War I, the Dillingham Commission found that a quarter of Slovenian migrants in industrial cities owned their apartments or houses, which was the highest percentage after the migrants from northern Italy and significantly above the average for the "new immigrants" (Reports of the Immigration Commission, 1911e, 468–469). The circumstances improved further during the war and in the first half of the 1920s. They were also linked to the progression on the ladder of working qualifications and the move from the working class ranks to employment in the service industry or independent activities (cf. Bodnar, 1977, 73; Klemenčič, 1995). In the 1930s, 50 percent of Slovenian families in Cleveland were owner-occupiers (Grill, 1979). Many migrants who settled in the United States remained closely connected to the old place and their activities became a transatlantic extension of the home economy. The remittances, particularly from the United States, were not limited to the packages for the families but also materialized as savings and investment accounts in banking institutions. Material gain and prosperity of the "American Slovenia" were linked differently to the old place. When it came to the use of funds from abroad (and the concrete numerical data and analysis are missing on this, too), we can say that on the microeconomic level, they were generally used to pay off debt, purchase land, renovate or build houses, purchase farming and other equipment, maintain and improve the life of relatives, and increase consumption.

The proceeds from the growing economy of "American Slovenia," of which fraternal organizations, insurance, savings, and lending companies, and other business institutions were an important part, helped to renovate churches, construct communal buildings, and provide for other needs of their communities of origin, which habitually turned to migrants overseas for all kinds of material aid (Bernik, 1939, 3, 54, 82–88). However, this economy did not become a contributing factor to the changes and development of macroeconomic structures. There was a lack of actors and instruments in Slovenia that could mobilize such funds and steer them toward a more long-term economic project that would benefit the migrants' places of origin. The Slovenian space and its migration pool remained mostly economically marginalized and excluded from infrastructural and economic modernization processes. Slovenian politicians were well aware of this problem. Fran Šuklje, a Slovenian deputy in the Reichsrat (Parliament in Vienna, or the Imperial Council), stressed that the majority of remittances and savings remained "unproductive" or unused because they were mainly used for fixing or building residential houses and savings. At the same time, returnees' intent to "multiply" the money they earned in the United States through the economy mostly turned out to be illusory. Šuklje illustrated that with the case of Lower Carniola, which lacked infrastructure and other conditions and initiatives that would allow the region to take full advantage of its potential (Šuklje, 1905).

In his socioeconomic, national, and political articles and appearances on the migration question, theologian and representative of the Christian socialist movement Janez Evangelist Krek warned of the need to strengthen the role of the local savings associations and cooperatives so that the funds that migrants sent to their homeland would contribute to the development of the migrants' regions of origin, particularly via investment in the modernization of the agricultural economy (1913). Migrants' money—from the United States and elsewhere—mainly contributed to the persistence of the countryside's economic structure and the rural population's social dynamics. The fragmentation of property did not change significantly and remained the main obstacle to the shift from subsistence to market farming. Remittances further hindered this process; they often accelerated fragmentation and increased the price of land and the agrarian workforce. Those leaving for the United States to earn money wanted primarily to strengthen and improve the situation of the existing farming estate. Only a small proportion of the returnees transformed the money earned and experience gained into new or fundamentally changed economic initiatives. In the Prekmurje region, thanks to the seasonal migration, migrants introduced improvements into farming operations, which they learned while working on large estates. However, in general, the "migrant economy" in the place of origin complemented or replaced the economy and thus hindered more radical changes, particularly structural shifts, toward modernization.

Nonmaterial Incentives and "Ideals"

Economic and social dynamics functioned as direct and indirect factors of mass migration. Migrants set off into the world to fulfill material needs, make money, provide a livelihood, and ideally improve their material standard of living. To understand the impulses and the meaning of migration as a phenomenon and in terms of agency, we must not neglect the processes of general transformation that marked the social development and ways of life of this period in history. Along with the concrete material needs and pressures that emerged from the economic imbalances and social distress, individual and collective migration choices were born from the desire for change, emancipation, and the adoption of new lifestyles, which entered the imagination, mentality, and ambitions of young generations through modernization. Juro Adlešič, a lawyer and an observer of the migration phenomenon, saw the new way of thinking as a primary reason for migration, which, like a "secret force," pushed young people to cities and abroad, convinced that they would implement and fulfill their life plans more easily there (Dražumerič, 1987, 182). In 1913, the newspaper *Gorenjec* (Upper Carniolan) reported: "America muddles the brains of our men and boys terribly. There is enough work and bread here, but they prefer to go and seek it abroad. It's not just the yearning for income; it's a disease, akin to madness, pulling people away from their homeland" (quoted in Drnovšek, 1989, 208). Priest and politician Anton Korošec stated at a Catholic meet-up in 1906 that it was not hunger and desperation that prompted migrants to leave but their vigor and the desire to fulfill their potential in a way they could not at home (Lampe, 1907, 20).

Although the cultural and psychological reasons for emigration are less often studied than the economic and social ones, they are essential, particularly in implementing alternative visions of personal development and life choices. The United States attracted young people not only with its offer of work and material goods but also with the magnificence of its development; it was perceived as a country of permanent progress. Emigrant authors wrote about the traits of the American world that left a deep impression on them: relaxed interactions, the culture of familiarity in address, the position of women, rich food, the sense of freedom, as well as the living conditions. Migrants, particularly immediately after their arrival, may have often shared abysmal accommodation and experienced disappointment about the real America, which was quite unlike the images they had conjured about it. However, by World War I, most members of the Slovenian communities (in Cleveland and elsewhere) were apartment-dwellers, residing in buildings with modern conveniences not common even in Slovenian urban environments (Molek, 1934; Prisland, 1950, 1968; Grill, 1979, 18, 33).

Young people who met returnees observed their American clothes and habits, listened to their stories, and absorbed those images that became part of the collective imagery (Adamič, 1983, 23; Melik, 1956, 249). In a public school in Bela Krajina, complained deputy Šuklje, the pupils seemed to know more about the geography of the United States than that of Slovenia and the Dual Monarchy (Šuklje, 1905). They grew up expecting that they, too, would leave.

In the image of the United States, the poise of the "Americans" stood out. Regardless of the success of their American experience, for the returnees, this poise expressed a self-confidence that was not a trait of either simple farmers or workers in Slovenian society. It shook off the humbleness that was adhered to and expected of ordinary anonymous persons. In the United States, writes Trunk, migrants learned different standards of justice and freedom. America turned "yokels" and "maids" into "misters" and "misses" who walk around "with their head held high" (Trunk, 1912, 399). Many also got used to entrepreneurship, initiative, and other virtues that the modern world encouraged and rewarded, which were not part of their thinking at home (Trunk, 1912, 398–399). Above all, they all encountered a "New World" in the United States and "tasted" a more relaxed way of life, different habits, and freer worldviews. Emigration to America meant a practical and symbolic step from the pre-modern into the modern world.

Slovenian press often described returnees as restless because they could no longer fit into the routine at home and were drawn to cross the ocean again, even if their life there was not easy. The awareness that, in the United States, you worked not only to survive but to improve your standard of living became commonplace. Likewise, despite the difficulties, there was a notion that this goal was attainable for all because its path was not strewn with so many social, legal, and class obstacles as in the old place. The Social Democrat politician Anton Kristan wrote about his visit to the United States in 1927, among other things: "[...] I could be happy for the majority of our people. I found the kind of prosperity in their houses that is one of the conditions for happiness. [...] Mostly, their situation is such that they can sit on velvet chairs when traveling [...]" (1927b, 866).

Suggestive images created by association with migration can also be discussed in connection to the European cities and centers of intensive economic development and social transformations. In Europe and Slovenia, large cities and even provincial towns represented a social threshold many villagers wanted to cross to achieve progress, whether material or spiritual. Their destinations included fast-developing cities from Trieste, Vienna, and Graz to Zagreb, Rijeka, and Ljubljana. This migration, too, was a consequence of searching for economic sources and ideas about a different kind of life, better-paid jobs, social change, and escaping from the clutches of the traditional family and village environments and their limiting social rules. These tendencies, along with the economic attractiveness of the labor market, became more prominent in the 20th century (cf. Fassmann, 1994). In her work about the establishment of the Triestine proletariat, Marina Cattaruzza finds that, in the 19th century, migration to Trieste allegedly developed out of harsh social circumstances. While migration initially meant searching for an exit to satisfy life needs in alternative environments, the phenomenon later increasingly intertwined with the desire to move from the rural world to more liberal urban environments and participate in social and cultural modernization processes (Cattaruzza, 1979, 6–7). External incentives notwithstanding, migrations in this phase also became a self-reproducing mechanism.

Migration Regimes and Policies

The State and Migration

The great migrations we have been discussing coincided with the liberal attitude of the states and politics toward spatial mobility within the monarchy and internationally. Imperial Austria legalized such an approach in the 1867 *Dezemberverfassung* (December Constitution) that guaranteed its citizens, among other rights, the freedom of emigration. The path to the liberalization of spatial movement in Austria was long. In the era of enlightened absolutism, the Theresian-Josephine reforms, which encouraged marriages, social and professional differentiation, productivity, and population growth, gradually removed the obstacles to mobility within the crown lands and the state. The absolutist state applied cameralist and populist principles to conduct a plan to colonize its own territory and offered privileges to attract experts and skilled workers from abroad. At the same time, the expatriation of its subjects with the intention of permanent emigration was strictly prohibited (Hahn, 2011, 88; Kalc, 2016a, 113).

The Emigration Patent of 1832 somewhat relaxed the regulations from the repressive Emigration Patent of 1784 but continued to allow emigration only with the state's consent (*Emigrazionskonsens*). Such emigration was conditioned with the loss of citizenship and, following the absolutist notion of subjects as a country's wealth, the payment of an emigration fee that would offset the damage that the migrant caused to the state by leaving, and, in the case of illegal migration, the confiscation of property (Kalc, 1997, 13). In the 1850s, the regime was slightly mollified and allowed free movement within the empire's borders and into the German states with which Austria formed a free movement zone. In the 1860s, with reforms that eventually brought the shift from neo-absolutism to liberalism and the establishment of the constitutional monarchy, the principle of freedom of emigration was finally implemented into legislation (Komlosy, 2003, 160, 165; Becker, 2010, 40; Burger, 2000, 19–20). The freedom of emigration allowed the citizen to choose, at will, temporary or permanent residence and living environment anywhere in the country or abroad. This right was explicitly limited only by military duty but could also be conditioned by other duties that citizens had to fulfill in their relationship with others and society. A passport remained obligatory to travel abroad, even after citizens received the freedom of migration. However, other identification documents were sufficient for traveling abroad—for example, work papers (known as *Arbeitsbuch*)—as long as the competent political and administrative authorities confirmed their validity (F. K., 1907; cf. Burger, 2000, 19–20).

In the era of constitutionalism, the proponents of economic liberalism—who considered freedom and self-initiative to be cornerstones of development and progress—encouraged leniency when it came to the movement of people and that showed in the Austrian laissez-faire attitude toward migration. Unlike some countries that limited emigration by law and tried to regulate and direct migration according to economic and other interests, Austria developed neither positive migration legislation nor planned and active migration policies. By guaranteeing freedoms, the state allowed migration but did not encourage or favor it. It guaranteed its citizens an unobstructed realization of their fundamental rights as long as leaving the country did not endanger other citizens or national interests (Kalc, 1997, 13–14). Generally, the state apparatus limited its role to observing and supervising the legality of the process. Until the 1890s, this policing approach reflected the disinterest of politics and the state apparatus. However, once migration abroad became a mass phenomenon and consequently a question of national importance, this attitude became a reflection of the divergent views on the issue and related interests, as well as the technical and organizational—even systemic—inability to deal with migration. All attempts to legislate the "emigration question" hence failed. However, emigration was a fixture on the political agenda in Imperial Austria, just as it was in the Kingdom of Hungary and in other European countries.

Two opposing views formed. The shipping and commercial circles, backed by the Ministry of Commerce, emphasized emigration's economic benefits for migrants, their families, and the state treasury and its indirect benefits for the Austrian navy and international commercial interests. In line with the constitutional principles, they supported complete freedom of migration abroad, with the state's regulatory intervention only to protect migrants and provide them with organized travel and entry into the international labor market. Industrialists and big landowners, who were worried about the lack of workforce and the increase in the price of labor, supported a more restrictive policy. Their allies were the Ministry of Interior and the Ministry of War; the latter was losing conscripts to migration. This side advocated a more active role of the state in managing migrations, curtailing the freedom of emigration, keeping the phenomenon within the boundaries of the real economic needs of migrants, and establishing a stricter control system to suppress illegal migration at home and in foreign ports. Political parties were similarly divided, so the endeavors to legislate the question of emigration faced a demand for a compromise between the opposing views and solutions. Topics related to migration also filled newspaper pages, and legislation was a subject of lively political and expert debate of different points of view—administrative, economic, security, and police—of those wishing to protect migrants and other proposals.

The Parliament in Vienna debated emigration several times. Two extensive surveys were carried out in 1905 and 1912 among experts who either researched and worked on the practical aspect of migration issues in Austria or were otherwise acquainted with them. In many published papers, lawyers, economists, and other experts discussed the issue, its effects, and the state's role. Some proposed setting up a specialized state office that would track the phenomenon and solve all the aspects of transatlantic and continental migration for the benefit of migrants and the state itself.[21] However, the opinions on migration differed so greatly that it was impossible to achieve compromise legislation that would comprehensively regulate the issue and define the guidelines for migration policy. Any attempt to target certain groups and either limit or actively manage migration clashed with the constitutional principle of freedom of emigration. Like the legislation of other countries (particularly Italy and Germany) that proposed regulating migration's legal, economic, and social aspects and allowing the state to take an active role in regulating them, individual drafts of laws failed. The approach that let migrations be led by events and influences of the liberal economy factors prevailed. The state only interfered with occasional police measures, using them to enforce specific benefits or curtail the most detrimental misdemeanors and excesses (cf. Chmelar, 1974, 140–159; Kalc, 1996a, 2016; Brunnbauer, 2016, 152; Bader-Zaar, 2003, 148). The national question also carried much weight regarding the phenomenon of migration. In the multinational monarchy with

21 One of the core questions was also keeping reliable statistics on migration. Despite the elaborate plan for an innovative system of statistics, it remained unsolved (Englisch, 1911, 106–120; Kalc, 1996b, 59).

growing national tensions, migration was at the core of clashing national interests. These interests influenced power struggles over regulating migration by law and migration policies, as these could benefit or damage the position of individual national groups (Bader-Zaar, 2003, 149).

The Position of the Authorities and Illegal Emigration

The authorities, however, did have the possibility to influence migration. When issuing passports, they checked if the applicants fulfilled the conditions to enjoy the right to migrate or if there were any limitations to this right. In addition to military duty, the freedom of migration was conditioned by other elements and circumstances. The authorities could refuse to issue a passport to citizens who, in their opinion, were not able to sustain themselves abroad or whose dependents might become a burden to society. When assessing the applications, other factors were also at play, from the age, gender, and health status of the applicant, the type of migration (temporary or permanent), the purpose and destination of the move, possession of the necessary funds for departure, to the entry requirements in the destination country and other factors from which the citizen needed protection or which could compromise the purpose of migration. The authorities' position could differ from region to region and period to period, depending on the economic, social, and other circumstances. When assessing an individual's position, the authorities could decide more or less restrictively within the regulations. There were also conflicts among different political sectors and levels of administration. Generally, the central authorities focused on national interests (military, security, protecting citizens from danger and the state from the cost and international bureaucratic complications). Local authorities were more sensitive to the benefits for their economic and social municipal budgets (cf. Becker, 2010, 34; Kalc, 2021b, 42). There were also deviations from the legislation.

When, in the 1870s and 1880s, the first groups from the Austrian Littoral wished to depart to Argentina and Brazil, the authorities rejected many passport applications. They held back conscripts and prohibited the departure of those families and individuals who were thought to have decided to migrate under the influence of the agitation from migration agents and those whose reasons to emigrate were not serious enough because they did not live in poverty. In such cases, the freedom of emigration was limited based on an arbitrary and restrictive evaluation of the applicants' social and economic situation and their justification for departure. The authorities used bureaucratic practices to restrain migration, such as delaying the issuing of passports (Cecotti, 2003, 18). Such delays had many reasons: prejudice against permanent migration, which was seen as an expression of mistrust and ungratefulness toward one's homeland, the fear that South America would turn out to be an unpleasant adventure, and, above all, pressure from big landowners. Fearing they would lose the workforce, the latter demanded that emigration be prevented and branded it as a phenomenon that damaged the social order and the homeland. They limited temporary labor migration for tenants in contracts (known locally as *coloni*) so that the productivity and income from agrarian property would not suffer (Benvenuti, 1982). Some municipalities opposed such an attitude and demanded that people be permitted to leave while they still had funds and before the deepening agrarian crisis made them a burden on the public social network. Local clergy also backed migrants, as they believed they deserved less constricting living conditions than they enjoyed under the rule of local landowners (Cecotti, 2003, 18–20).

In the phase of mass migrations to the United States, from the 1890s onwards, the authorities no longer stopped people if they met the conditions to leave for work abroad. The passport applications from the Logatec district (Carniola) reveal that the municipalities tasked with providing evaluations on the applicants' character and property status and the right to assert impediments against migration did not only consider the prevention of the negative consequences of migration. They were more concerned with its benefits because entering the North American labor market represented an important source of income and alleviated social problems within the municipality. The municipalities recommended applicants for passports as long as there was a possibility that they would make better money in the United States. Migration to the United States was a financially self-sustainable system.[22] A contributing factor to this self-sustainability was the improving organization of transatlantic transportation; the climate and the attitude toward the phenomenon that permeated the social and economic dynamics also changed. In the 1890s, even landowners in Goriška no longer opposed the migration of their tenants quite as forcefully as before because it turned out that migration did not threaten their economic interest but rather helped them get rid of many a troublemaker (Kalc, 2021b, 52–53; Cecotti, 2003, 18–20).

---

22 In their reports to the higher authorities, the municipalities explained that there is no concern for the migrants to the United States because they had the money for their return at their disposal, should inauspicious circumstances force them to return. This became obvious during the 1908 crisis when many were forced to return but the social situation in the district did not worsen. Arhiv Republike Slovenije, 136 Okrajno glavarstvo Postojna, box 34, January 1, 1901; box 27, 7013/1908; Kalc, 2021b, 41–42.

The authorities' attitude was particularly obvious regarding men subject to military service who could only exercise their right to migrate under certain conditions. Men under the obligation to serve in the army represented the bulk of male migration since the migration age coincided with the age of compulsory military service. During the time of military duty, from age 17 to 36, men were only banned from going abroad during presence service, which started at age 21 after a successful draft and lasted from two to four years. Before and after, those who exhibited economic need and guaranteed they would return to either enlist for active service or attend the reserve force training could expatriate with permission from military authorities. Over the years, the Logatec district (and other areas of transatlantic migration in Slovenia) increasingly issued passports to conscripts before their presence service. Prospective migrants' guarantees and oaths that they were not leaving to avoid military duty played an important role in this process, as did the lenience of the authorities. People who favored limiting migration reproached the authorities for believing conscripts' promises, even though so many broke them and remained abroad (Šuklje, 1905; Kalc, 2016a, 2021b, 53). A telling example was the case of the renowned community organizer in the United States, Frank Česen, who obtained a two-year passport in 1906 from the Logatec district authority, after which he was supposed to return and serve his military duty. Something, as he explains, neither he nor the district head who issued the document believed (Česen, 1970, 196).

Like elsewhere in the monarchy, migration in the Slovenian ethnic territory often took illegal forms. Most breaches of legislation were linked to restrictive military regulations. In 1913, 11,000 of Carniola's military conscripts resided in the United States (Drnovšek, 2007b, 18), with only Dalmatia recording a larger ratio. However, many of the Dalmatian absentees were seamen and thus absent because they were at sea (Kalc, 2014b, 34–35). The most pronounced migration areas in Carniola also recorded the highest numbers of absent conscripts. From Bela Krajina, for example, there were 2,863 draftees in the United States in 1913, while in 1905, the districts of Kočevje, Novo Mesto, and Črnomelj recorded a lack of 5,240 military conscripts between 21 and 36 years of age. Among the cohorts of 1882–1884 alone, 510 of the 1,370 conscripts in the Kočevje district failed to appear for the draft. In Novo Mesto, the number was 580 of 1,380, and in Črnomelj, 670 of 1,030. The Črnomelj district head suggested, somewhat sarcastically, sending the draft commission to carry out their work in the United States (Izseljevanje, 1905; Drnovšek, 2007b, 18).

The Ljubljana police—which, considering Ljubljana's position in international migration traffic, had a lot of work surveilling and intercepting illegal migrants from Austrian and Hungarian regions—annually arrested around 70 persons from other municipalities in the Slovenian ethnic territory who set off to the Americas without correct papers; most were military conscripts (Pogačnik, 2000, 17). There were control points in other places and at the border with Italy. Surveillance duties were carried out by the regional police authorities and the municipalities themselves. They had to report all presumably illegal departures to the higher authorities and prevent passport misuse. Nevertheless, the surveillance system was rather lax and permeable, so many easily circumvented it. For example, it was known what was going on in the border area between Austria, Liechtenstein, and Switzerland, through which the route toward Le Havre passed: illegal emigration took place in front of the eyes of Austrian authorities, which simply could not check all passengers because trains, according to international agreements, only stopped for a few minutes (Markitan, 1912, 11–12; Chmelar, 1974, 88–89). Crossing the border to Germany was even easier. The monarchy fought for direct control of Austrian citizens in the German ports of Hamburg and Bremerhaven, from where most Austrian migrants traveled. German authorities, defending their shipping companies' interests and territorial sovereignty, did not concede (Brunnbauer, 2016, 178). In the Slovenian territory, secret routes led migrants on foot across the Austrian–Italian state border and onward to Genoa and other ports by rail (Caro, 1909, 61–66). Migrants easily passed through monitored border points, too. The only consistent control Austrian authorities exerted over migrants overseas was in Trieste and Rijeka. Even there, many could avoid them with the help of shipping companies.

Authorities of transit and destination countries only asked for identification papers from migrants, as they were not interested in migrants' obligations to their countries of origin. Many conscripts migrated illegally to avoid military service. As Austria did not have treaties of extradition of military conscripts with the destination countries, they were "safe" abroad (Rahten, 2018, 48–51). Many left for economic reasons with appropriate travel documents approved by military authorities. They became illegal migrants once abroad for failing to appear at consulates for the draft. They either did not apply to waive or postpone their military training or failed to return home to complete their military service. This phenomenon was particularly widespread in transatlantic migration. At the inter-ministerial consultation conference to prevent illegal migration in Vienna in October 1913, the attendees were told that the Tolmin district (Goriška), where continental destinations prevailed, had almost no illegal migration. Although around 1,400 passports were issued to workers annually, migrants knew their rights and military limitations well and there were almost no absences from drafts (Kalc, 2014b, 40).

Many migrated without appropriate papers and permits for other reasons, such as to avoid family, criminal, legal, or other obligations. Military conscripts searched for illegal paths, erroneously believing they were not allowed to leave the country because they could not obtain passports or because they doubted that the authorities would issue them one (Martinčič, 1960, 232). Newspapers thus recommended that conscripts not leave in secrecy because they could also emigrate if they met certain conditions. They were advised to turn to the authorities with well-founded and convincing arguments and proof about leaving abroad in good faith to earn income. Guarantees from parents or other trustworthy persons were helpful (Goršič, 1911; Kalc, 2014b, 40). The crucial factor in proving a violation of military service duty was intent to escape, which was proved by suspecting or catching a conscript during an attempt to leave without papers, with invalid travel documents, or if he failed to return to perform his civic duty without a reasonable excuse (Kalc, 2014b, 38). The Military Service Act prescribed hefty fines and one to twelve months imprisonment for those who wanted to expatriate to avoid military duty. The actual fines were far lower and, in cases of irregularities, only administrative. This was yet another reason why many paid little heed to the risk.

The Control of Commercial Actors and Limits to the Freedom of Migration

Illegal migration came hand in hand with soliciting and the assistance that official and unlicensed agents offered to people wishing to leave the state territory illegally. Processing and transporting migrants, in particular, were prone to legislation violations that worked against the migrants and the state. The state banned migration agencies in the 1850s and later—to stop them from actively soliciting migrants—intervened several times against shipping companies' representatives and travel agencies: in 1897, by issuing stricter provisions against prohibited forms of business and, in 1907, with new trade legislation (Kalc, 1995, 49, 1997, 17). Agents, however, were extremely crafty at what they did and skillfully avoided control (Kalc, 2014b, 39).

Police, district, and municipal archives in Slovenia keep records and proof of the unauthorized operations of agencies and their unlicensed collaborators in the field. Confiscated from persons caught trying to leave their homeland illegally, these records range from the shipping companies' brochures and literature about the situation in the Americas to letters from agents with detailed instructions on how to avoid border control and cross the state border. Travel agencies were not allowed to offer trips with shipping companies that had no concession, so passengers who were illegally solicited to travel with them only bought tickets during their journey at one of the transit stops abroad or at emigration ports (Becker, 2010, 46–47). The concessioner companies and agencies used an underground organizational network and banned forms of operation that were difficult to eradicate.[23] Notaries were also involved, among others. There were known cases of their issuing false papers on parenthood or kinship to either minors or senior citizens, without which US authorities would not allow them on US soil (Kalc, 2021b, 49).

Authorities and the public were particularly attentive and untrusting toward migration to South America. On the one hand, the political situation there was precarious and often in a state of war. On the other hand, migrants to Brazil, in particular, frequently found themselves in difficult economic circumstances and an unenviable position vis-à-vis their employers or landowners. Their country of origin often could not help them when in distress or lost contact with them because of insufficient legations and unfamiliarity with the legal situation. Foreign agencies, particularly from Friuli, which solicited migrants from their base in Italy and were thus not risking any sanctioning from Austrian bodies, were especially problematic. Among the most notorious was the Nodari Agency from Udine/Videm/Udin, which extended its network of collaborators as far as Galicia and Bukovina. One name that often appears among migration agents who recruited migrants for South America in the Austrian Littoral and Carniola is the Slovenian agent Anton Gergolet from Doberdò/Doberdob near Gorizia. Both Austrian and Italian authorities had him on the list of the most aggressive and unscrupulous agents, but he had the trust of people and was deemed the person to turn to if one wanted to migrate to South America (Umek, 1979; Kalc, 1995, 48–49; Grossutti, 2013, 61–62).

---

23 Jožef Paulina, an entrepreneur who was, among other things, an official agent for the Hamburg-Amerika Linie shipping company in Ljubljana is an illustrative case. When trying to edge out the competition, he went against the rules and led an active propaganda campaign throughout the countryside by distributing posters, brochures, and even dictionaries, sent people personalized letters to convince them to emigrate, used the connections he had in the field as an insurance agent, bribed railway employees, porters at the stations, toll collectors, and others to direct people to his agency, he denounced competing agents to the police, provided migrants with false passports, etc. Once he lost his license as an agent, he continued his activities under the guise of a travel office, acting as an unofficial agent for other European agencies, until, after countless reports, fines, and sentences, he lost his trade license forever (Umek, 1967).

The authorities were additionally skeptical toward migration to Brazil from the Trieste port after the negative experience with the initiative of Triestine bankers and honorary consuls of Brazil, the Morpurgos, to organize migration in 1888 and 1889. Because the migrants from the first of the three organized trips returned home as early as two months after departure, having realized they had been cheated by false promises, the land and local authorities demanded the prohibition of emigration. The Morpurgos were forced to close down their enterprise (Umek, 1979; Kalc, 2001; Cecotti, 2003). The affair, which made waves among the public and filled newspaper columns with returnees' testimonies (Kalc, 1889), coincided with similar events and resonant scandals and affairs elsewhere in the country. For this reason, 1889 brought a renewal of the ban to establish migration agencies on the state level and in individual crown lands (Grossutti, 2013, 57). The measure did not work; the activity of foreign agents continued. Migration to Brazil continued to elicit opposition because of many repatriations and costs for home municipalities (Brunnbauer, 2016, 185–190). Because the Morpurgos' plan to set up routes for migrants to Brazil from Trieste failed, migration continued via Italian and other foreign migration ports: controls of Austrian authorities were thus eschewed, and migrants were even more exposed to the danger of being scammed and exploited (Kalc, 2001).

The second affair that resulted in restrictions on the freedom of migration broke out after the Canadian Pacific Railway Company moored in Trieste in 1912 with its own route for migrants and threatened the interests of the North Atlantic shipping cartel. The government in Vienna gave a concession to the Canadian company to increase the role of the home port in international migration transportation and in its attempts to "nationalize" Austrian emigration.[24] The cartel and Austro-Americana set up a competing line to Canada in response. A price war started, accompanied by a newspaper propaganda campaign. The cartel accused the Canadian company of systematically targeting military recruits and undermining the foundations of the Austro-Hungarian Army. Such accusations intensified the questions of emigration and the mass absence of conscripts, upsetting the opponents of free migration for over a decade. In October 1913, the government banned the operations of the Canadian company and closed down all its agencies. However, the investigation showed that all the companies and travel agencies, for economic reasons, systematically violated legislation and sought military conscripts and other illegal migrants. The national Austro–Americana

24 The Slovenian consul Ivan Švegel, who worked in Montreal in 1909, strove for this (Rahten, 2018, 61).

did the same and consequently suffered the imprisonment of several of its managers and closed down some of its offices in Galicia (Chmelar, 1974, 144–152; Kalc, 2014b, 40).

The pressure from the Ministry of War and other opponents of the liberal attitude to migration within the state apparatus and political circles led the state to tighten control over migration. It did not change the legislation but rather reinforced and reorganized the control apparatus that controlled the movement of the population and the agencies and ensured it implemented the existing regulations throughout the monarchy's territory and its borders. The tightened regime, which, in addition to the police, involved the collaboration of the financial guard and railway personnel, was also implemented in the Slovenian ethnic territory. The measure was contested, and it divided the government circles themselves. It was a political maneuver by the forces aiming to get migration under immediate state control and regulate it according to their conservative principles and interests (Chmelar, 1974, 156). Liberal and leftist political forces saw no real benefit of this measure to the state. They chided the state that it did not care for the citizen's actual needs but rather arbitrarily curtailed fundamental civic rights and the freedom of accessing economic sources that the homeland could not provide to its population (Kalc, 2014b, 41).

After the "Canadian affair," the Austrian government finally took the role of a strong actor in regulating relationships between shipping companies and protecting national interests. It took advantage of the dispute that the scandal caused between the historical allies and the leading partners in the North Atlantic cartel, the Norddeutscher Lloyd from Bremen and the Hamburg–Amerika Linie based in Hamburg. The Austrian Ministry of Commerce directly intervened in negotiations between shipping companies, used the help of Viennese banks to free Austro-Americana from its dependency on German capital, and won a larger traffic quota for Trieste. This condition mostly met the economic interest of the Austrian shippers, who were also favorites of the Ministry of Commerce. All other questions—from the most burning one, the protection of migrants, to the planned migration policies, which many championed—remained unsolved. The quota of migrants embarking for North America from Eastern and South-Eastern Europe via Trieste was expected to increase from 4 to 7 percent in 1914 and reach 10 percent in 1919 (Chmelar, 1974, 140–159; cf. Kalc, 1992, 493–495).

The Aspects of Migration Politics

In the last decades before World War I, migration and dealing with migrants demanded more and more work from the state and its administrators. From ministerial down to municipal levels, the political and administrative apparatus faced, in line with their competencies, new tasks and administration methods linked to organizing migration issues and asserting state sovereignty over citizens outside the state's territory. To do this, it had to coordinate with the legal systems of destination countries and carry out their legislation and certain administrative procedures. Political authorities, for example, demanded medical certificates about the physical state of applicants and proof of other conditions to enter the country of destination to issue a passport. In Trieste, under the criteria of acceptance to US soil, port authorities, police, and medical professionals, together with consular officers of the United States, carried out preliminary checks of documents, health status, political, moral, and other characteristics of passengers (Kalc, 2016a, 114–115, 2015b, 83–84; Brunnbauer, 2016, 191).

In the destination countries, consulates had to provide legal protection and help to their citizens (Phelps, 2013, 109–110). In the United States, solving their civilian law issues and representing their interests was the primary area of the Austrian consulates' activities. Private institutions, for example, the Österreichische Gesellschaft von New York (Austrian Society of New York), also served this purpose. Among its founding members were also the renowned inventor Nikola Tesla and Frank Sakser, the most prominent Slovenian businessman in the United States (*10. Jahresbericht der Oesterreichischen Gesellschaft von New York*, 1908). Avgust Jakopič, a co-founder of the Slovenian St. Raphael Society in New York (with Rev. Father Kazimir Zakrajšek), managed the hostel for newly arrived migrants from Europe, known in the community as the "Austrian home" (Friš, 1995, 310).

It was in the interest of the Austrian authorities for the migrants to see themselves as subjects of and remain loyal to their Habsburg homeland, retain citizenship, and not resettle abroad. The Austro-Hungarian government favored temporary entry into the US labor market, which provided a larger influx of money to the national economy. It antagonized migrant communities that promoted permanent settling, naturalization, and practices that allowed migrants to strengthen their social position in the migrant environment. Migrant organizations ensured that ties were retained with the homeland but predominantly as a part of individual national groups and with the mediation of their organizations (Cf. Agstner, 2012, 111–113; Phelps, 2013, 178).

Diaspora organizations often filled the void created by the "absent" or "insufficient" interest of the country of origin for their citizens abroad, and Slovenian migrant organizations were no exception. While such organizations did play a role in maintaining migrants' connection to Austria-Hungary, they were far more successful in encouraging regional cohesion, that is, organizing migrants according to their ethnicity. At the same time, they divided migrants along their philosophical and political beliefs, as the Chapter 2 describes. The relationship between Slovenian migrants and Austria-Hungary is a complex question as are the Slovenian diaspora's relationships with other national groups, which sometimes mirrored the traditional relationships between them, but sometimes created new alliances, affinities, or conflicts. These questions have not been well studied. Historiography has long followed the national principle when studying migration from Austria-Hungary and, under the influence of methodological nationalism, only studied migrations by individual national groups separately. Nevertheless, newer (and some older) studies (Steidl et al., 2017; Steidl, 2021; Phelps, 2013; Bodnar, 1977) show that migrations have to be observed not only from the perspective of the country of origin and the destination country (the "national" approach) but also from the perspective of interethnic relationships in specific regional setting in the countries of origin and in immigrant communities of the receiving countries alike. In the United States, these interethnic relationships could be highly conflictual, as the United States, based on its racial criteria or the concept of "nations and peoples," gravitated toward differentiating migrants from Austria-Hungary on ethnic grounds. At the same time, the Habsburg Monarchy attempted to implement its sovereignty abroad on the concepts of territoriality and citizenship. In the Slovenian case, one must also consider that migrants originated from both parts of the monarchy, as well as from Benechia in the Kingdom of Italy. They moved under different migration regimes and were, at home and abroad, subjected to different national policies. These particularities were reflected in separately organized migrant communities, different ethnic identifications, and levels of emotional engagement with places of origin and reference countries. Strong local or regional identity added to the complexity of this issue.

Migration policies and regimes covered migration abroad and—albeit in different forms—movements within the same monarchy and the Slovenian Crown Lands. Important factors of migration politics were, once again, local administrations, which intervened in managing migrations through social care and residency rights competencies. They found a basis for this in the *Heimatrecht* legal principle, which stipulated that citizens could only enjoy full civic rights in the municipality where they had the right of domicile (municipality of their permanent residency). Elsewhere in the country, regarding social and legal affairs, they were, administratively speaking, aliens.[25] Municipalities could thus return

25 Austria knew two types of aliens: citizens of other countries (*Ausländer*) and foreigners

citizens to their municipalities of domicile if they were left without income or had no means to support themselves and became dependent on social support. They also denied applications to grant newcomers the right of domicile. This practice made it possible (at least until the 1880s) to exploit migrants as labor capital while loading the social burden on their original municipalities. The domicile and social care legislation system was, via ordering expulsions, a factor of forced migrations that also included wives and descendants of migrants. Because both gained the right of domicile (the same was true for citizenship) from their husband or father, this "return" to the domicile municipality could mean leaving for alien environments.[26]

The bourgeoisie often used this instrument and other forms of exclusion to regulate migration and protect their communities' economic and social interests (Kalc, 2016a, 117–118). Once the right of domicile legislation became more democratic and the change of domicile easier, moving to richer environments became an economic and social security choice for many. This practice was particularly noticeable in the city of Trieste, which needed considerable productive forces. It could accept a large population, so the authorities initially granted the right of domicile to migrants quite liberally. However, toward the end of the century, poverty started its unstoppable spread among the weakest strata of society. The economic and political elites voiced loud opposition to the immigration of those supposedly coming to abuse the city's social benefits. They defended the interests of the "locals" before the "newcomers." Municipalities from which the migrants originated often denied the claims of the municipality of Trieste to refund the cost of their medical treatment or social care. Trieste's annual balance sheets thus recorded chronic deficits because of unrecovered costs for foreign domicile holders while they lived in legal and social insecurity (cf. Scartabellati, 2009).

Internal migration policies were also linked to the national question and emphasized in the Slovenian ethnic territory. This aspect is a well-known theme in the history of the Habsburg Monarchy in the 19th and 20th centuries. Internal migrations impacted the numeric relationships and ratios between the individual national groups in many areas of the multinational empire. These

---

(*Fremde*), that is, Austrian citizens who did not have the right of domicile in the municipality in which they resided (Becker, 2010, 34).

26 The system started changing in the 1880s, and, since 1901, migrants and their family members could obtain right of domicile without restrictions in the place where they resided for 10 years (Komlosy, 2003, 161–165).

processes were not only the effects of spontaneous development but also of economic and other politics that directly or indirectly intertwined with national interests. For example, we can mention the German-speaking population's migration and its influence in Lower Styria or emigration from the Slovenian territory to industrialized German-speaking Styria, where Slovenian-speaking inhabitants were encouraged to declare themselves German during population censuses; a large number of them did assimilate (Klemenčič, 1979). As mentioned before, in Trieste and Gorizia, immigration was at the core of the national confrontation and related politics; the revision of the 1910 census found 50 and 63 percent more Slovenian speakers, respectively, than the Italian municipality authorities originally showed, as they abused their statistical competencies to benefit their national interest (Čermelj, 1958; Marušič, 2012, 83; Bufon, 2021). They also made good use of the mechanisms described above. Municipal authorities in Trieste also favored migration from the Kingdom of Italy and the employment of Italian citizens in public institutions. This migration was supposed to be a measure by Italian national liberals to retain the Roman image of Trieste vis-à-vis the Austrian government, which, as an instrument of anti-Italianization politics, encouraged the migration of Slovenians and other Slavs. There were harsh confrontations between the municipality and the state authorities about recognizing or refusing the right of residence and expulsing Italian citizens (cf. Silvestri, 2017).

Migrations as a Socioeconomic, Moral, and National Problem

Around the mid-19th century, migrations in (and from) the Slovenian ethnic territory entered public discourse. Public opinion was critical of the phenomenon, yet, as long as it was kept in a moderate scope and took place in familiar forms, it did not trigger too much concern. Once it acquired epic dimensions, it became a political issue worthy of debate. As migrations coincided with the period of nation-building and intense implementation of national ideology, mass migration gained its place among the core national questions and became a topic of patriotic discourse. Regardless of their philosophical, social, or political orientation, the Slovenian thinkers and public opinion makers were consumed by the idea that the phenomenon opposed the nation's interest and threatened its very existence. Newspapers and journals regularly reported on migrations and the growing numbers of Slovenians abroad. Negatively labeled in any public speech, migration was veiled in derogatory metaphors that underlined its dramatic dimensions and fatal consequences. The most widespread and telling were the metaphors of "a nation bleeding" and "a cancerous wound on the body of the Slovenian nation" (Drnovšek, 2010a).

In 1911, Janko Mačkovšek summarized the core of the problem in his discussion on demographic development and Slovenians' position concerning neighboring nations in the monarchy. He claimed that as late as the mid-19th century, Slovenian losses were minimal and mostly occurred due to assimilation on the edges of the Slovenian ethnic territory. In the second half of the century, particularly in the last decades, the known "enemy" was joined by emigration, which took the best life forces from the center of the Slovenian ethnic territory. Mačkovšek claimed that foreign lands kept the "blossom," the most intelligent part of the Slovenian people, for itself and returned "the old, sick, and helpless." Despite the natural vitality and the growth of the Slavic population, Slovenians had regressed in numbers relative to the other nations in the monarchy. They were the nation with the lowest number of members in the best age for procreation and work and the highest level of the aged population. Slovenehood disappeared quickly in non-Slovenian Austrian Lands. On the left bank of the Drava River, the Slovenian ethnic territory was shrinking due to the growth of the German-speaking population (Mačkovšek, 1911, 245–250). These theses, which the author supported with numbers from population censuses, appeared in one form or another in prominent social and political places and the wider discourse in the nation. Generally speaking, principles expressed toward migration showed more or less an understanding of the problem and different views of solving migration's social and moral themes. The Catholic press paid the most attention to it, the liberal and social-democrat press slightly less.

In the zeal of the debate, demands started spreading to interrupt or even prohibit migration. Nevertheless, considering the socioeconomic situation, such a ban would be neither realistic nor morally acceptable because people could not be prevented from leaving if there was no possibility of surviving at home. Jurij Trunk, who knew the subject well, felt migration was not entirely necessary but also not unjustifiable or evil in itself (Trunk, 1912, 412). Franc Šušteršič, the spiritual pastor among the Slovenians in the United States, claimed that while people did not become millionaires in the "New World," they lived better than they could ever have at home (Šušteršič, 1891). Slovenian bishops, too, defined migration as a natural phenomenon that helps humanity bridge material needs and difficult social situations and contributes to commerce's progress, spreading technical knowledge, behavior, and self-confidence (Prečastiti duhovščini in vernikom ilirske cerkvene pokrajine pozdrav in blagoslov v Gospodu!, 1913, 137–138). However, they all agreed that migration needed to be regulated in light of migrants' "justifiable" reasons and those of their homeland.

Public discourse put individual factors of migration in the dock and called for greater accountability of the state and politics on the central, regional, and local levels. By publishing a "dramatic" statistic on migration from Carniola, Ivan Mulaček, a returnee from the United States and a highly educated and perceptive contemporary observer of the Slovenian diaspora, scolded the "fanatically indifferent" attitude of politics to one of the most important "among the social and national questions" (1913, 16–17). Slovenian representatives occasionally spoke in the Reichsrat in Vienna to warn about the implications of migrations and voice their demands to legally regulate the migration question (Caro, 1909, 180; Izseljevanje, 1905; Prečastiti duhovščini in vernikom ilirske cerkvene pokrajine pozdrav in blagoslov v Gospodu!, 1913). The small Slovenian ethnic territory, divided between different regional administrative units and heavily decentralized, only featured marginally in the relevant state forums where the debate on migration took place. It is quite telling, for example, that the 1912 state-wide survey by the Ministry of Commerce about emigration from Austria did not include a single expert from Carniola or Lower Styria (Protokoll der im k. k. Handelsministerium durchgeführten Vernehmung von Auskunftspersonen über die Auswanderung aus Österreich. Wien: K. k. Hof- und Staatsdruckerei, 1912).

Public discourse also targeted migration agents in its criticism, with critical Slovenian voices blaming them not only for their fraudulent practices but also for the exceedingly mass nature of the phenomenon. Also under attack was the state's laissez-faire attitude, which left the migrations to self-regulation and the organization of migration to greedy private entities rather than supporting migrants and regulating the phenomenon in their—and the national economy's—favor (Adlešič, 1909, 170–173). The state was blamed for partiality and supporting capitalist and elitist interests at the expense of migrants and their places of origin (Krek, 1913). The reproach to the state was that its (non-)policy let the United States have the human capital without compensation, thus yielding to the mighty shipping cartel and letting money bleed into foreign ports rather than favoring the strengthening of the national port and controlling emigration (Šašek, 1912; Kalc, 2015b, 80). The critics warned that when calculating the worth of migrants and their work, only one-third of the capital returned home; that because of the migration, labor was scarce at home; that because of the rising price of labor, the agrarian economy suffered; and that the very existence of larger homesteads was in question (Izseljevanje, 1905; Trunk, 1912, 412–415).

Complaints were also aimed at the insufficient protection of migrants worldwide, where they needed daily assistance and protection in different cases of civilian law, from asserting their rights in cases of accidents and damages to repatriation and help in cases of unemployment. State offices abroad could not carry out their duties because there were not enough consuls to efficiently cover the territories where migrants lived. Besides, among the consular staff, few spoke Slavic languages (Die Auswanderungsfrage im österreichischen Abgeordnetenhause, 1912). Most showed little interest in migrants' needs and cared primarily about their own status concerning the US authorities because they were, as representatives of "new immigrants" themselves, often subjected to prejudice (Caro, 1907, 551; Phelps, 2013, 150, 161).

The reproach was aimed at German and Hungarian nationalist policies, too, which were said to covertly use migration as an instrument for limiting the socio-economic development of the Slavic regions and the political establishment of the Slavic people (Zavertnik, 1925, 258). This politics also reached abroad, where the Kingdom of Hungary strove for close sovereignty over its subjects and the expansion of Magyarization (Puskás, 1982, 96; Phelps, 2013, 184–185; Brunnbauer, 2016, 156). Migrants from Prekmurje to the United States were subjected to it, and "Wendish" origin was forced onto them via clergy and other organizations. Their separation from "American Slovenia" was encouraged so that they could identify as a part of the Hungarian nation (Kuzmič, 2001, 100–139). As far as the legal regulation of migration was concerned, Slovenian writers and thinkers suggested Italy as a model. With the law from 1901, Italy dropped the police approach. It assumed a more active role in organizing migration and its economic, organizational, socio-protective, and national-political aspects. When it came to the very attitude to the migration question, they often referred to the Kingdom of Hungary, which traced a part of transatlantic migration through Rijeka and used legislation from 1903 and 1909, like Italy, to reserve itself the possibility of regulating the migration currents (Chmelar, 1974, 140–141; Brunnbauer, 2016, 154–157).

In Search of Solutions

Along with the criticism, the ideas of necessary interventions to limit the negative effects of migration and use its benefits also appeared. The need for investments was emphasized, as it would guarantee permanent income to the population with only a few months' worth of work per year available at home. Catholic circles, with the Slovenska ljudska stranka (Slovene People's Party) at the forefront, defended "strategic agrarian economic policy" as a recipe to stop migration. The more radical thinkers saw the solution to the migration question as a conditio sine qua non for economic development and national perseverance in industrialization. However, because the industrialization primarily took place at the edges of the Slovenian ethnic territory and functioned as a centrifugal force for the nation—dispersing rather than concentrating it—from the national point of view, industrial growth had to be encouraged in the central Slovenian space and thus improve the circumstances for the nation's defense and its political, cultural, and all-round progress (Cf. Rostohar, 1912; Drnovšek, 2010a, 124–125, 127). Another demand was the improvement of vocational training so that "Slovenian sons" would not leave to work in mines and factories abroad, while Slovenian factories had to hire skilled foreign workers (Drnovšek, 1989, 207).

In discussions about where Slovenian migrants should be going, several voices looked to Egypt. Physician Karol Pečnik defended the thesis that male migration should follow women and shift to Egypt. There, migrants could establish themselves in trade and other professions and join the middle classes rather than toil as the exploited proletariat in the United States. The gender structure would be balanced and strong Slovenian colonies could be established, benefiting migrants and the Slovenian economic and national interests (Pečnik, 1897; Mlekuž, 2016). Henrik Tuma, a lawyer and socialist leader who spent most of his adult life in Gorizia, saw the Slovenians scattered around the world as potential intermediaries to promote and develop Slovenian export and trade (Marušič, 1995, 63–65). Andrej Gabrščček, the editor of the *Soča* newspaper from Gorizia, advocated a rationalized transfer of Slovenian savings to the homeland and personally intervened with state authorities to improve the situation of Slovenian women in Egypt and establish appropriate and secure banking channels (Gabrščček, 1934, 260).

Like all advocates of positive migration legislation and "sensible" migration politics, Jurij Trunk believed that the Habsburg government should limit migration according to the needs of the people and steer them toward the most auspicious countries. Together with Italy and Russia—the three countries that supplied the majority of all transatlantic migration—Austria-Hungary was to negotiate better circumstances for its people. He supported the reorientation toward South American countries, where Slovenian artisans would have better opportunities to establish themselves. In Argentina, he also saw possibilities for seasonal migration of agrarian laborers, like the type that Italy knew (Trunk, 1912, 415–416). Farm work was supposed to be more aligned with the very nature of the Slovenian people, and knowledge gained in Argentina would benefit agriculture at home (Adlešič, 1909, 177).

All the political sides demanded that the state provide better protection for migrants and regulate and monitor the shipping companies and their agents. With the strong revival of transatlantic migration after the crisis of 1907 and 1908, amidst the state's unresponsiveness, voices emerged that promoted self-initiative. Lawyer Juro Adlešič advocated complete subordination of migration to public control and suggested an organization under the auspices of the regional government with a central committee in Ljubljana, which would branch out to individual villages via local sub-committees and municipal divisions. It would be a public interest organization, working as a charity for the nation's interest and not for private economic benefit. It would advise migrants, protect them en route, direct them to destination countries according to their needs and the interests of their homeland, find work for them, negotiate favorable prices for the journey and reduce the negative influence of agencies. Such "preventative work" would direct migrants into the "right hands" and retain at home those not driven abroad by a genuine need. The organization should include priests, mayors, and other trustworthy persons. Adlešič's proposal aligned with the most illiberal views of and conservative approaches to the migration issue. In his opinion, migration should be "limited, curtailed and barraged," as it was often spurred by false assumptions about the foreign world and agitation (Adlešič, 1909; Drnovšek, 2010a, 119).

Janez Evangelist Krek, a representative of Christian social thought and the vice-president of the Austrian St. Raphael Society, worked with his allies in the Viennese parliament to pass protective legislation for migrants. In 1912, he was the first signatory of the parliamentary petition to achieve international regulation of migrant protection (Antrag des Abgeordneten Dr. Krek und Genossen betreffend das Auswanderungsgesetz, 1912, 99–100). He emphasized the role of cooperatives, which he promoted as an economic and social strategy to protect the position of peasants from the invasive influence of capitalist relationships and competition in the international market. Migrants' savings, kept in urban monetary institutions and invested in speculative and non-agrarian economic activities, should be diverted to agrarian saving and lending societies. The money the migrants earned abroad would thus return to villages in the form of loans to help the progress of rural communities (Krek, 1913). In 1913, Krek and the Slovene People's Party members of parliament promoted establishing a consulting service in the Carniolan *Landtag* (assembly), which would help agrarian laborers and artisans find work at home and abroad. The consulting service would work with municipalities, cooperatives, and educational and professional associations through local administrative bodies. Such a land-based public organization would be better than a private one. Existing city agencies would help to level the labor market, which caused unemployment in industrial centers and cities and the lack of workforce in the countryside (Krek, 1913; Obravnave deželnega zbora kranjskega v Ljubljani, 1914).

Between Moralistic Discourses and Reality

Considering the negative attitude to which it was subjected, the public discourse in Slovenia usually treated migration pejoratively and often with contempt. Newspapers and other publications, which otherwise did not fail to mention the success Slovenians had abroad, only rarely had something positive to say about it. The leitmotif was the scolding, warning about the dangers of foreign countries and negative consequences for the nation and migrants themselves. As early as the 1860s, the *Kmetijske novice* warned migrants about losing the Slovenian language, homeland, faith, and Catholic upbringing (Drnovšek, 2010a, 54–63). In 1883, a short story by Jakob Alešovec, *Ne v Ameriko!* (Not to America!), was published. Other writings also followed that showed Slovenians as rural people not fit for work in industry and mines or life in cities. Migration was associated with exploitation, the destructive influences of unhealthy habits, moral decay, adopting wrong ideas, and inappropriate modern lifestyles. Compared to the praiseworthy rural homeland, foreign countries were generally painted in dark and gloomy colors. Trunk's book *Amerika in Amerikanci* met with criticism because it allegedly presented America too favorably (Klemenčič, 1999; Drnovšek, 2010a, 121). Migration was also a topic in some literary works, usually with a negative tone. Slovenian writers always presented protagonists as unhappy or negative characters and migration and life abroad as sources of painful problems and bitter fates (Hladnik, 1985; Drnovšek, 2010a, 129–139; Lazarević, 2015, 166–168). There was no shortage of moralistic censure, either. Such censure was often reserved for independent female migrants. At the top of the list of targets were aleksandrinke, particularly those among them who were mothers and thus an easy target for moralist musings (they left their children behind, they traveled by themselves, and they worked in a non-Christian environment). However, none of the women, even those who were unmarried and responded to invitations from the Americas or simply left to work in cities and towns in the monarchy, escaped unscratched. The latter were thought to be particularly in danger of their national consciousness being corrupted by coming into contact with the Italian or German environment. Ironically, Karol Pečnik thus advised that girls should come to work in Alexandria rather than Trieste or Gorizia because in Alexandria they remained Slovenians (Mlekuž, 2016; Verginella, 2006, 73–93; Testen, 2013).

Migration—whether abroad, to other Austrian Lands or Slovenian towns—was thus at the core of patriotic rhetoric, which on the Catholic and liberal sides linked the nationally ideal to the rural. Slovenian social elites were mostly linked to the traditional social system in their ideology. They supported economic progress but opposed the uncontrollable intrusion of economic and social modernization, not least because this modernization appeared in the form of foreign (German and Italian) economic and national influence. The elites understood progress more as strengthening and preserving agriculture and less as strengthening other industries (cf. Lazarević, 2015, 79, 82). With its social, cultural, and political implications, migration threatened the foundation of national thought, the traditional social order, and the position of power. Regardless of the rhetoric, Slovenian elites did not resist migration because the economic reality did not leave any room to limit the phenomenon; they could only curb the enthusiasm for the foreign world and implore people not to indulge in new lifestyles.

The critical discourse on migration was one side of the attitude toward migration. The other side was concrete care for migrants, in which the role of the Church stood out. With the state and its administration so reserved, the Church—the institution anchored most deeply in the nation and with a capillary network in place—took the leading role in caring for migration issues. Mass migrations gave it new tasks—namely, offering concrete help to migrants on route and pastoral and charity work in the destination countries—aimed at preserving migrants' religious and moral integrity. The Slovenian Church and clergy followed the guidelines of the Roman Catholic Church, which, in the period of liberalism, busied itself with solutions to the problems of migration and migrants. In the Slovenian ethnic territory, too, the Catholic Church—given the lack of interest of the liberal and socialist camps in the migration question—filled the void and became more active. However, it also lagged behind compared to the churches in countries like Italy and Germany and Imperial Austria itself. In the early 1890s, the Slovenian clergy came up with the idea of setting up an association to help migrants on their way abroad. The idea was only realized when the St. Raphael Society opened in Ljubljana in 1907 and, soon after, in New York. However, a committee to protect migrants had been working, as a part of the Roman Catholic Archdiocese of Ljubljana, since 1903. The St. Raphael Society worked, among other things, on the incentives "for employment in the Slovenian territory and the colonization of Bosnia and Herzegovina" (Kalc, 2021b, 56). It developed well as an organization and quickly tried to take over the Slovenian migration agency in New York. Still, few migrants used its services; it failed to dissuade people from migrating.

The question of emigration was a constant topic of discussion at Catholic meetings and in Episcopal pastoral letters and programs. The Church saw its role not only in fulfilling its pastoral and social mission in its work with migrants: its task was also to preserve the Slovenian language and ethnic identity in Slovenian migrant communities around the world. This role was particularly expressed in the United States where a string of Slovenian parishes and churches was established. The separation between the Church and the State enshrined in the US legislation required a different relationship between the clergy—of whom some advanced quite high on the ecclesiastical hierarchy—and their congregation, particularly as parishes depended on parishioners' financial contributions. Slovenian migrant communities developed other forms of social organization, support activities, and cultural and political life around the Church and other religious institutions. Together with liberal and socialist organizations, they co-created the rich social life of "American Slovenia" (cf. Friš, 1995).

World War I: A Turning Point

World War I broke the great cycle of European—and Slovenian—mass migrations from prior decades. With the declaration of war and the start of mobilization, Austria-Hungary banned the departure of persons fit for military service and limited migration rights within the state (Kalc, 2016a, 118; Svoljšak and Antoličič, 2018, 247–248). Immediately, transatlantic migration ceased almost completely. Thirty-seven thousand Slovenian and Croatian nationals had entered the United States just a year before the war. In the first year of the war, this number dropped to 2,000, then remained at a couple hundred per year (Willcox and Ferenczi, 1929, 384).

In Trieste, where according to the newly negotiated quotas between European ports, an increasing number of Eastern and Southeastern European transatlantic migrants would be processed, traffic stopped completely in July. The British Cunard Line Company withdrew from the Adriatic. The domestic Austro-Americana laid off its staff and signed over the building where migrants were processed to the Red Cross, which turned it into a military hospital (Bertuzzi and Fait, 2010, 82). The international transit of migrants who crossed the Slovenian territory on their way to and from other European ports also stopped (Kalc, 1992, 482–483, 2015b). The war surprised many permanent or temporary continental migrants far from home, including the many seasonal laborers from Benechia, Goriška, and Prekmurje.

A small number of Slovenian migrants (according to US statistics, 2,500 Slovenian and Croatian nationals) returned home from the United States at the beginning of the war. Consular appeals to the Austrian army reserve and draftsmen to fulfill their duty to their homeland were published in the US-based Slovenian press, as were appeals for Slavic migrants not to join the fraternal war against Serbians and other Slavic nations (Klemenčič, 1987, 43). Austro-Hungarian authorities did not seriously count on absent conscripts: the United States became their "foster mother," so they were, in terms of the military, lost to their old homeland (Kalc, 2014b). Patriotic appeals failed to mobilize them, as did threats of criminal prosecution for deserters and even general amnesty. One concrete reason the Habsburg Monarchy could not count on them was logistics: most who responded to the mobilization could not be brought home from the United States.

Some Slovenians, Croatians, and Serbians who supported the Yugoslav idea joined the army of Serbia as volunteers. They trained in different camps, including Bizerte/Binzart in present-day Tunisia, from where they were deployed to the Salonica Front in 1917 and 1918. Migrants from South America, Australia, and New Zealand volunteered as well (Turk, 1938, 58; Klemenčič, 2002, 330; Svoljšak and Antoličič, 2018, 129–130). Others participated in the war in Europe as American soldiers, including Louis Adamic, who talks about it in his novel *Laughing in the Jungle* (1932). According to some estimates, around 2,000 Slovenian immigrants from all the destination countries and many of their children took part in the Entente Powers (Svoljšak and Antoličič, 2018, 130). The Cleveland community saw around 400 sons of Slovenian parents serve in the US Army or Navy during World War I (Ledbetter, 1918, 15).

Most Slovenian migrants from Benechia, who were expected to join the Italian Army, remained in the United States. Many returned after the war after the amnesty cleared them of the accusation of deserting. Migrants also remained in the United States for economic reasons. With the economic growth, the position of the working class in the United States improved during the war, and Italy was rather lenient toward these "deserters" so as not to decrease the inflow of valuable remittances from across the Atlantic (Kalc, 2018a, 113). Migrants in Germany found themselves at the center of the war operations and, as Austrian citizens, could not avoid mobilization into the Austro-Hungarian Army. Many were sent to the Russian Front at the war's beginning; some were later sent to the Soška fronta/Isonzo Front. Some women from the Slovenian communities in Germany, particularly those left without income after their husbands were drafted, returned home with their children; others spent the war in the Ruhr Valley (Wörsdörfer, 2017, 180–181). Many aleksandrinke remained in Egypt, while male citizens of Austria-Hungary and Germany were interned or expelled by the British authorities at the beginning of the war (Ismail and Gabriel, 2011, 58).

Slovenians in the United States took different and changing positions toward the war and the political issues it created in the "old country." Many factors influenced their opinions: their legal situation (most of them were Austrian or Hungarian citizens); pressure from the Austro-Hungarian authorities; their views; the US entry into

World War I; and the political forces that shaped the postwar organizational destiny of Slovenian ethnic territory and the Yugoslav question. Many Slovenian migrants and organizations in the United States, whether Catholic or liberal-socialist, harbored patriotic feelings and expressed bias for Austria-Hungary, at least until the US entry into the war. In the Catholic ranks, reservations toward Serbian and Russian Orthodoxy played a role. Because of suspicions that they sympathized with the Central Powers, some Slovenian migrants shared the fate of "hostile aliens" and were imprisoned or persecuted (Molek, 1979, 192–193). Others strongly supported the anti-war stance from the very beginning and condemned the war against Serbia as unjust. They were warming up to the Yugoslav idea and joined the movement for a new postwar order and establishing a Yugoslav state as early as 1915.

The resulting points of view, forged through their attitude to the war, did not simply reflect traditional philosophical and political divisions. Within the camps themselves, they also caused deep fissures. The deepest ones were created in the Yugoslav socialist movement. Attitudes toward anti-militarism and internationalism principles, the class struggle against the bourgeoisie, capitalism, and US entry into the war tested their unity. Slovenian and Croatian socialists disagreed on the last point. Regarding the Yugoslav question, the "Ententephiles" from the Slovenian League and the Yugoslav Committee in London, the supporters of the "loyalty to the Austrian homeland," and the socialists engaged in critical debates. All sides offered more or less radical forms of unification and independence of the Yugoslav nations, some within a restructured Habsburg Monarchy, but most within an independent state (Klemenčič, 2001c, 53–54).

When discussing this complex and intense chapter of the political activity of Slovenian (and Yugoslav) migrant communities, a particularly relevant historic act must be mentioned: the Chicago Declaration of June 29, 1917. This declaration, signed by the representatives of Slovenian liberals and socialists at a meeting in the political, economic, and organizational center of Slovenians in the United States, supported the establishment of a federal Yugoslav republic as an independent state. In such a state, the constitutive nations—Slovenians, Croatians, and Serbians—should enjoy autonomy in all those areas in which they were culturally different and in all the matters they did not share. This declaration was the foundation for the Slovensko republičansko združenje (Slovenian Republican Alliance), which later, together with its Croatian counterpart, united into the Jugoslovansko republičansko združenje (Yugoslav Republican Alliance, or JRZ) (cf. Čizmić, 1974; Drnovšek, 1978; Kodrič, 1983). During the war, Slovenian migrants in South America also organized to support the Yugoslav cause. As members of the Jugoslovanska narodna odbrana (Yugoslav National Defense, or JNO), they supported the Yugoslav Committee in London and, at the end of the war, established the first pro-Yugoslav Slovenian organization in South America (Mislej, 1989b).

When the war started, the Slovenian ethnic territory became a stage for forced migrations. On the one hand, these migrations were triggered by steps taken against citizens of hostile countries who had to leave Austrian soil or face arrests, expulsions, confinements, and deportations (Cecotti, 2001b; Kalc, 2016, 119). The problem intensified when Italy entered the war in 1915. Restrictions against citizens of hostile countries and the Italian mobilization stopped mass seasonal migration from Benechia and the entire Friuli to Imperial Austria and the German Empire. The impact of the war on the "economy of migrations" was obvious in the first few months of the conflict: the mass return of seasonal workers resulted in no income and worsened the situation of their families. Therefore, many continued returning to Austria until May 1915, while Friuli witnessed increased unemployment and, consequently, social unrest and protests (Di Caporiacco, 1969, 260).

Austrian measures against the citizens of hostile countries—they had to leave the country by a certain deadline while many conscripts were interned—also affected most citizens of the Kingdom of Italy in Gorizia and, particularly, in Trieste, where around 50,000 lived before the war. They had to retreat to their legal "homelands" or were placed under strict surveillance as hostile or suspicious aliens. In Trieste, the authorities interned or confined 5,000 *regincoli* (i.e., citizens of the Kingdom of Italy) who did not retreat and expulsed more than 8,000. In the hostile environment, many Italians became a target of intolerance—as did Austro-Hungarian citizens and their property in Italy, for example, in a series of attacks in Milan (Biondi, 2001, 52, 56; Cecotti, 2001a, 30). Austro-Hungarian authorities also monitored their citizens at home and abroad because certain nations of the monarchy were deemed less loyal, patriotic, and trustworthy than others, which was particularly obvious when it came to the Italians (Stergar, 2004).

The second forced migration included the population of wider areas along the Soška fronta/Isonzo Front; they, too, had to leave their homes and go into exile. One hundred seventy thousand civilians were displaced. Among them, 90,000 Slovenians, mostly from the left bank of the Soča/Isonzo River, and around 12,000 from the right bank, which was occupied by the Italian Army. Habsburg authorities moved those from the left bank to Carniola, Lower Styria, and Carinthia and to refugee camps in Lower Austria. Around the same time, Italian authorities sent the latter to different towns and camps around the Apennine Peninsula. Most were moved during planned evacuations or moved away from the frontline voluntarily in the first months after Italy entered the war in 1915. Subsequent evacuations followed in August 1916, before the Sixth Battle and the Italian occupation of Gorizia, and after the Italian success in the Eleventh Battle in August 1917 on the Banjšice

Plateau. Around 50,000 Slovenian refugees remained in Carniola, Lower Styria, and Carinthia. Those who were not able to support themselves were sent to different refugee camps, such as Steinklamm near St. Pölten, Gmünd on the Czech border, Wagna near Leibnitz, Bruck an der Leitha (the largest one with 4,200 Slovenian refugees), and Strnišče near Ptuj (Svoljšak and Antoličič, 2018, 280–282).

Picture 1.8: Children Evacuated from the Front Line around the Soča/Isonzo River in the Steinklamm Refugee Camp in Lower Austria. Author: unknown. Courtesy of: Muzej novejše in sodobne zgodovine Slovenije (Museum of Contemporary History of Slovenia).

While Italian authorities expelled the population from occupied Austrian territory, they interned the non-enrolled Austrian conscripts, clerks, and clergy as potentially dangerous persons. Unlike Austria-Hungary, Italy did not create large refugee camps—the "wooden cities." Most refugees lived in mid-sized or small colonies scattered throughout the country (Svoljšak, 1991; Malni, 2001). As the frontline moved during the war, refugees started returning from exile in Austria and Italy. Still, most returns occurred after the war and continued until 1922.

The war triggered other migrations linked to the extraordinary circumstances. In September 1914, different towns around Slovenia received thousands of refugees from Galicia and Bukovina (Svoljšak and Antoličič, 2018, 278). By 1915, the number of people living in Trieste had dropped from 245,000 to 155,000. Among departees were employees of different public and private institutions who were moved to the hinterland for security reasons. One of the largest groups was railway workers. These were mostly Slovenians reassigned to the internal railway network, many of them because of the proximity of the frontline and because they feared an Italian breakthrough. At the same time, Trieste received refugees and other migrants. Departures, arrivals, and shorter and longer stays of old and new inhabitants followed each other, depending on the events and turns on the front (Cecotti, 2001c).

Internal mobility retained its economic function despite strict surveillance in the war circumstances. Changes in production and the labor market caused migrations dictated by the needs of the population and wartime economy (cf. Strle, 2015a). Authorities implemented special measures to solve the workforce shortages and relocate workers to different strategic fields. Refugees were a reserve workforce for the war industry as well as agrarian and other industries. The Urad za begunce (Agency for Refugees) from Goriška helped refugees find employment in Ljubljana and other towns. It was mostly the socially disadvantaged that were sent to camps, while skilled people and those capable of working were relocated throughout the inland areas—depending on possibilities for life and work. For example, an entire shoemaking cooperative with 1,200 villagers moved from Miren (near Gorizia) to Rečica, Mozirje, and nearby villages in the Savinja Valley (Svoljšak and Antoličič, 2018, 284).

Refugee camps were centers of production for different goods. Three thousand refugees from Galicia and Bukovina were employed in building a railway, while between 20,000 and 30,000 Russian prisoners of war were mostly engaged in constructing military infrastructure in Slovenia (Svoljšak and Antoličič, 2018, 278, 313). Slovenian refugees in Italy, too, supplemented the social support received from the Italian state by working, since factories, workshops, and farms were left without a workforce due to the draft. Because of that, refugees often moved (Svoljšak, 1991, 25–26). Thus, both the migrations that the governments managed and spontaneous moves remained important economic and social factors even in wartime.

Emigration from the Slovenian Ethnic Territory between the Two World Wars: A Demographic Frame

World War I introduced a new chapter in the history of European migrations. The war changed the political geography of Europe and other parts of the world. Its echoes also influenced the scope, typology, and directions of international migrations and migration regimes and policies toward migrations and migrants.

New states were created from the embers of the German, Austro-Hungarian, and Russian empires. Their borders divided previous territorial units or unified previously divided ones. New political divisions cut into the national geography. At the war's end, this cut—in addition to the return of soldiers, prisoners of war, and refugees—caused new migrations of people who found themselves outside the borders of their homelands or without homelands altogether. It created millions of members of national minorities who became targets of discriminatory policies and a source of further migrations as foreign elements in national states (Ther, 2016, 33–35). Many established prewar routes between places of origin and end destinations for labor migrations closed down, and many internal moves became transnational. For the same reason, other migration routes lost their international character and became internal.

The Slovenian ethnic territory became one of the complex grounds of these changes and processes. While it was almost completely united in a single state before the war, it was divided between four nation-states by its end. The central part, which we will call Yugoslav Slovenia, became part of the newly formed Kingdom of Serbs, Croats and Slovenes. It covered Carniola, Lower Styria, a small part of Carinthia, and Prekmurje, which, following the 1919 Treaty of Trianon, was unified with Slovenia, while Slovenians living along the Rába/Raba continued to be Hungarian citizens. Most of Carinthia, including the ethnically predominantly Slovenian part, became part of the new Republic of Austria. Most of its Slovenian population voted for this solution at the 1920 Carinthian plebiscite, partly because of historical and sentimental ties, partly because of the real and artificially created skepticism or opposition toward belonging to the Yugoslav state. No less important for the rejection were economic reasons and being separated from the regional capital Klagenfurt. In the west, the Kingdom of Italy occupied the parts of the Slovenian ethnic territory that used to be a part of the Austrian Littoral and parts of western Carniola in 1918. Two years later, the 1920 Treaty of Rapallo between the Kingdom of Italy and the Kingdom of Serbs, Croats and Slovenes formally annexed the territory, henceforth known as Primorska, to the Kingdom of Italy, making it part of the new eastern Italian province of the Julian March/Venezia Giulia/Julijska Krajina. Over a quarter of the Slovenian ethnic territory thus remained outside the borders of the Kingdom of Serbs, Croats and Slovenes. Consequently, almost a third of the Slovenian population found themselves in the position of national minorities in the neighboring countries.[27] Eight percent of the population in Yugoslav Slovenia were national minorities, the largest being German and Hungarian speakers. Such division of the Slovenian ethnic territory and its inhabitants also meant belonging to different state economies, migration regimes, and policies tailored to serve different national interests and coordinated with different internal needs and positions of these states on the international scene.

The war and its shocks left visible traces on the population. Compared to the last Austrian census from 1910, calculations based on the first postwar census—indeed, censuses, as Slovenians now lived in four different countries—showed a population drop throughout the Slovenian ethnic territory except in Prekmurje. Decreasing numbers affected all districts; the only population growth occurred in Ljubljana, Maribor, and Celje in Slovenia and, in Italy, Trieste and some individual towns where, because of their particular economic, infrastructure, or administrative function in the postwar regional configuration, the population increased. In Goriška (now in Italy), which was most directly impacted by the war, the population decreased significantly in the town of Gorizia and its Slovenian northern and eastern hinterland, while the south-western industrial district of Monfalcone/Tržič, where the shipyards were restored immediately after the war, recorded substantial growth by 1921. These dynamics and structural changes resulted from several factors: the mid-war drop in natural growth due to increased mortality and decreased fertility; in many places, migrations; the population turnovers caused by war; and the new political division that immediately followed. In some places, these factors also persisted in the years following the war (Vogelnik, 1965, 71).

In the 1920s, the re-established economy, the strengthening of industrialization, and the revitalized natural demographic increase brought about renewed population growth. By 1931, the number of inhabitants across the entire Slovenian ethnic territory was increasing in almost all districts, including those that, before the war, recorded a permanent population deficit, for example, the Črnomelj district in Lower Carniola. The only places that regressed were the towns of Celje and Ptuj (Yugoslavia) and the former districts of the Gorizia-environs and Sežana in Goriška (Italy). The natural

---

27 According to the last Austrian census from 1910, around 326,000 persons with Slovenian as their *Umgangsprache* lived in the territory that was annexed to the Kingdom of Italy, and a further 52,000 lived in Benechia according to the Italian statistics. Around 100,000 Slovenians remained in Austria (in Carinthia and partly in the border regions of Styria), and 6,000 in the Rába region in Hungary. The 1921 census in Yugoslav Slovenia counted 1,055,000 inhabitants (Kresal and Perovšek, 2005, 177–178).

growth in Yugoslav Slovenia, which (except in Styrian towns) was close to the level in the decade before the war (on average, 11.5 percent of the population in 1921), contributed to this positive development. In Primorska (Italy), population growth was slower. Population growth—also because of immigration from other Italian provinces—was concentrated in the bigger economic, administrative, or commercial centers, while in the Slovenian countryside the population stagnated. In the 1930s, natural population growth in Yugoslav Slovenia also slowed, as did actual population growth. Like before the war, the share of the agrarian population continued to shrink. In the 1930s, the only two regions where it remained above the 60 percent mark were Styria and Prekmurje—in the latter, which was still totally rural, it stood at 87 percent (Vogelnik, 1965, 71; Blaznik et al., 1970, 106–111). Likewise, the population continued to move from the countryside to urban areas and centers of intensive industrial progress.

Effects of the natural demographic growth in this period were further reduced by emigration. According to the census data for the period 1921 to 1931, the net emigration was 35,000 in Yugoslav Slovenia (Vogelnik, 1965, 60), almost a third of the natural demographic increase. It is impossible to be more precise in calculations for the 1930s because the next census only occurred in 1948. In Carinthia (Austria), where migration decreased significantly in the 1920s, negative net migration for 1923–1934 was 5,213 persons, or 12.6 percent of the natural increase. Events at the end of World War I and the first few postwar years triggered strong migrations. Authors estimate that in the entire interwar period around 25,000 persons left the Slovenian parts of (Austrian) Carinthia, and substantial numbers went to Yugoslavia (Klemenčič and Maver, 2017, 179). Primorska, a part of the Julian March, witnessed even greater population shifts due to its annexation to the Kingdom of Italy and turbulent political circumstances. On the one hand, this time was marked by a relatively intense emigration of the Slovenian (and in Istria, also Croatian) population and German and other nationalities from the erstwhile Habsburg Monarchy. Estimates reveal that over 100,000 Slovenians and Croatians emigrated from the Julian March in the interwar period (Čermelj, 1974, 174). At the same time, people from other Italian regions immigrated in high numbers to the Julian March for economic reasons or to carry out administrative or other functions.

Reconstructions of the demographic development show that the territory of today's Slovenia lost around 130,000 inhabitants through emigration between the two wars. Demographer Živko Šifrer calculated that the net emigration in the territory of present-day Slovenia between 1919 and 1941 was 132,830. There was an 8,858-person deficit in 1919–1921; 67,241 in the decade 1921–1931; and another 56,731 by World War II (Šifrer, 1974, 16–19). Demographer Dolfe Vogelnik corroborates these estimates, stating that the average annual net emigration between 1921 and 1940 was 6,200 people, amounting to almost 118,000 persons for the entire period (1965, 71).

The Slovenian space, whose central part (Carniola) had been, in relative terms, the most emigration-prone region in Imperial Austria, was also at the very top of all European countries when it came to interwar emigration. The average annual negative net migration rate within the borders of present-day Slovenia was 4.6 per thousand persons, followed by Norway (4.1 per 1,000) and Ireland (3.2 per 1,000). With its net emigration rate, Slovenia also stood out heavily compared to the rest of Yugoslavia, from which, on average, one in every thousand inhabitants emigrated yearly (Vogelnik, 1965, 59). It is also worth emphasizing that, at the same time, immigration to Yugoslav Slovenia increased sharply as well: prewar migrants returned from different former Austrian Crown Lands; refugees and other migrants arrived from Carinthia and Primorska; many wartime refugees did not return to their homes in Goriška; political refugees from Russia arrived; returnees came from overseas; and migrants entered from the rest of the Yugoslav space. Vogelnik estimates that this influx reduced the net emigration from Yugoslav Slovenia between 1921 and 1931 by one-third (1965, 60–61).

Slovenia was among the most economically developed regions in the new Yugoslav state because of its industrialization and general economic advancement. At the same time, it was also the region with the lowest rate of demographic growth and the highest rate of emigration. The slow population growth was mainly caused by the low rate of natural growth, which was, among other things, a consequence of the strong prewar emigration and the weakening of the demographically active segment of the population. Diverse cultural patterns, aspirations for modernization, and related ambitions and lifestyles brought various levels and characteristics of demographic development or greater or lesser intensity of migration movements and directions in different Yugoslav regions. In Yugoslavia's southern regions, the population remained firmly anchored in the countryside and bound to rural traditions. In Slovenia and Croatia, however, the changing socioeconomic model was strengthened through the expansion of the non-agrarian economy and popularization of urban social patterns and lifestyles (cf. Melik, 1938; Lazarević, 2015).

Like before World War I, the biggest emigration epicenters in Yugoslav Slovenia were the Lower Carniolan district of Kočevje, where between 1921 and 1931 the net emigration (–4,739) completely annulled natural growth, and the Prekmurje districts Dolnja Lendava/Alsólendva and Murska Sobota, where in the same period their combined net emigration (–10,470 persons) surpassed natural growth, thus causing their population numbers to decrease. Such high levels of migration in these districts also resulted from the departure of the German-speaking and

Hungarian-speaking populations, which became, like Slovenians in neighboring countries, a subject of national discrimination and homogenization policies. More pronounced emigration regions were also the districts of Krško (80 percent of the natural growth), Ljutomer (58 percent), a wider area between Ptuj, Celje, and Brežice (57 percent), Črnomelj (55 percent), and Litija (50 percent). The positive net migration—a result of the "flight to the cities"—was observed in Ljubljana and its surroundings, Maribor, Kranj, Celje, and Ptuj. Everywhere else, emigration surpassed immigration. Regionally, the phenomenon was stronger in Styria, where the negative net migration was 3.6 percent of the initial population. In Carniola, it was 2.3, but at the top was Prekmurje with 11.3 percent. In Primorska, emigration affected the entire region. In large numbers, the Slovenian population was leaving rural areas, Trieste, and Gorizia, for Yugoslavia, other countries, and inland Italy. Nevertheless, migration from the Slovenian countryside of Primorska to Trieste, Gorizia, and other important regional economic centers continued.

Map 1.2: The Slovenian ethnic territory in the period 1919–1929.

A Turn in Migration Geography

The geography and scope of Slovenian migration completely changed between the two world wars. The new configuration's three distinguishing traits were an overall decrease in migration abroad as a consequence of the drop in transatlantic migration, an increase in migration to European countries with the number of Slovenians migrating within Europe at least three times that of those migrating overseas, and the redirection of migration routes to various parts of the newly established Yugoslav homeland. The statistics of the Jugoslavanski iseljenićki komesarijat (Yugoslav Emigration Commissariat) (see Table 1.6) show that between 1921 and 1939, other continents were a destination for 23,000 migrants from Yugoslav Slovenia. However, between 1927 and 1939 (the years for which data of all migrations abroad are available), 14,000 persons crossed a sea, while 63,000, or 82 percent of all international migrations, went to one of the European countries. Migration to previously established and new European destinations started in the first postwar years. For example, before World War I, France was only a marginal goal for Slovenian migration. In 1929, it was home to around 16,000 migrants from Yugoslav Slovenia (Lipoglavšek-Rakovec, 1950, 43).

Table 1.6: Yugoslav Slovenia: overseas and continental migration, 1921–1939. Source: Statistički godišnjak 1929–1939.

	Overseas		Continental	
Year	**Emigrants**	**Returnees**	**Emigrants**	**Returnees**
1921	2,489	1,914		
1922	440	1,523		
1923	1,312	458		
1924	1,197	611		
1925	1,579	697		
1926	2,134	649		
1927	2,727	677	1,403	
1928	3,162	674	2,307	
1929	1,976	719	5,075	

	Overseas		Continental	
Year	**Emigrants**	**Returnees**	**Emigrants**	**Returnees**
1930	1,671	1,264	6,585	3,022
1931	703	507	4,032	3,402
1932	339	708	2,227	3,021
1933	307	258	2,111	1,506
1934	420	232	2,935	2,352
1935	408	188	2,920	1,942
1936	434	220	3,590	1,782
1937	557	319	6,865	2,886
1938	648	319	7,636	9,873
1939	435	164	15,345	11,353
Total	22,938	12,101	63,031	41,139
1927–39	13,787	6,249	63,031	41,139

Another characteristic of interwar international migration was the high number of returnees because of the repatriation of migrants who had lived in the receiving countries for a long time and the change in migration patterns, more precisely, the increase in seasonal and other short-term contract migrations. The net emigration of Yugoslav Slovenia between 1921 and 1939 was approximately 35,000 migrants (see Table 1.7). Added to this number must be those Slovenian migrants who migrated from the territories that, after World War I, became parts of Italy and Austria. Based on rough calculations, the net emigration from the territory of present-day Slovenia would include around 75,000 people.

The End of the Great Transatlantic Migration

As an extension of the prewar cycle, transatlantic migrations continued in substantial numbers until the early 1930s, when they significantly diminished (see Tables 1.6 and 1.7). In the 1920s, it included 18,700 persons, or 81 percent of the entire interwar transatlantic migration from Yugoslav Slovenia. It was still predominantly oriented toward the United States. However, immigration quotas to the country significantly limited the access for Yugoslav and, therefore, also Slovenian migrants: in 1921, when the quota was calculated as 3 percent of the immigrants from individual countries who were included in the 1910 census as permanent residents of the United States, the Yugoslav contingent was limited to 6,426 arrivals per year, with 700 of those arrivals allocated to immigrants from Yugoslav Slovenia (Izseljevanje v Zedinjene države ameriške, 1922). The Immigration Act of 1924, which reduced the share to 2 percent and chose the 1890 census as a point of reference, when the presence of Slovenian and other "new immigrants" was significantly lower, reduced the quota for Yugoslav immigrants to 671 arrivals per year and effectively shut that country's door to them. Increasing the quota to 845 arrivals per year in 1929 did not significantly change arrival possibilities. The legislation allowed immigration beyond annual quotas to include family members of migrants who had become naturalized US citizens (Klemenčič, 2011, 113). After the quota restrictions, family members represented the majority of new Slovenian and other Yugoslav immigrants to the United States. In addition to the legal routes, some immigrants avoided the quotas using illegal routes through neighboring countries (Đikanović, 2012).

Many Slovenian migrants in the United States postponed their planned return home because of the beginning of World War I. However, the war and the changes in Europe and the homeland often contributed to migrants opting for permanent or extended stay in the United States, with many deciding to become naturalized US citizens. Vatroslav Grill, a judge, editor, and political leader, states in his memoirs that the war was a watershed moment for the identification with the "new homeland" and that immigrants "took American citizenship in scores and were proud of it" (Grill, 1979, 107). Naturalization was, of course, also a way for immigrants to declare their loyalty to the United States and protect themselves from pressure on the citizens of hostile countries (cf. Bodnar, 1977, 121). In the postwar years, permanent settling was also accelerated by quotas preventing returnees without US citizenship from re-entering the United States after a longer than six months' absence. This restriction changed the nature of migration because the United States stopped being a transnational labor market and no longer allowed the iterative temporary coming and going. Many moves that were initially meant to be temporary thus changed into permanent ones. The postwar migration pattern to the United States—including family reunification and arrivals and departures connected to the organization of life across the ocean—primarily reflects the stabilization phase of the Slovenian immigrant communities in the United States.

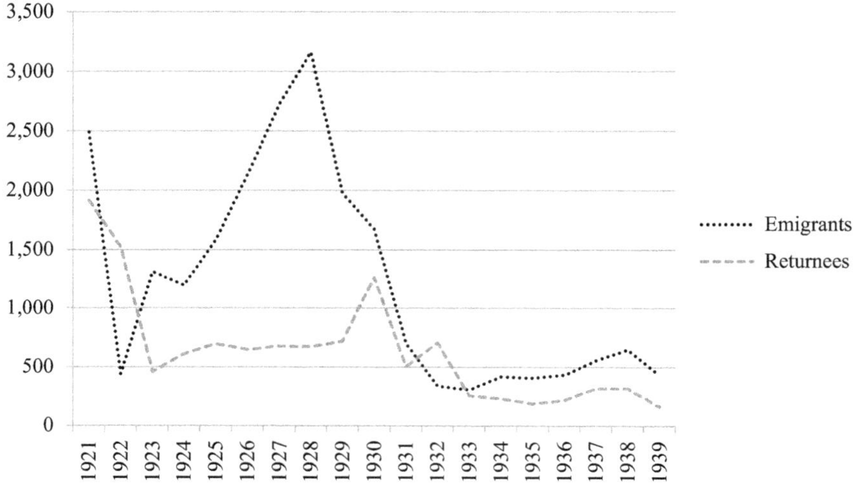

Graph 1.4: Yugoslav Slovenia: Overseas migration, 1921–1939. Source: Statistički godišnjak 1929–1939.

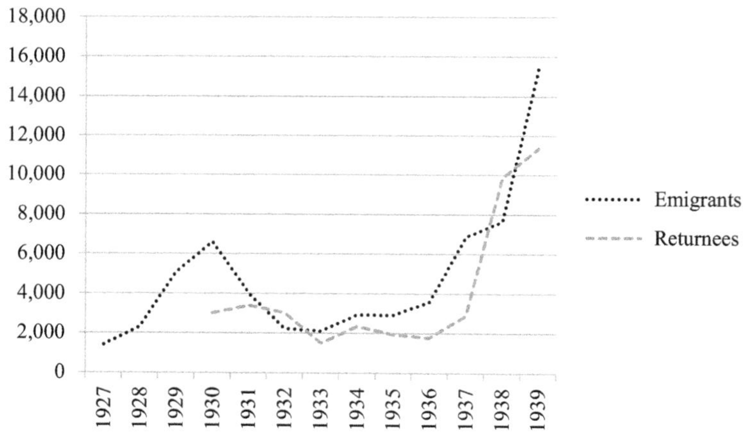

Graph 1.5: Yugoslav Slovenia: Continental migration, 1921–1939. Source: Statistički godišnjak 1929–1939.

However, even in this phase, transatlantic migration was not unidirectional. If some had irrevocably chosen the *frej kontra* (free country), others decided to return home. Between 1921 and 1939, just over 12,000 persons moved to Yugoslav Slovenia from overseas, which amounted to 53 percent of emigration in the same period; mostly, these were returnees and their descendants. Many migrants returned immediately after the war when the transatlantic return migration was the greatest and significantly surpassed the number of emigrants. By 1921, over 40,000 persons had moved to the Kingdom of Serbs, Croats and Slovenes from overseas (Izveštaj, 1925/26). The crisis at the end of the wartime economy cycle in the United States and a favorable dollar exchange rate contributed to the return of prewar migrants (Karanović, 1978, 357–358), as did the expectations upon the establishment of the Yugoslav state, which for a certain period even offered to pay migrants' journey home. However, the fact that, in the following years, about 30 percent of the emigrants from Yugoslav Slovenia were foreign citizens (between 1927 and 1935, there were 3,965) shows that many returnees from the United States (and partly from other countries) who returned home as foreign citizens, eventually decided to re-emigrate, many because they were disillusioned with the "new old country" (Kolar Dimitrijević, 2012).

One consequence of the system of quotas and increased family migration or reunification was the changed age and gender structure of migrants to the United States, which now included a larger share of women and children (Jezeršek, 1929–1930, 177–178). Between 1918 and 1924, women emigrants from Yugoslav Slovenia outnumbered men significantly. Many postwar women migrants and their children traveled to reunite with their husbands and fathers; unmarried women often followed their fiancés. Some migrants traveled to join their other relatives. Ship passenger lists confirm similar trends for the migrants from Benechia (Kalc, 2018a, 115–116). The transatlantic migrants were still overwhelmingly employed in the agricultural sector, but there were significantly more skilled workers and independent professionals than before the war. This shift testifies to the changed migration due to the socially structured "American Slovenia" across the ocean. Another telling fact related to the community in the United States is that more than half of the postwar migrants departed for the United States to join their family members and relatives, who, in many cases, covered their travel costs (Valenčič, 1990b, 66).

The second most popular destination of overseas migration from Yugoslav Slovenia was Canada. In the interwar period, 4,359 people migrated to Canada, and 1,234 persons returned, bringing the net emigration to Canada to 3,125 (Genorio, 1989, 72). Emigration to Canada shrank by the end of the 1920s since, due to the Great Depression and unemployment, the country severely limited immigration in 1930. Since 1924, Canada had already categorized Yugoslav migrants as "undesired" but continued to accept them because it needed a workforce. It favored migrants who had the means to purchase land as well as agricultural and forest laborers. Laborers, whose physical abilities were checked upon arrival, had to commit to working on farms for a fixed period and only later seek employment in the non-agrarian sector, mainly mining, forestry, and infrastructural construction (Dražumerič, 1987, 174; Genorio, 1989, 60–61; Miletić, 2009, 102–105).

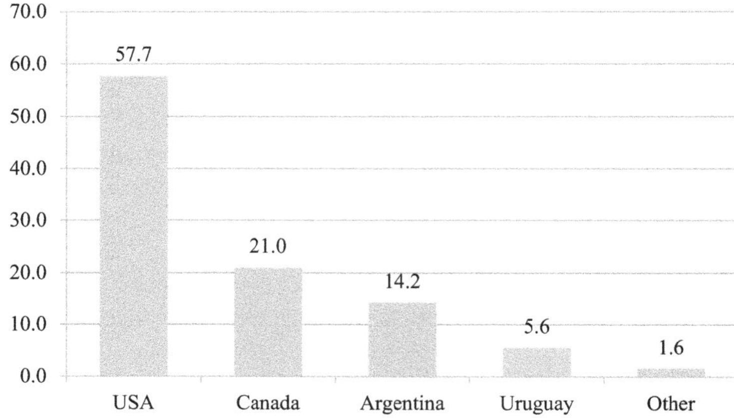

Graph 1.6: Yugoslav Slovenia: Overseas migration by country of destination, 1929–1939 (in percent; N=7,898). Source: Statistički godišnjak 1929–1939.

Argentina followed Canada with 3,024 migrants and Brazil with 525. The numbers for other countries, from Uruguay to Australia, which introduced immigration quotas following the US example, were lower. South American countries also tightened immigration conditions and required a medical and professional certificate and a clean criminal record. They checked newcomers' political orientation to prevent admission to agitators and avoid social conflict (Miletić, 2009, 105–107; Devoto, 1989). South American countries became prominent migration destinations once the United States introduced immigration quotas. Before the Great Depression, it was primarily a destination for labor migration and, as such, predominantly male. By the late 1920s, however, more and more countries began to impose immigration quotas, and the migration pattern to South American countries shifted from predominantly labor migration to migration for the purpose of family reunification. As a result, female migration surpassed male migration in all overseas destinations in the 1930s. This structural shift denoted the process of migrants settling in the Americas and brought the long migration cycle to a close.

Picture 1.9: Slovenian and Croatian construction workers in Cacheuta, Mendoza, 1934. Author: unknown. Courtesy of: Slovene Ethnographic Museum, Ljubljana.

In the interwar period, transatlantic migration almost completely lost its prewar role in the Slovenian social and economic context. Its function as an economic resource diminished, as did its effect on alleviating demographic pressure on existing economic sources in the places of origin. When we consider the high number of returnees, the negative balance of the migration exchange between Yugoslav Slovenia and overseas destinations was just under 11,000 persons (see Tables 1.6 and 1.7). Emigration from Yugoslav Slovenia represented about 10 percent of all overseas migration from Yugoslavia, which amounted to 202,000 people between 1920 and 1939. Of the emigrants, 79,000 moved to the United States, 34,000 to Canada, just under 8,000 to Australia, and the rest to South America, mainly Argentina and Brazil. The state gained significant economic benefits from Yugoslavs in the United States, and an important share of remittances still came from Slovenians. The otherwise decreasing inflow of money kept coming until the start of the Great Depression and substantially impacted the Yugoslav balance of payments; on a smaller scale, it continued in the following years. From the war's end until 1933, emigrants were said to transfer 250 million dollars home,[28] while the Yugoslav debt was 280 million dollars (Hranilović, 1987, 327–329). Based on official data from the Ministry of Finance, the National Bank, and other state institutions, the net inflow of remittances from migrants fell to one-eighth in 1933 and 1934 compared with the 1920s. In the second half of the 1930s, the amount increased to two- or three-eighths (Brunnbauer, 2016, 227).

Old and New European Routes

Continental migrations show completely different dynamics. As a reaction to the closure of the traditional North American labor market, migration to European countries in the 1920s surpassed the transatlantic one and doubled in the 1930s. However, Table 1.7 shows that a significant part of migrations was temporary or seasonal. This trait was reflected in the number of immigrants from the European countries in the 1930s, most of whom were returnees: over 47,000 persons emigrated between 1930 and 1939, and 37,000 immigrated. Almost half of all migrations between 1935 and 1938 were seasonal.

Fluctuations in emigrations and returns were generally also influenced by the economic situation and changing possibilities of employment, immigration policies, bilateral agreements on work contracts, and, of course, the political atmosphere in some countries and changed attitudes toward foreigners. The general economic crisis—which reached its peak in 1934—also impacted this dynamic. The economy started recovering in the middle of the decade, and in the following years the workforce demand increased again, particularly in Germany, Austria, and Czechoslovakia (Lazarević, 2005, 473–479; Drnovšek, 2012, 30–31).

28 Estimates vary significantly. Some claim that between 1919 and 1923, the inflow was between 26 and 27 million dollars annually (Miletić, 2009, 158). Other sources cite significantly higher numbers, namely 60 million dollars for the year 1919, and between 20 and 40 million dollars annually after 1924 when migration quotas were introduced in the United States, while before World War I, it was said to reach 160 million dollars (Govorichin, 1961, 107).

Table 1.7: Yugoslav Slovenia: migration abroad, 1921–1939 (in thousands). Source: Vogelnik, 1965, 62.

Type of migration	**Period**		**Year average**	
	1921–1931	1931–1939	1921–1931	1931–1939
Overseas				
Immigration	9,2	2,9	0,9	0,3
Emigration	18,7	4,3	1,9	0,5
Net migration	−9,5	−1,4	−1,0	−0,2
Continental				
Immigration	11,2	37,0	1,1	4,1
Emigration	24,4	47,7	2,4	5,3
Net migration	−13,2	−10,7	−1,3	−1,2
Total				
Immigration	20,4	39,9	2,0	4,4
Emigration	43,1	52,0	4,3	5,8
Net migration	−22,7	−12,1	−2,3	−1,4
Total Per 1,000 Inhabitants				
Immigration			1,8	3,7
Emigration			3,9	4,9
Net migration			−2,1	−1,2

The net emigration from Yugoslav Slovenia to European countries was just under 24,000 persons in the interwar period. Because of the economic depression in the early 1930s, restrictive migration policies, and the beginning of World War II, more migrants settled abroad in the 1920s than in the following decade. Migrants from Yugoslav Slovenia represented around 30 percent, and in the second half even surpassed 50 percent of all Yugoslav migration to European countries. In Yugoslavia, continental migration became an increasingly

Slovenian characteristic in absolute and even more so in relative terms. This characteristic is true for traditional destinations, such as Germany and Austria (in the 1930s representing 67 and 80 percent of all the Yugoslav emigration, respectively), France (50 percent), and the Netherlands (58 percent). Emigration from the rest of the Yugoslav territory was also directed to other countries, particularly Turkey, Romania, Albania, and Greece. The non-Slavic ethnic groups emigrated there from Yugoslavia for political and national reasons.

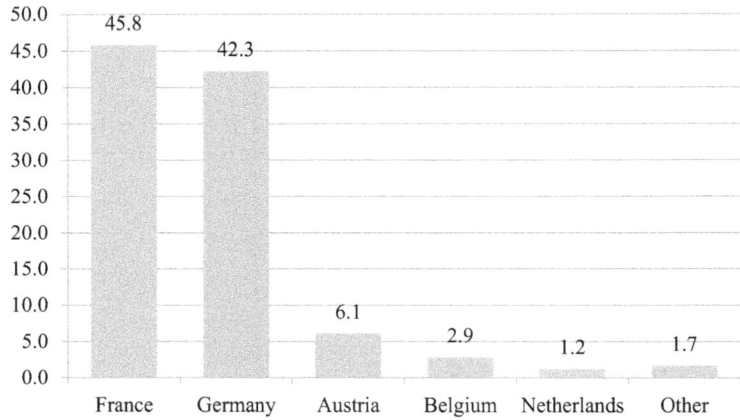

Graph 1.7: Yugoslav Slovenia: migration to European countries, 1929–1939 (in percent; N=59,321). Source: Statistični godišnjak 1929–1939.

Germany

After the war, the moves from the Slovenian ethnic territory to the rest of Europe reconfigured. They were influenced by the postwar economic situations, needs, and migration policies in European countries and changes in migration regimes that occurred on the international level with the implementation of economic protectionism and state interventionism. Although the German Empire had been the largest workforce importer before the war, Weimar Germany significantly limited immigration and subjected it to strict state management. Economic and social circumstances were different from those in the neighboring countries, which immediately opened to migration due to the severe workforce deficit. The country had to face the difficult repatriation of two million foreign civilians and prisoners of war, the social reintegration of its own soldiers, and incoming refugees from Russia and Poland. The population moved abroad due to economic problems and high unemployment in cities and industrial centers.

At the same time, the agrarian economy felt shortages of its seasonal workforce, which had to be imported just as before the war. Based on demands and growth in individual production sectors, the centralized office for employment and labor agency hired and regulated the influx and professional profile of the foreign workforce using criteria negotiated with employers and union organizations. Foreign workers were only hired if there was a deficit in the local workforce. They had to have previously arranged formal employment contracts that stipulated equal wages and social rights; these contracts were initially available for the agrarian sector and only later for other economic sectors. Work and residency permits in Weimar Germany were limited to one year, with permanent residency and access to all employment sectors only available to those who had resided in Germany for a certain number of years. Previously only for agricultural workers, this policy expanded the legislated obligation of return to all foreigners (Herbert, 1990, 121–126).

Difficult circumstances in the German economy prompted many Slovenians to return from Westphalia in the first years after the war or move to France, Belgium, the Netherlands, or the United States. Their returns to Slovenia continued until the mid-1920s when the economic situation in Weimar Germany improved. Slovenian presence in Germany thus shrank. The Yugoslav Emigration Commissariat put it at 20,000 in the middle of the decade (Valenčič, 1990b, 67), while the official German statistics showed 14,000 Slovenian migrants, of whom around 5,500 were in the Rhine Valley and 4,800 in Westphalia (Drnovšek, 2012, 145; Wörsdörfer, 2017, 142). In 1928, it is estimated that about 20,000 to 30,000 Slovenian immigrants and their descendants resided in Germany (Lipoglavšek-Rakovec, 1950, 40–41; Drnovšek, 2012, 145). Most Slovenian immigrants in Westphalia had firmly settled in Germany by then. Many took German citizenship when the Yugoslav state was established. Some opted for it in 1928, when the Yugoslav law no longer permitted dual citizenship. Others acquired it by marrying German citizens, some quite opportunely and not least because, despite the formal equality with German nationals, they were considered second-class residents compared to Germans. Often, their children partly or fully assimilated, as did some of those Slovenian immigrants who had arrived before the war (Herbert, 1990, 126; Drnovšek, 2012, 144–5; Wörsdörfer, 2017, 127). Throughout the interwar period, deciding on citizenship was among the factors that influenced the decision to remain, return, or migrate to other countries (Wörsdörfer, 2017, 119). The number of people taking German citizenship increased with the economic crisis, growing concerns regarding the ability to retain jobs and social security benefits, and the political pressure on foreigners. Tendencies for naturalization were alleviated by an agreement on labor insurance and an agreement on paying annuities to miners who returned to Yugoslavia (Drnovšek, 2012, 148–9).

In the late 1920s and early 1930s, migration from Slovenia to Weimar Germany briefly increased. The increase was linked to a 1929 German–Yugoslav agreement on the mobility of agricultural laborers. Migration again became lively after 1937 because of seasonal and more permanent migrations of agricultural laborers (Drnovšek, 2012, 146–7). Nazi Germany, which limited migration—despite workforce shortages—for autarkic and monetary reasons, opened the door to foreign agricultural workers to fill needs created by the increasing mobilization of the German workforce for the military industry. Nazi Germany, therefore, signed agreements with several countries and, between 1937 and 1939, Yugoslavia was one of them. In three years, over 20,000 laborers from Yugoslav Slovenia left for Germany; in 1939 alone, there were 13,000. Even though it wanted to procure more people from Yugoslavia, Germany was quite hesitant about a larger influx of foreign workers. Adhering to its racial and cultural paradigm, it partly solved the labor shortage by recruiting people from Czechoslovakia and Austria. The Nazi regime also tightened security measures and selective criteria regarding foreign workers. It set up an apparatus of supervision and specific regulations aimed at individual ethnic groups. It repressed and deported immigrant workers who, with their behavior, might in any way deviate from the regime's interests (Herbert, 1990, 127–131).

Estimates of the size of the permanent Slovenian immigrant community in Germany in the 1930s vary, probably due to very different census criteria or definitions of a community. In 1933, German statistics recorded just over 17,000 Yugoslav citizens (mostly Slovenians). Yugoslav authorities list 25,000 because many had kept Yugoslav citizenship after becoming naturalized German citizens (Lipoglavšek-Rakovec, 1950, 40). The higher estimates, as a rule, also include completely assimilated descendants. There were approximately 60 Slovenian immigrant communities near factories and mines in the Ruhr region, while in the entire country, there were approximately 140 (Lipoglavšek-Rakovec, 1950, 41; Wörsdörfer, 2017, 143–146).

France

After World War I, France became the most frequent destination of Slovenian migrations abroad. After the United States restricted access, France became the metaphorical "new America" in the eyes of Slovenian labor migrants. Already in the decades before World War I, France was a significant importer of laborers. With the increased population deficit caused by the war, the need for a foreign workforce to help restart the economy also considerably increased. It was spurred, on the one hand, by social changes, internal migrations, professional mobility, and the restructuring of the labor market. On the other hand, the acquisition of the predominantly mining and industrial regions of Alsace and Lorraine also influenced it because a part of the German population, which had previously represented an essential share of manpower, had retreated. The French mining, industry, construction, and agrarian economy thus became attractive labor markets for immigrants from Eastern and Southeastern European countries. Since 1919, France signed contracts to acquire workers with countries that traditionally supplied workers, such as Italy, and the newly formed European countries, the most important of which was Poland (Amar and Milza, 1990, 7–12).

Yugoslavia was among the friendly countries France turned to for help to fill the labor gaps and workforce shortage. Yugoslavia contributed incomparably smaller immigrant contingents than other countries. Slovenian laborers first left for France from Germany, with supposedly around 3,000 miners moving from Westphalia to Pas-de-Calais in 1921. In the following years, more moved to the Moselle department (Bogovič and Cajnko, 1983, 34; Drnovšek, 2012, 190). Soon, they were joined by migrants from Slovenia, whose numbers grew after 1923. Migration from Slovenia to France peaked in the early 1930s and again in 1936 and 1937 when seasonal agricultural workers also went there. In 1929, according to the data from the Yugoslav Emigration Commissioner in Paris, 17,000 Slovenians from Yugoslav Slovenia resided in France (Pislar Fernandez, 2004, 274). Data from the Foreigners' Bureau in Paris have prompted some authors to state the number 25,382, which includes temporary migrants (Bogovič and Cajnko, 1983, 122; Pislar Fernandez, 2004, 78).

At the end of the 1930s, 23,000 Slovenians were living permanently in France. The Emigration Commissariat estimated that, together with those from the Julian March, there were 34,000 (Lipoglavšek-Rakovec, 1950, 43) living in larger and smaller communities throughout the country. Most Slovenians, however, lived at both ends of the northeastern border of the French hexagon: Lens in Pas-de-Calais and Aumetz and Merlebach in Moselle (Lorraine). The Slovenians in Pas-de-Calais mostly resettled there from Westphalia, while new migrants joined the existing prewar communities in Moselle. Many also lived in the Centre Region, in the departments Loire, Puy-de-Dôme, Nièvre, and Saône-et-Loire, as well as in Paris and Île-de-France, where the postwar migration created the smallest Slovenian immigrant colony (Pislar Fernandez, 2004, 276, 305; Drnovšek, 2012, 192–193). During the economic crisis between 1932 and 1936, French authorities allowed Slovenian workers to seek employment in the agricultural sector instead of mining and other industries.

Picture 1.10: Miners' neighborhood in Merlebach, France, in the 1930s. Author: unknown. Courtesy of: Research Centre of the Slovenian Academy of Sciences and Arts, Slovenian Migration Institute.

The Netherlands, Belgium, and Luxembourg

Mining in the Netherlands only started developing when private and, later, state mines opened at the end of the 19th century, but the real growth occurred after the war. In the Limburg mining district, the number of inhabitants jumped from 40,000 at the beginning of the century to 170,000 at the end of the 1930s. In the same period, the number of miners increased from 6,500 to 25,000. The response to the growing demand for workers first came from neighboring Germany, and many of those coming from Westphalia and the Rhine Valley were Slovenians. They moved because their situation in Germany was not ideal and were welcomed because they agreed to the mining wages in the Netherlands, which were considered unacceptably low by Dutch miners. Later, miners also left for the Netherlands from Belgium and France. When miners began arriving from Yugoslavia in 1926, they were all Slovenians.

Immigration from Slovenia peaked at the end of the 1920s when an estimated 4,000 Yugoslav immigrants and their families lived in the Netherlands. Most Slovenians came from Ljubljana-environs and Celje-environs; some were from Austria and Italy. They mostly lived in the Limburg province along the borders with Germany and Belgium. By 1939, the number of Slovenians in the Netherlands had halved due to the economic crisis and unemployment. Many returned to Yugoslavia; some left for other countries. Those who stayed found work in factories or in other sectors. Women followed the market's demand for domestic service and other jobs and moved from settlements in Westphalia to Dutch cities (Mulders, 2010, 21; Drnovšek, 2012, 264–268).

The migration structure was similar in Belgium and Luxemburg, whose economies advanced considerably in the 20th century. After the war, Belgium attracted many migrants from neighboring countries and the countries in the European east and south. Slovenians came to Belgium from Germany and Austria in the first postwar years. From 1923, they also came from Slovenia and, with Italian contingents, from the Julian March. Slovenian migration to Belgium peaked at the end of the 1920s and then shrunk significantly with the economic crisis. At the beginning of the 1930s, Belgian authorities estimated that around 3,000 Slovenians were living in the Belgian Limbourg. Official statistics from 1936 show 5,500 Yugoslavs in the entire country. More than half of them were Slovenians. By 1939 their number dropped to around 2,500; these numbers do not include Slovenian migrants from Italy. Slovenian migrants were more scattered in Belgium than in the Netherlands or France. Most of them worked in the coal mines of Limbourg and Liège and were steel industry laborers in Charleroi in central Belgium (Drnovšek, 2012, 300–301).

Predominantly Organized Emigration and Contract Work

As before the war, the postwar labor migration to European countries was partly spontaneous but increasingly organized via direct recruitment by state or other official work agencies. This intermediary work was done in accordance with the migration policies and agreements which destination countries used to regulate workers' arrivals and rights. Until the beginning of the economic crisis, destination countries and employers, particularly mining companies, competed for migrant laborers, as they were crucial for their economic interests. Hence, they recruited workers directly from their countries of origin. In Slovenia and elsewhere in Yugoslavia, for example, France initially recruited workers even without the permission of the Yugoslav authorities (Pislar Fernandez, 2004, 74). Workers who left for France independently had to report in Toul (Moselle), where a processing and dispatching center for foreign workers was set up. From 1924 onward, the workers were recruited through the Société générale d'immigration, which worked together with the Yugoslav Emigration Commissariat of the Ministry of Social Policy. In 1926, it opened its office in Zagreb. Together with employment offices in Yugoslavia, it scouted interested workers, checked their physical and professional abilities, officiated labor contracts, and organized group transportation. French immigration authorities directed Slovenian and Yugoslav immigrants primarily to coal and iron mines (Pislar Fernandez, 2004, 75).

Dutch mining companies were also actively scouting for workers in Yugoslavia. By recruiting them directly in their homelands, they tried to find employees who were more manageable than the unionized and socially aware laborers in the Netherlands. They favored migrants without mining experience who simply had to meet certain physical and moral criteria to be selected. Agents searched among the rural Catholic population, convinced that such a workforce would associate along the ethnic and religious lines under the patronage of the clergy and then return home. Dutch agents operated in the field, without intermediaries, recruiting, for example, people in taverns on Sundays (Mulders, 2010, 29–30).

Migration to Germany was also primarily based on recruitment and under the supervision of relevant state bodies. German mining companies worked hard to lure laborers from competitive mines elsewhere. In all European countries, as a rule, migrant workers had to obtain work permits, sign contracts with their employers, get residency permits, and, in some places, visas in advance. Occasionally, they were required to produce a certificate of their work experience and good behavior. Migration based on recruitment and a pre-existing agreement was better for migrants because their transportation was paid for, and they were offered accommodation where they worked. When people left independently, without a contract or other arrangements, they had to deal with all of this on their own. Despite strict control and regulation of migration processes, characteristic of countries of origin and even more so in receiving countries, some left Slovenia illegally and entered their countries of destination without work permits. Employers also took on workers who arrived in their countries in such a way (Mulders, 2010, 35).

In the entire interwar period, the typical employment sectors for the Slovenian immigration communities in the European countries were mining and, to a lesser extent, other industries. Migration abroad thus continued to retain a working-class character and was professionally part of the lowest segments of the labor market. Workers took piecework in mines because they could earn more. The unions fought against it because it weakened workers' negotiating power and worked against their efforts to reduce exploitation, physical exhaustion, and numerous work accidents (Pislar Fernandez, 2004, 100–101). Like other immigrant workers, Slovenians were also characteristically mobile. That was connected to the general economic situation in individual countries and the search for better working conditions and higher wages. Changing occupations and employment sectors or combining economic sources were also widespread. Some immigrants entered trades and services; these two industries often developed as a segment of the "migrants' economy," that is, migrants providing services to meet the needs of their communities.

In France and Belgium, migrants opened taverns or took over managing canteens for single workers. Taking tenants and providing services for single migrants were, as before the war in the German Empire and the United States, relatively widespread sources of income, particularly once migrants were joined by their families and settled in separate apartments in working-class neighborhoods. Large working-class houses—often surrounded by garden allotments—allowed families and extended households to grow produce and thus reproduce the combined agricultural and wage labor economy they had known in their places of origin. Renting additional allotments for market production was also relatively widespread, as was wage work for large landowners, employment of women in canteens, domestic service for the senior employees of the mine's administration, and other permanent or temporary forms of creating income, which also worked as a compensation system when economic crises hit the mining sector (Pislar Fernandez, 2004, 74, 97–100; Mulders, 2010, 21; Drnovšek, 2012, 147, 196, 298–299). Professional advancement and social mobility were low and only slightly improved in the generation of their children through vocational education and professional training. That was particularly true for Germany where the Slovenian presence was older and the immigrants were better integrated into the social fabric. However, in the mining regions—as before the war—the sons mostly followed in their fathers' footsteps, professionally speaking.

Theoretical Equality and the Reality of the Great Depression

When European countries intensified their concern for social welfare, they also increased their concern for the regulation of the status of migrant workers. The receiving countries were forced to do so because they needed foreign workers, and labor organizations also influenced the attitude toward the immigrants and their position. Labor organizations played a significant role as social partners with the state and employers in negotiating labor and social legislation and influencing migration policy. Germany, France, Belgium, and the Netherlands treated foreign workers the same as the local workers and guaranteed them equal positions regarding payment and social rights. Immigrants were provided with accommodation in either hostels, if they were single, or in residential houses, which only in Belgium were of lower quality, while in France and particularly in the Netherlands and Germany they were relatively spacious and had modern equipment (Lipoglavšek-Rakovec, 1950, 48–49; Drnovšek, 2012, 194, 268).

Like before the war, housing policy was part of the strategies for migration policy. Its purpose was not only to provide adequate living conditions but also to stabilize migrants and limit their mobility. This question was particularly pertinent in coal-mining regions, as the cycle of coal production was extremely unstable and the demand for laborers changed a lot. Mines tried to keep the laborers from departing to other production sectors during low production spells, when miners were unemployable, so that they were available as soon as the production cycle started picking up again (Sori, 1979, 365). In addition to employment possibilities, migrants were also lured by the social security policy. Slovenian migrants' economic and living standards in European countries were generally better than at home. Once their contracts ended, they strove to remain; many returnees decided to leave again. In the 1920s, women with children, fiancées, and other relatives soon followed men, particularly to France as well as the Netherlands and Belgium. Many seasonal workers remained in France as well. Slovenian colonies thus grew and solidified their character as structured communities (Drnovšek, 2012, 194, 196).

However, theoretical and declarative equality did not guarantee actual equality. The position of migrant laborers in everyday practice often differed from that of the national workforce. It also depended on agreements between receiving countries and countries of origin, which used conventions to recognize specific rights and harmonized reciprocal social pension and insurance systems. Such agreements influenced the general attitude of employers, state bodies, and society toward immigrants. In this respect, Yugoslav laborers were worse off than migrants from other countries, as Yugoslavia's government was late in solving such issues and representing its migrants. The competent ministries did not put forward clear positions and guidelines on migration policy. The country's negotiating position was also weak so, in bilateral economic relations, it could not advocate for its workforce as efficiently as other countries did.

The Treaty on Social Insurance between Germany and the Kingdom of Serbs, Croats and Slovenes was only concluded in 1928, and the Agreement on Providing Social Equality for agricultural, mining, and factory workers a year later. In 1928 and again in 1939, the countries concluded the Arrangement concerning Serbian, Croat and Slovenian agricultural seasonal workers, which, considering pre-defined wage and insurance conditions, allowed specific contingents of Yugoslav laborers to integrate into the German industrialized agrarian economy. In 1933, a new agreement confirmed that the status of the Yugoslavs who worked in Germany for a certain amount of time was equal to that of German laborers. The same agreement also guaranteed Yugoslav workers benefits in case of unemployment and social transfers for their family members (Drnovšek, 2012, 32–33).

Despite friendly relations and mass emigration from Yugoslavia, the Yugoslav government did not sign a convention with France until 1932. Because of delays by France and the economy's slow recovery after the crisis, the convention was only ratified in 1939, so it was never implemented (Drnovšek, 2012, 189). Yugoslavia did not sign similar agreements with the Netherlands and Belgium. Yugoslav migrants in Belgium had no right to unemployment benefits, and workers in Belgium were thus only eligible for pension if they retired after full length of service (and not if they took early retirement). In the case of coal miners, however, few reached retirement age because hard work and silicosis—known as "black lung disease" or coal workers' pneumoconiosis (CWP)—killed them en masse (Drnovšek, 2012, 297–298). The position of immigrants from Yugoslav Slovenia thus differed from that of immigrants from many other countries, particularly their fellow Slovenians who migrated from the Julian March, who, as Italian citizens, enjoyed greater protection and social rights (Pislar Fernandez, 2004, 102; Drnovšek, 2012, 36). Even before World War I, Italy and France negotiated a convention on reciprocal recognition of the rights of labor migrants. After the war, Italy was at the forefront in its endeavors to implement international labor law criteria as part of the International Labour Organization, at international conferences on migrations, and on the level of bilateral agreements between countries (Tosi, 2002).

If, in the 1920s, the differences in levels of protection were buffered by the favorable economic situation and liberties, during the Great Depression, the unresolved issues regarding the status of immigrant workers, caused by the absence of international agreements, became quite dramatic. It did not help that economies were shifting into national molds and protectionist logic. Foreign workers in France felt the consequences most acutely: the country, which in the 1920s was known for its openness to immigration and the acceptance of newcomers into the French society, now fostered a prevalent negative policy toward foreigners and an intolerant, xenophobic atmosphere. Throughout the 1920s, the French government had already considered abandoning its liberal immigration policy and was debating issues such as desirable and undesirable immigrants and the French national identity (Noiriel, 2007, 187). The state passed a series of discriminatory measures. In 1932, it passed the Law for the Protection of National Labor. In 1935, it limited the right to family reunification. In 1936, it limited the immigration of certain ethnic groups and social categories and, in 1939, it abolished the rights of free association for foreigners (cf. Cross, 1983, 188–207). Slovenian and other Yugoslav immigrants found themselves in a particularly precarious position because of the negotiated but unratified convention between France and Yugoslavia to guarantee them a certain level of workplace protection and social security. In addition to dismissals and the aggravated economic situation that pushed many into poverty, they were subjected to bullying, were not paid unemployment benefits or other social transfers, and lost work permits. They were thus forced to leave the country and return home where they could not claim the years of pensionable service earned abroad (Drnovšek, 2012, 187, 195).

Picture 1.11: Children of Slovenian immigrants in Pas-de-Calais, France. Author: unknown. Courtesy of: Združenje Slovenska izseljenska matica (SIM), Ljubljana.

The economic crisis first and most strongly hit foreigners everywhere and prompted many to return home. France, Belgium, and the Netherlands encouraged these departures by paying the travel fees to the border, or even all the way to Ljubljana, and by paying incentives for those who decided to leave. Despite the 1933 agreement, Slovenian workers in Nazi Germany faced discrimination during the hiring process. They were also more easily dismissed from their jobs, which caused high unemployment among Slovenian immigrants (up to 39 percent) compared to German workers (Wörsdörfer, 2021, 181). The logic—one also adopted outside of Germany—favoring domestic workers' employment made immigrants among the first to be dismissed. In Germany and the

Netherlands, unemployment benefits and social transfers were also guaranteed to foreign laborers, so the St. Raphael Society advised Slovenian migrants not to return home, even though they were interested in doing so. In Yugoslavia, the St. Raphael Society could not guarantee any aid, and the living conditions awaiting them were even worse. In all the countries at the time of the economic crisis different political activists fared the worst, particularly communists. Communists, and all those foreigners who had brushes with the law, faced expulsion. In Germany in the 1930s, immigrants felt the increasing pressure of Nazism. As early as 1933, the authorities pondered the expulsion of all Slovenians in Westphalia, ostensibly in response to the post-World War I expulsion of German-speaking nationals from Yugoslav Slovenia. The expulsion of Slovenians did not happen for two reasons: technical difficulties and, in principle, because German authorities recognized Slovenians as people integrated into German society (Wörsdörfer, 2021, 181). However, they were subjected to different forms of discrimination and repression, from revising the already acquired German citizenship, limiting religious rights and the use of the Slovenian language in church rituals, and Nazi interference in organizations to arrests, prohibition to associate, house searches, arbitrary dismissals from the workplace, and accusations of anti-German activities. Consequently, many Slovenian immigrants declared German as their mother tongue in the census, Slovenian organizations expressed loyalty to the regime, membership in organizations declined, and assimilation strengthened. The latter was already characteristic for the period before World War I, but it now accelerated, particularly among younger generations (Werner, 1985; Drnovšek, 2012, 250–251). Many members of Slovenian communities also adhered to the ideology of Nazism even before 1933; they advocated the principles of anti-Yugoslavism, Germanophilia, and the "German authenticity" of the Slovenian ethnic territory. At the end of the 1930s, German-born community members were pressured to join the German Army (which would secure their citizenship) and/or profess German loyalty and identity (Wörsdörfer, 2021, 184, 188–9). Some Slovenian migrants left the Netherlands for the USSR, which was inviting Western workers and experts. Although the USSR did not experience the economic crisis on the same level as the capitalist world, the experience of Soviet-style socialism soon led Slovenians to return to their homeland (Drnovšek, 2010a, 215, 2012, 35–41, 267).

The economic crisis also hit migrants in the Americas hard. Despite hardships and unemployment, most managed to take care of themselves and survive; consequently, however, their remittances decreased significantly. Many contemplated returning due to the unclear economic future. In the mid-1930s, around 5,000 members of the Slovenian communities in the United States required solidarity benefits to survive after several years of unemployment. Most of them were not "unsuccessful" or poor in the narrow definition of the word. The collapse of US banks had wiped out their savings and property; moreover, they had no access to their savings and securities deposited in banks and cooperatives in Yugoslavia because payments were frozen due to bureaucratic or other complications. Migrants in European countries experienced similar situations. The mistrust reduced the money flow from immigrants to the old place even further. Migrants asked the Yugoslav state for help, calling for it to abolish bureaucratic problems and to conclude contracts with governments of receiving countries to solve problems of protection and equality for the Yugoslav migrants. Nevertheless, Yugoslavia was slow to respond to these demands. The insufficient Yugoslav labor legislation further prevented migrants and returnees from enjoying the rights they gained from working abroad (*Prvi slovenski izseljenski kongres v Ljubljani dne 1. julija 1935*, 1936).

Migrations to the Yugoslav Territory

Another characteristic of Slovenian migration in the interwar period was its redirection toward the Yugoslav territory. We do not have direct statistics showing its dynamic traits, exact scope, structural characteristics, and more detailed geographical directions. However, censuses show that between 1921 and 1931, around 34,900 persons moved from Yugoslav Slovenia to other parts of the country, while around 22,600 moved to Slovenia. The territory thus recorded a 12,300-person deficit due to migration to other parts of the country (Vogelnik, 1965, 60). The Yugoslav census of 1931 found 58,000 persons outside the borders of Slovenia who declared themselves Slovenians. Strong assimilation among prewar migrants must be considered here; many no longer declared themselves Slovenians (Valenčič, 1990a, 68–9). The number of people born in Slovenia was significantly larger. Šifrer estimated that the net emigration from the territory of present-day Slovenia in the interwar period stood at around 60,000 persons (cf. Šifrer, 1962, 1974; Vogelnik, 1965). Just before World War II, around 100,000 Slovenians lived elsewhere in Yugoslavia, including immigrants from territories beyond the state borders (Austria and Italy) (Kuhar, 1939, 533).

Slovenian migrations to other parts of the Yugoslav state were, in many cases, a continuation of traditional prewar movements. These were, for the most part, interregional and remained within Austria-Hungary only reaching to a small extent beyond the state borders into Serbia. Because these migrations—such as other internal migrations—were not limited to the wage-laborer, artisan, and rural strata, they were structurally different from the great transatlantic and continental migrations. They included other social groups and professional profiles, from merchants and workers in technical and independent professions to clerks, educators, gendarmerie, military, entrepreneurs, and others. They occurred in connection to entrepreneurial opportunities, public service, career advancement, and other dynamics of the internal labor market. In addition to permanent or long-term migrants, there were seasonal agricultural, forest, and sawmill workers, hard laborers, rafters, peddlers, spinners, and other artisans (Kržišnik Bukić, 2006, 421).

Map 1.3: The Slovenian ethnic territory and Drava Banovina as part of the Kingdom of Yugoslavia.

After the war and the creation of the Kingdom of Serbs, Croats and Slovenes, certain routes into the territory of the erstwhile empire, previously important for Slovenian migrants, were interrupted. However, the Yugoslav state territory became a new area of opportunity for economic and social ambitions, compensating for the loss of prewar migration destinations. The migration routes toward the southern regions of the new state must be seen through the optics of the development dynamics in Slovenia and other parts of the Yugoslav territory. Their scope increased, new destinations were added to traditional ones, and migrants themselves were professionally and socially more diverse. Census statistics in 1931 show that more than half of the people who were born in the Drava Banovina (that is, Slovenia) but resided elsewhere lived in the Sava Banovina (Croatia) and the rest of them were dispersed through other banovine (Valenčič, 1990a, 69).

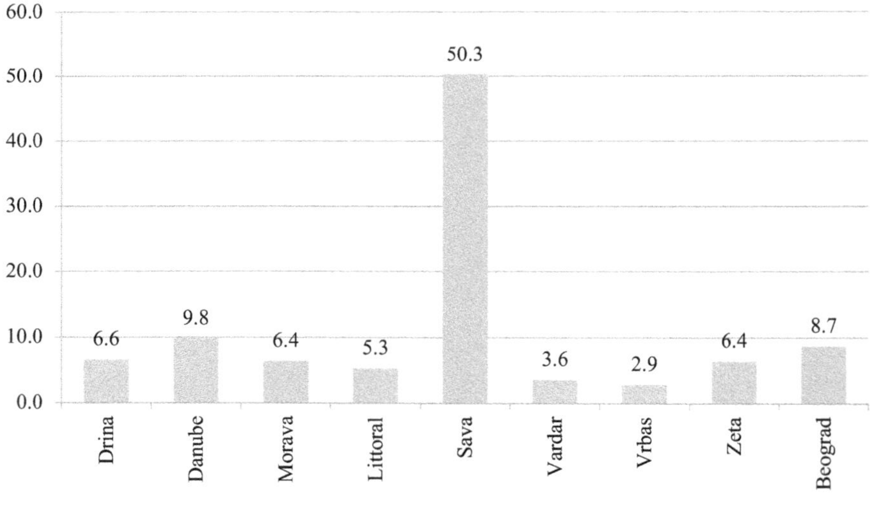

Graph 1.8: Persons born in Drava Banovina residing in other banovine by their banovina of residence in 1931 (in percent; N=57,251). Source: Valenčič, 1990a, 69.

Regarding the gender structure of the migrants to the Yugoslav territory (see Table 1.8), men only had a slight advantage over women: the gender balance characterized this migration. Significant differences in the geographical distribution of genders were linked to different types of work and reflected the labor market. In the Sava Banovina, women were more prevalent due to the attraction of Zagreb and other cities, and the same was true for Belgrade. In other regions, there were more male migrants. In some banovine, the share of men was double or triple that of women, which can be partly explained by the stationed military population. The age structure shows that the interwar period migration (like the prewar emigration abroad) directed people in their most productive age toward the country's inland. Among the migrant population, the 1931 census found almost 85 percent of persons in gainful employment, fewer than 8 percent were children, and only a few percent more were seniors. This internal migration was more diverse age-wise than migration abroad. It included the working class and a wider professional and social spectrum of the population (Valenčič, 1990a, 70).

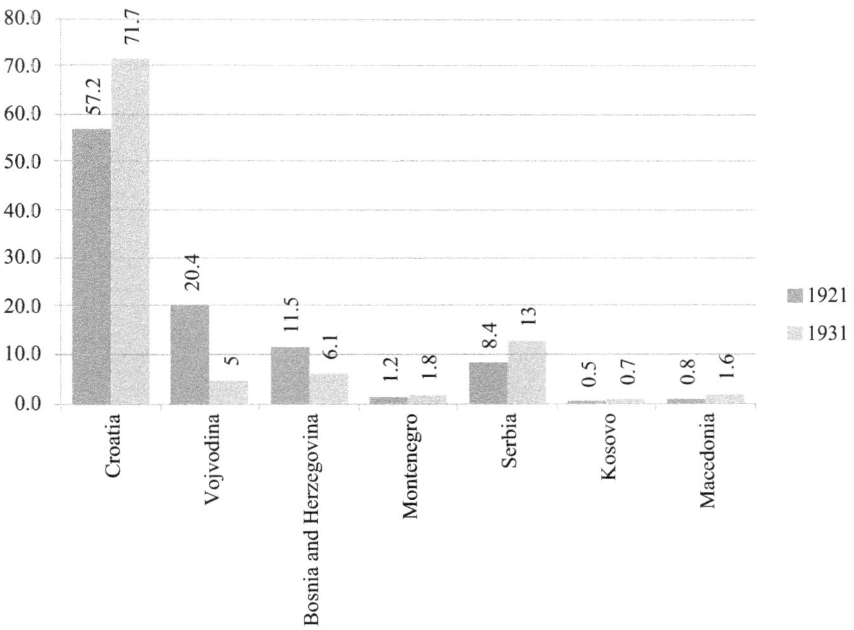

Graph 1.9: Persons with Slovenian as their mother tongue in the Yugoslav territory in 1921 and 1931 (in percent; N=40,616 (1921); N=72,857 (1931)). Source: Josipović, 2014.

In 1938, on the 20th anniversary of the establishment of Yugoslavia, Anton Melik wrote how in the new country, the Slovenian prewar "flight from the countryside" continued and was even reinforced, as—besides the economic motivations—it was fueled by the desire for better material and other advantages of the urban environment. Zagreb, where the largest Slovenian community lived even before the war, became a real magnet for Slovenians in the interwar period. The multifaceted economic development of the Croatian capital was accompanied by massive demographic growth and an expansion of agglomeration. The population, of which 70 percent were immigrants, doubled by the mid-1920s compared to the prewar numbers. By the 1940s, it doubled again to 260,000. In 1921, 8,699 persons with Slovenian as their mother tongue lived in Zagreb. In 1931, 17,627 inhabitants born in the Drava Banovina were residents of Zagreb (9.4 percent of its population). To these, we must add Slovenian migrants from Austria and Italy, so the total number was over 20,000 persons (Valenčič, 1990a, 69–70; Kržišnik Bukić, 1994–1995, 1995).

Table 1.8: Persons born in Drava Banovina residing in other banovine, 1931. Source: Valenčič, 1990a, 69.

Banovina	**Male**	**Female**	**Total**	**%**
Drina	1,915	1,322	3,237	6.6
Danube	2,814	1,557	4,371	9.8
Morava	1,843	811	2,654	6.4
Littoral	1,519	597	2,116	5.3
Sava	14,530	19,500	34,030	50.3
Vardar	1,051	442	1,493	3.6
Vrbas	842	693	1,535	2.9
Zeta	1,835	470	2,305	6.4
Beograd	2,511	3,099	5,610	8.7
Total	28,860	28,491	57,351	100.0

As many as two-thirds of Slovenians in Zagreb were originally from the region along the Sotla and Krka rivers. The last third originated from the wider territory historically inhabited by Slovenians, all the way to Trieste, Gorizia, and Istria. The least represented were Carinthia, Upper Carniola, and Inner Carniola. The professional spectrum of the Slovenian community in Zagreb was rather broad, although most migrants still came from the lower social strata. The higher echelons of the community consisted of intellectuals of different profiles—university professors, lawyers, doctors, school principals, engineers, entrepreneurs, and high-ranking public servants. Many secondary school teachers and university students only lived in Zagreb during the academic year. There were many merchants and independent artisans of all types, from carpenters and cobblers to tailors and locksmiths. Quite a few were tavern keepers, and Slovenian dairies were also quite known. Several hundred were railway workers, and many were waiters, but the largest professional group were domestic workers (Melik, 1938, 255–256; Kržišnik Bukić, 1993). This market was the largest for female employment and the most important factor in their migration to the city. In the mid-1930s, around 4,000 Slovenian domestic, retail, and hospitality workers lived in Zagreb. The economic crisis in the early 1930s rendered many of them unemployed. Like the women, male Slovenian immigrants who had come to Zagreb in the 1920s in search of better working conditions and to enjoy the city's good standard of living, also became unemployed (Izseljenci, 1935, 63).

Belgrade, like Zagreb, attracted many Slovenians. The country's capital also experienced expansion: in 1921, it had 110,000 inhabitants, and in the early 1940s, 420,000. Between the establishment of the Kingdom of Serbs, Croats and Slovenes and the early 1930s, the number of Slovenians in Belgrade grew from 1,000 to over 6,000. It is estimated that by the end of the decade, their number had increased from 18,000 to 20,000 (Kuhar, 1939, 533; Valenčič, 1990a, 69). In Belgrade, Slovenians of different professions also found employment in the private sector and public administration. Many were artisans and merchants. Most were still workers employed in factories and companies, laborers in warehouses, sailors on the Danube and the Sava rivers, and so on. This group lived primarily in the suburbs and was hit hard by the early 1930s economic crisis. As the St. Raphael Society reported, many struggled with unemployment and suffered poor living and material conditions (Kordin, 1933). Again, domestic servants were the largest professional group among Belgrade Slovenians. Like in Zagreb, where the stereotypical Slovenian was a domestic servant (Kolar, 1995, 116), they were, it appears, very highly regarded, and many worked for highly respected bourgeois families (Ravnik, 2010, 171).

Old and new migration routes also led to Karlovac, Varaždin, and Osijek (Sava Banovina) and Split in Dalmatia (Primorska Banovina). In Bosnia and Herzegovina (Drina Banovina), the main destination was Sarajevo, then Banja Luka, Mostar, Bihać, and Travnik as well as the mining centers of Semizovac, Vareš, Tuzla, Zenica, Breza, and Kakanj. Similar to before the war, Slovenian miners continued to move to Bosnia, many with professional and training experience from the United States and European mining countries. The Slovenian–Bosnian prewar mining collaboration continued and, in many mines, Slovenian miners and mining experts were part of the management (Koblar, 2008, 17–18; Hašimbegović, 2008, 39–41). Many doctors, engineers, veterinarians, and agronomists contributed to the modernization of agriculture (Papež Adamič, 2009, 25).

Public servants and experts of different profiles migrated to work for the banovine or political districts' administrations, hospitals, public infrastructure construction, company boards, railways, and other services (Koblar, 2008, 17–18). Educators, mostly teachers in technical and vocational schools, were another category of migrant workers. Their postings were sometimes politically motivated (Lilek, 1933).

As destinations that attracted a concentration of workers from many places, mining centers generally registered a more or less numerous Slovenian presence. One such colony, the largest Slovenian settlement in Kosovo, was Kosovska Mitrovica and the nearby Trepča Mines. Sources report that one could meet a Slovenian, from a digger to an engineer, on every corner. After 1928, they were all employees of Selection Trust, the British concessioner of the mine. Along with miners, the colony had members of other professions such as artisans, small merchants, and workers in the factory in Zvečan. In 1937, there were around 700 of them altogether (Mak, 1937, 81–83). Another settlement of Slovenian miners, including administrators and military personnel, was in Obilić near Pristina/Priština/Prishtinë (Slovenska kri v Južni Srbiji, 1937). Around 400 Slovenian miners and their families lived in the Serbian coal-mining district Ravna Reka. There were hundreds more in Kragujevac, Paraćin, Niš, Zaječar, Blagovica, Novo Brdo, Stanje, Rakita, Bor, Majdanpek, and Aleksinac. The copper mine in Bor employed several thousand Slovenians until World War II; the Rtanj coal mine was said to have employed around 1,500. The more than ten mines in the Timok Valley had many Slovenians among their technical personnel (Piljušić, 2013, 5–9). In the late 1920s, around 300 Slovenians worked in the Knežev Arsenal in Kragujevac, which needed toolmakers, mechanics, and other metalworkers. Slovenians also lived in small groups scattered throughout smaller places where they searched for work on their own initiative. Many, however, were posted there for a shorter or longer period, among them gendarmes, financial officers, military officers, public administrators, railway workers, and educators.

Migrations of National Minorities

Circumstances, Causes, Reasons

With the reconfiguration of the European political space at the end of World War I, national minorities became a central problem of the old and new "nationalizing states"—states which prioritized the titular nation (at the expense of others) and aimed at imbuing every corner of their territory with the dominant national(ist) ideology. Despite the protection of minorities enshrined in the Treaty of

Versailles, minorities were discriminated against; they became causes of international disputes and protagonists of more or less forcibly induced migrations and population transfers (Brubaker, 1996, 63; Ther, 2016, 29–38; Palko and Foster, 2021). Interwar migrations from the regions that, at the end of World War I, came under Italian or Austrian rule—where the Slovenian population was a national minority—gained clear national and political connotations. In addition to the economic factors, migrations from Carinthia and particularly the Julian March were urged by political pressure and systematic denationalization policies intended to decrease the "foreign" population and force it to assimilate into the majority German or Italian culture. This migration affected all social groups in the Julian March and Carinthia, particularly weakening the higher social strata and intelligentsia. On the one hand, migration resulted from the hostile environment, direct repression, and violence against the ethnic minority population. On the other, it was a result of the economic situation, which, in the case of the Julian March, was linked to the slow integration of the former Habsburg regions into the Italian economic system and the intentional eroding of the Slovenian and Croatian communities' economic foundation (something that Italy could carry out because, as a winner of the war, it was not bound by the international agreements protecting national minorities). The project *bonifica etnica* (ethnic homogenization, that is, deslavicizing the newly incorporated eastern provinces and reinforcing their Italian character) included the emigration of the Slavic population from the territories near the Italian-Yugoslav border and the immigration of ethnic Italians to them (cf. Kalc, 1995).

People were already leaving the Julian March during the Italian military occupation: not just the newly unemployed Austrian public servants and recent arrivals into the region—the autochthonous population that had lived there for generations was leaving, too. German, Slovenian, and Croatian speakers of all social classes left for several reasons: to escape the region's uncertain future, Italian rule, and, often, pressure and violence. Nearly 1,500 railway workers from Trieste moved to the Kingdom of Serbs, Croats and Slovenes in 1919, most after being fired for participating in union activities and the strike wave that shook Italy that year. Italian authorities either expulsed or prevented the return of all who were not born in the Julian March or did not have the right of domicile there (Purini, 2010, 42–53). It is estimated that between 30,000 and 40,000 inhabitants of Slovenian or Croatian nationality left the occupied territory in 1919, along with most nationals of the former Austro-Hungarian Empire who had previously resided in Trieste (Kacin Wohinz, 1972, 115–116). In 1919, refugee camps in Yugoslav Slovenia hosted over 20,000 refugees from the Julian March (Nećak, 1973, 123–124).

Around 15,000 war refugees who lost their homes in the Soča Valley remained in Carniola and Styria, as did former Austrian soldiers and returnees from Russian captivity who faced internment in the Julian March as members of the enemy army and potential revolutionaries. At the same time, around 40,000 *regnicoli* (prewar immigrants from the Kingdom of Italy) returned and new Italian migrants arrived in the Julian March—from administrative staff to people searching for new opportunities in life. To this number, we must add the military and the members of the police forces who represented a large share of the Italian presence in this militarized border region. The ethnic structure perceptibly changed in a very short period. At the last Austrian population census before the Great War, Slovenians and Croatians (467,000) represented half of the Julian March population, and Italians (355,000) represented 38 percent. In 1921, the Slovenian and Croatian share was down to 38 percent (350,000), while the Italian increased to 52 percent (467,000). In Trieste, where almost 60,000 Slovenians and Croatians lived before World War I, their numbers halved in 1921 (Novak and Zwitter, 1945; Svoljšak, 1991; Purini, 2010, 34–55, 84–85).

Emigration continued with the arrival of Fascism and its systematic politics of denationalization. With the reform and Italianization of schools between 1923 and 1927, at least half of the Slovenian and Croatian teachers were either posted to inland Italy or emigrated to Yugoslavia (Vovko, 1992, 88; Lavrenčič Pahor, 1994; Verginella, 2021). Railway workers and other public servants had a similar experience: they were posted all over the Italian peninsula with their families. Numerous political representatives and activists in Slovenian social and political organizations left the Julian March because of Fascist pressure, intimidation, and persecution. Italian authorities were keen to suppress any revolutionary activity, so they often prevented students from returning from Yugoslavia in order to curtail their national and political influence. "Abyssinians" were another category of political émigrés. These were around 500 military conscripts who escaped to Yugoslavia in 1935 to avoid being recruited and participating in the Second Italo-Abyssinian War (Vovko, 1978, 449–451; Kalc, 1996a, 24–28; Purini, 2010).

Along with migration for political or national motives, Slovenians migrated for economic reasons throughout the 1920s (and, to a lesser extent, the 1930s). This migration was directly or indirectly triggered by the politics of deslavicization, which deprived the Slovenian community of banks, cooperatives, and other economic institutions that enabled economic and social growth before the war. The regime used tax policy to weaken agricultural estates and offered transatlantic migration as a possible escape from the economic distress—all at a time when Italy changed its migration policy. Until 1927, Italy buffered the social imbalances by "exporting" the workforce, advocating for the mollification of the

US immigration restrictions, and searching for new destinations for its migrants, even in the Soviet Union. It encouraged demographic growth and retained the population to include their labor potential in developing the national, corporative, and autarkic economy (Nobile, 1974).

This change occurred when, in the mid-1920s, it became obvious that the existing Italian economic system, which relied heavily on mass emigration and its benefits, was unsustainable. The new starting points of the economic policy—whose primary goal was to reduce the dependency on imports, stop inflation, and improve the conditions for foreign investments—were based on strengthening the rural economy and stabilizing the rural population. Following these principles, the Italian authorities also tried different methods to stop migrations to urban areas and carried out a policy of colonizing and meliorating agrarian land (Treves, 1976; Sori, 1979, 406, 427–429). Migration policies were a part of the new Italian demographic politics that abandoned the previously held doctrine of reducing demographic pressure and conceded that the scope and growth of the population proved the strength of the nation and state, which was used as an argument for Italy's foreign policy and its "right" to territorial expansionism. In this renovated ideology, migration abroad (except for short-term migrations and selected groups) was forbidden and declared unpatriotic (Nobile, 1974; Sori, 1979; Cannistraro and Rosoli, 1979).

These limitations were not placed on Julian March's "foreign" populations—known to the Fascist regime as the *alloglotti* (speakers of foreign languages) or *allogeni* (population of foreign extraction)—which obtained passports for going overseas without problems. Believing that it was in the Italian national interest and the interest of the *bonifica etnica* policies of the Julian March, Mussolini himself sent a circular to the local authorities urging them to simplify the emigration process as much as possible for the *allogeni* (Kacin Wohinz, 1995, 24). Shipping companies and agencies were allowed to advertise overseas emigration in the Julian March and approach potential recruits. This policy did not openly expulse the Slavic population; it simply opened the door and hinted they leave. People applying for a passport had to prove their non-Italian origin because citizens of Italian origin were not allowed to leave the country. This regulation created many conundrums in practice because it was at odds with the practice of forcibly Italianizing Slavic names (or, as the official line claimed: "names, reversed to their original, Italian form") designed to prove people's "Italianness" (Grossutti, 2013, 249).

Migration to Yugoslavia was quite a different story. Migrants to Yugoslavia were often politically very active and worked to influence the relationships between the two countries, and Italy monitored them closely. Even before the emigration regulations changed in 1927, many who emigrated to Yugoslavia chose (or were forced) to do so illegally to avoid the complicated bureaucratic procedure of acquiring a passport. Applicants had to guarantee that their families would be able to support themselves without them. The authorities used this condition to hinder emigration if they considered it opposed to Italian interests. They tried to suppress illegal departures by imposing fines on the families and prohibiting people from returning (Grossutti, 2013, 119–127). It is estimated that since the rise of Fascism in the 1920s, on average, 500 people annually migrated to Yugoslavia illegally, and in the 1930s, around 1,000 people annually (Kacin Wohinz, 1990, 86, 326). The programs of the *bonifica etnica* assumed the colonization of the Slovenian territory with the Italian rural population, which was only realized on a small scale (Vanello, 1981). Public servants from all over Italy were attracted to the Julian March because of the higher wages and benefits for patriotic work in this border province. Many Italian migrants came voluntarily and with special concessions to open their own small enterprises. Public works, particularly military barracks, were an important field of employment (Kalc, 1996a, 26–27; Purini, 2010, 160; Bearzatto, 2012).

Other migrations directly triggered by political and national interests were those linked to the situation of the German minority now living under Italian rule in South Tyrol/Südtirol/Alto Adige and Val Canale/Kanaltaal/Kanalska dolina. The question about the status of these two regions surfaced upon the German annexation of Austria and Sudetenland. Because of friendly relations with Italy, Hitler did not consider the annexation of South Tyrol and Val Canale. Nevertheless, he wanted the state border between the two countries drawn and consolidated according to national belonging. For this reason, an agreement was concluded in June 1939 based on which the German population in South Tyrol could opt to take German citizenship and migrate to the Reich. The agreement was primarily an opportunity for Mussolini to eliminate the German irredentist elements (Ferenc, 1968, 14–16).

The same regime was instituted in Val Canale, which also had a Slovenian population. When faced with a choice that did not consider their national identity, most Slovenians (1,600 of the 1,750) opted for Germany, while a smaller part abstained. More than political affinities for either country, the decisive factors for Slovenians were their historical and cultural links to Carinthia, the belief in a better economic situation, the propaganda they were exposed to, which promoted the move to Germany, and the fear of forced displacement to other parts of Italy (Steinicke, 1984, 45). While the Germans did move, Slovenians largely avoided doing so because the war stopped the process. Only about 100 Slovenians left the valley and settled, similar to the German "optants" in Carinthia and Styria. Most of them were miners and laborers who got jobs and apartments, while the Slovenian rural population mostly persisted in their estates. Although the choice of German or Italian citizenship was formally voluntary, such options—considering the circumstances and the accompanying political pressure—were forced migrations, even more so for the Slovenian population. With most of the Germans gone and the settlement of the Italian population, the ethnic structure of Val Canale was radically changed (Pahor, 1987).

There were many similarities between the emigration of the Slovenian population from the Julian March and the Republic of Austria, both from Carinthia and other Austrian regions, to the Yugoslav territory. The 1931 population census in Yugoslavia counted almost 30,000 people born in Austria, of whom 17,000 lived in Slovenia (Drava Banovina) and more than 6,000 in Croatia (Sava Banovina) (Valenčič, 1990a, 68–70). They left the Austrian territory during the postwar armed conflict between Slovenian and Austrian units about the future of the ethnically mixed Carinthia and Styria (known in Slovenian historiography as *boj za severno mejo*, or the Struggle for the Northern Border, and in German-language historiography as *Kärntner* Abwehrkampf, or the Carinthian Defensive Struggle) and the ensuing plebiscite in October 1920; the causes were political and national as well as economic. The national question and anti-Slovenian atmosphere indirectly or directly caused the latter. In mere months after the plebiscite, which determined that southern Carinthia be a part of the Republic of Austria, the first group of Slovenians (said to be between 2,000 and 3,000 strong) left Carinthia: factory workers, foresters, land lessees, artisans, merchants, railway workers, skilled workers, teachers, and other public servants who were dismissed from their jobs because they voted for the annexation to Yugoslavia. People of liberal professions left, as did intellectuals and priests; students left to study in the Slovenian language. Emigration increased once again after the Annexation of Austria to the German Reich in 1938 and after the beginning of World War II. In 1938 and 1939, around 400 people fled to Yugoslavia (Grafenauer, 2010).

Minorities in Yugoslav Slovenia

Almost one-fifth (18.8 percent) of the population of the newly created Kingdom of Serbs, Croats and Slovenes were national minorities (Janjetović, 2012, 61). In Yugoslav Slovenia, this share was lower. As mentioned before, 7.9 percent of the population listed a mother tongue other than Slovenian in the 1921 census—3.9 percent chose German and 1.9 percent Hungarian. Although the peace treaties bound the newly established Kingdom of Serbs, Croats and Slovenes to respect national minorities, these minorities, just like minorities in other "nationalizing" states of the interwar period, found themselves in opposition to the Yugoslav nation- and-state-building process: in the state's attempts to homogenize the Yugoslav nation from the three constitutive ethnic groups (or "tribes") and integrate the state territory, members of ethnic minorities—just like the Yugoslav minorities in the neighboring countries—faced discriminatory policies and measures that encouraged assimilation and emigration (cf. Brubaker, 1996; Janjetović, 2012; De Azcárate, 1945, 49–51). In Yugoslav Slovenia, this was the case particularly for ethnic Germans and Hungarians. The German and Hungarian domination and the national conflicts in the decades before World War I conditioned their relations with them. They were additionally contaminated by the postwar events, territorial disputes, and negative attitudes toward national minorities.

German speakers were present in large numbers throughout Yugoslav Slovenia: two-thirds of them lived in Styria, one-fourth in Carniola, mostly members of the centuries-old Gottscheer community in the Kočevje district, and a small number in Prekmurje and Slovenian Carinthia. They were present in towns (particularly Maribor, Celje, Ptuj, and Ljubljana) and the countryside. Their social scope ranged from peasants and landowners to workers and wealthier middle and upper-middle classes, from artisans and merchants to industrialists and those practicing independent professions, as well as intellectuals. No matter where they lived, they were well organized and, in certain places, also retained political influence and dominant positions. Despite the National Minorities Protection Act of 1920, they were subjected to much of the same policies as Slovenians in Austria. In schools, the German language as a means of instruction was limited to classes with a certain number of German-speaking students—public use of German decreased. The toponomastic was Slovenized. Associations were dissolved, and estates were expropriated. The German-speaking population was forbidden to purchase real estate in the area near the newly established border, which, in the 19th century, was subject to Germanization with the arrival of Südmark settlers (Judson, 2007, 100–102; Klemenčič, 1979). Assimilation was encouraged, as was the return to the Slovenian identity for those who were Germanized before the war. Slovenian authorities were, in short, in favor of returning to the situation before the denationalization processes before the war (Cvirn, 1998). The consequence of this politics of Slovenization and emigration was that the number of the German-speaking minority dropped from 45,500 in 1921 to 29,000 in 1931, or from 4 to 2.5 percent (Ferenc and Repe, 2004, 149). The reduction in their numbers did not significantly decrease their economic power. This economic power as well as the aspirations for the autonomy of the historic German linguistic island in the Kočevje region, the adherence of numerous members of the different German-speaking communities to Nazi ideology, and the propagation of ideas about joining the German Reich intensified the national conflict, tensions between the German community and the state, and with Germany (Komac, 2015; Nećak, 2004).

A part of the German-speaking population left the territory of Yugoslav Slovenia immediately after the war because of the mass layoffs from public service jobs. The emigration also continued after the 1921 deadline for opting for Yugoslav or other citizenship. Most moved to Austria, where, in the middle of 1919, there were 1,870 refugees from Styria, and two Gottscheer associations were created. Many Gottscheers joined the Gottscheer émigrés in the United States (Kosi and Cvirn, 2012, 62; Čuček, 2021).

Unlike the German speakers, who were scattered throughout the Slovenian ethnic territory, the Hungarians lived in contiguous communities in Prekmurje, along the border with Hungary. The national question here was intertwined with religion because both Slovenians and Hungarians followed the Roman Catholic rite, and a minority followed the Protestant/Evangelical rite. The Hungarian community also had a far more complex class structure than the Slovenian. With the Treaty of Trianon, Prekmurje was "cut" from the newly established Hungarian state and annexed to the Kingdom of Serbs, Croats and Slovenes, but it was distant socially, economically, culturally, and ethnically from the rest of Slovenia. The integration policy of the Slovenian authorities thus had a distinct anti-Hungarian charge; it tried to remove the upper Hungarian echelons of the society and disable anyone who did not profess loyalty to the Yugoslav state (Kosi, 2018, 2020). It operated in the educational, religious, and socioeconomic fields to imprint the space and the social structure with a more prominent Slovenian seal. Under this pressure, intellectuals and big Hungarian landowners emigrated, mostly to Hungary, as did numerous smallholders, because the agrarian reform served as an instrument of national politics. Hungarian smallholders were exempt from the land redistribution or, more precisely, they had the right to obtain the land expropriated from the big Hungarian landowners on the condition that they identified themselves as Yugoslavs. Nevertheless, the reform administrators favored impeccable and nationally conscious citizens, such as refugees from the Julian March. Slovenian national politics was particularly pronounced here because of the settlement patterns and the influence of the Magyarization that still lingered after the war (Komac, 2015, 63; Kovàcs, 2002, 2007).

The Migration Destinations of Stranded Slovenian Minorities

Home to Yugoslavia

The extent of interwar migrations from and to the Julian March is difficult to evaluate because the available statistics do not provide reliable data, and the data available do not allow for distinguishing the population and migrants based on their national identity. The complex processes included permanent emigration as well as temporary and iterative moves and returns to different destinations. The main destination of the emigration from the Julian March was certainly the Kingdom of Serbs, Croats and Slovenes, or Yugoslavia. Rough estimates claim that around 70 percent of all Slovenian and Croatian emigrants from the Julian March moved there (Čermelj, 1974, 174). Many Slovenians and Croatians in Italy perceived the newly established state as their homeland, so it is not surprising that it became the immediate and—particularly when thinking in national terms—ostensibly natural exit for the first massive postwar migration and subsequent moves. Migrants usually remained in Slovenian and Croatian territories yet also settled elsewhere in Yugoslavia in search of employment and opportunities for a better life. Most lived in urban areas and economic centers, some in the countryside. The Ljubljana city census of 1928 revealed that more than 6,000 inhabitants—or 10 percent of the city's population—were born in the Julian March and that most of them had moved to Ljubljana after the war (Krašovec, 2016).

Maribor, which, once the Kingdom of Serbs, Croats and Slovenes was established, saw its German-speaking population dwindle, was also an important destination for migrants from Italy and Austria. It became a new home for the largest community of exiles from Carinthia. In 1931, 4,000 out of its 25,000 inhabitants were from the Julian March. There would have been many more if the town could have absorbed them. Many newcomers thus looked for a home in other Slovenian and Yugoslav towns (Golouh, 1935). More numerous colonies of *Primorci* (Slovenians from the Julian March) were found in Ptuj, Celje, and Kranj; many lived in communities dispersed throughout the countryside (Hartman, 1976; Grašič, 1990; Kalc, 1996a, 34). In Croatia, many settled in Zagreb; in Bosnia and Herzegovina, they settled in Sarajevo and Banja Luka. There were large colonies in Belgrade and other Serbian cities. Large or small groups were scattered southward to Kosovo and Macedonia (Rusić and Novak, 1973). Most Slovenians from Austria headed to Yugoslavia and settled in the towns and their surroundings, from Jesenice, Kranj, and Škofja Loka, to Ljubljana, Celje, and Maribor, where the largest community from Carinthia lived. Many left for Zagreb, Belgrade, and other Yugoslav centers (Grafenauer, 2010, 477, 480).

Despite significant bureaucratic, organizational, and practical problems, by the mid-1920s, the migrants found relatively favorable opportunities for employment and integration into the society in Yugoslavia. From intellectuals and people in independent professions to former public servants, clerks, and employees of private enterprises, the middle class was considered a precious human resource for the newly established country, which was building its administrative apparatus and facing numerous economic and social challenges. They replaced the former Austro-Hungarian clerks with qualified and experienced public administration staff infused with a deep pro-Yugoslav sentiment. They were employed as clerks in companies that opened their own workshops, shops, and sometimes large enterprises. Some rose to prominent and responsible positions (Vovko, 1978, 451). Among them was Drago Marušič, the Ban of Drava Banovina. Others include Fran Vodopivec, who became the grand mayor of the Maribor Oblast (county); Vilko Baltič, a former officer at the Trieste police, who rose to be first the grand mayor of the Ljubljana Oblast and then a grand mayor of the Belgrade Oblast; Rudolf Golouh, one of the leading Triestine socialists, who became the deputy mayor of Maribor; lawyer and university professor Boris Furlan; the mayor of Ljubljana Dinko Puc; and many others (Kalc, 1996a, 32).

Even more prominent public functions went to the Slovenian immigrants from Austria. We must at least mention the members of parliament in Belgrade, Franc Grafenauer, Anton Brandner, and Fran Schaubach; the archbishop of Ljubljana Gregorij Rožman; the ethnologist, theologian, and politician Lambert Ehrlich; the president of the University of Ljubljana Radoslav Kušej; and many university professors, lawyers, judges, and high public administration officials (Grafenauer, 2010, 478).

In the first postwar years, skilled and specialized workers also found many good opportunities, particularly in certain industries such as shipbuilding and, for migrants from Austria, mining. The Tivat shipyard in Montenegro employed around 2,000 former shipyard workers, mostly Croatian, many of whom were previously employed in Pula. Many found work in the railway depots and workshops from Ljubljana and Zagreb to Drvar, Kraljevo, and Slavonski Brod (Peruško, 1953, 160; Tumpić, 1991, 91, 94). Particularly among the migrants from Goriška, many went into farming. Some cultural workers joined notable Slovenian cultural institutions, from theatres to music ensembles. In many places, the migrants from the Julian March and Austria contributed to the organization of social and cultural life. A typical case is Maribor, where the migrants from the Julian March, among other things, participated in the founding of a permanent theatre and Glasbena matica and contributed greatly to elevating the Slovenian national identity of the city (Hartman, 1976; Makuc, 1987; Grašič, 1990).

During the deepening economic crisis in the second half of the 1920s, employment opportunities for newcomers from the Julian March in Yugoslavia worsened. They swelled the ranks of the Yugoslav unemployed and those dependent on social services. In the early 1930s, at the peak of the economic crisis, many lived on the poverty line; searching for work, they left Slovenia and moved to other parts of the country, returned to Italy, or applied to work abroad. They had difficulties finding sponsorship to move to destinations outside Yugoslavia because they were Italian citizens for whom Yugoslav authorities were not competent when contracting international work agreements (Kalc, 1996a, 33).

Although they integrated into the social, economic, and cultural life, immigrants from the Julian March were somewhat extraordinary in Slovenia and elsewhere in Yugoslavia. Their rights and relationship with the authorities and the milieu depended on their citizenship; while some managed to obtain Yugoslav citizenship, the majority retained Italian because of bureaucratic obstacles. In many ways, their position and daily trials, in their formal and practical aspects, were similar to those of foreigners (Golouh, 1935; Wörsdörfer, 2004, 288). Considering the difficult economic situation that was troubling the Yugoslav society, they became, as competition for the work positions and social benefits, an unnecessary element in the eyes of the Slovenian public. The general population was bothered by their employment in public administration, at many prominent posts, and in the police, which was particularly unpopular in times of social distress and unrest. In the colonized territories, concretely in Prekmurje, they were targets of prejudice because they received privileged treatment in the land distribution during the agrarian reform (Makuc, 1985).

The nascent dissatisfaction and, at times, resentment was expressed with insults and derogatory slurs for the immigrants from the Julian March—they were often called "Lahoni" (a pejorative name for Italians) or even "fascists." The distrust and negative emotions were also present in the administrative apparatus itself, particularly at the local level, which often declined to grant them permanent residency—a requirement to obtain Yugoslav citizenship (Čermelj, 1972, 39). They were increasingly overlooked when it came to obtaining jobs and getting social benefits. Not infrequently, as "Italians," they were targets for the reciprocal discrimination that the Yugoslav citizens experienced in Italy (Makuc, 1985, 57; Kacin Wohinz, 1990, 328; Wörsdörfer, 2004).

From the beginning, one of the characteristics of the migrants from the Julian March was the objective to stay connected and maintain identity traits with the people who had a shared experience of exile. Immediately after the war, committees and associations were established to provide urgent social relief for the compatriots from the Julian March and helped them integrate into the new environment. Liberal and Catholic associations, similar to the ones existing in the prewar Littoral region, were reproduced. In 1928, the Organizacija jugoslovanskih emigrantov (Organization of Yugoslav Emigrants), or ORJEM, tried to unite migrants and represent their interests concerning the Yugoslav state and conduct the political struggle against Fascist Italy and its denationalization efforts in the Julian March (Vovko, 1978, 452–455).

ORJEM, although a non-party organization at its core, had an influential leftist faction and was critical of the Yugoslav establishment. This criticism—and the Italian pressure—convinced the authorities to dissolve it. In the early 1930s, ORJEM was replaced by the pro-unitarian (and closer to the government in Belgrade) Association of Yugoslav Emigrants from the Julian March (Zveza jugoslovanskih emigrantov iz Julijske krajine), which counted more than 40 emigrant associations throughout Yugoslavia as its affiliates. Over time, anywhere between 5,000 and 10,000 migrants from the Julian March were politically organized and very active. Their leaders were in contact with Italian and international anti-fascists. Internally, they were split into two factions. The bourgeois, pro-monarchy, and the mostly liberal faction supported Yugoslav centralism and championed the annexation of the Julian March and Istria to Yugoslavia. The leftist faction, which saw King Alexander I's dictatorship as a Yugoslav variant of fascism, linked the struggle for the Julian March to the struggle for a more just society and placed it in the international movement of popular anti-fascist fronts (Vovko, 1978, 456–472, 1979).

The émigrés also secretly worked with the Slovenian and Croatian anti-fascist organizations in Italy and the Yugoslav and foreign, particularly British, military intelligence. Throughout the 1920s and even more so in the 1930s, the activists supported the Slovenians and Croatians in Italy in their resistance against denationalization and assisted escaped Italian and other European anti-fascists. Regardless of their ideological differences, internal conflicts, and divergences in their attitude toward the Yugoslav regime, the migrants from the Julian March in Yugoslavia played a visible role in the anti-fascist and national liberation struggles. They were also active in the political endeavors to solve the territorial question of the Julian March and the annexation of Istria and Primorska to postwar Yugoslavia (Vovko, 1979; Kalc, 1996a, 36–45).

Slovenian migrants from Carinthia organized themselves in Slovenia to help Slovenians in Carinthia. Exiles from Carinthia were generally well-received in Slovenia and did not usually face the typical problems those from the Julian March faced. Before the plebiscite, the first Carinthian clubs and associations were formed. The central one was Klub koroških Slovencev (Club of Slovenians from Carinthia) in Ljubljana. Its program included "national-defensive" work, connecting to Slovenians in Carinthia, supporting their struggle to gain cultural rights, and advocating for what they called "a more just border," that is, a border drawn more congruently along ethnic lines that would assign the parts of southern Carinthia with a substantial Slovenian population to Yugoslavia. Until World War II, the club was the main protagonist of political work in favor of the Slovenian minority, which, because of the different situation and position of Slovenians in Austria and Austro-Yugoslav bilateral relations, could function differently than the work of the emigrant organizations from Primorska (Grafenauer, 2010, 480–483).

South America

An important segment of the emigration from the Julian March in the interwar period turned toward South America, mostly Argentina, which recorded economic growth in the 1920s and where the demand for workforce grew. This emigration peaked between 1927 and 1931, at the very same time when the number of departures abroad dropped drastically elsewhere in Italy because the country banned free emigration. Many socialist and communist activists and sympathizers also left for Argentina because the country allowed political activism until the 1930 coup d'état. They could continue their work in internationalist circles (Genorio, 1988; Mislej, 1995a; Kalc, 2016b; Zobec, 2019, 2021a).

According to the estimates, 22,000 Slovenians from the Julian March lived in Argentina in the late 1930s. Many were among the 5,000 Slovenians in Brazil and other South American countries (Lipoglavšek-Rakovec, 1950, 34, 37–38; Zobec, 2021b, 109). Primorci thus represented the majority of the interwar Slovenian presence in Argentina.

Half of them lived in Buenos Aires, where first La Paternal quarter and then Villa Devoto became true "Primorska villages," just as the Avellaneda quarter became a village for the people from Prekmurje. Most Slovenians from the Julian March stopped in La Paternal, at least temporarily; it was also the center of organized social life. Following work and living opportunities, they moved from here to other quarters and communities outside the city. Some settled in smaller communities in Córdoba and Rosario, and some dispersed across the vast country (Lipoglavšek-Rakovec, 1950, 34–37; Brecelj, 1997; Zobec, 2013; Molek, 2019, 93–100).

Like the prewar emigration to North America, the postwar migration to Argentina was motivated by the labor market. Most immigrants originated from rural areas, but many were skilled workers. They were employed in construction, factories, abattoirs, artisan workshops, trading companies, and transport companies as field workers and similar positions. Some started businesses that eventually became successful enterprises, while others dedicated themselves to farming, particularly fruit growing. Like back home, women found good posts as domestic workers for urban families (cf. Mislej, 1995a; Brecelj, 1997; Molek, 2019, 60–65). One prominent figure was the architect Viktor Sulčič, who moved to Argentina in the early 1920s and gained fame by designing several iconic Buenos Aires landmarks including La Bombonera Stadium and the central market, Mercado de Abasto (Mislej, 1989a).

Once the economic crisis began, many found themselves in difficult material and social situations. Like those in Yugoslavia, the migrants from the Julian March in Argentina were closely connected, socially organized, and engaged in social and cultural life. Because of their background and migration reasons, their ranks were filled with an anti-fascist spirit and a zeal for political work connected to the Julian March. However, regarding their worldviews, the communities were defined separately and, because of their antagonisms, also had separate organizations (Molek, 2019, 93–100). The leftist faction was engaged in the union movement and political and revolutionary activity, influencing their relationships with other immigrants and the Argentinean, Italian, and Yugoslav state authorities. Those most active in politics were, just like in Yugoslavia, under special surveillance by the Italian Fascist authorities, Argentinean political police, and the Yugoslav authorities. The latter considered them a part of the Yugoslav nation and quickly responded to their criticism of the Yugoslav political regime (Zobec, 2019, 2021a). Because of their union and communist activism, some were deported to Italy, where they were confined (cf. Genorio, 1988; Kalc, 2016b).

Other Destinations

Among the European migration destinations from the Julian March, France and Belgium were at the forefront, with a smaller contingent of migrants also going to the Netherlands and Luxemburg. This migration was a part of the broader Italian working-class migration to these countries. Italian migration to France spread from the northwestern Italian provinces to the northeast, where prewar migration routes led to the Austrian and German territories. Unlike Friuli and Veneto, Julian March was only marginally represented here. Most migrants left for these countries from Goriška, while many from the region of Trieste left for France. The Yugoslav Emigration Commissioner in Paris estimated that, in 1929, 7,000 Slovenians from the Julian March lived in France; before World War II, there were around 11,000 Slovenians living in France. For Belgium, the numbers reach 1,000 (Lipoglavšek-Rakovec, 1950, 50). Most migrants worked in mines, fewer in industrial plants and construction; a part of migration had a temporary character and was seasonal (Lipoglavšek-Rakovec, 1950, 43–51). The Italian state institutions managed the migration from the Julian March. The national migration system broadly worked to improve the competitive edge of the Italian workforce and endorsed its inclusion into the increasingly closed off and protected foreign labor markets. The government concluded agreements with other countries and employers to secure the work and protection of Italian workers. It seems that after the implementation of the restrictive policy in 1927, the Italian authorities facilitated permanent emigration and family reunification on the continent for Slovenian migrants for national reasons as well (Grossutti, 2013, 217–218).

Migrations from the Julian March went in other directions, too. Because of the harsh economic situation and the devastation caused by war, Goriška saw the renewal of the mostly female migration to Egypt, which continued until World War II (Lipoglavšek-Rakovec, 1950, 53). Slovenian migrants in Egypt were now formally members of the Italian migrant community, which between 1917 and 1927 grew from 40,000 to 51,000 members. This increase was also the consequence of the inclusion of Slovenian immigrants from the newly incorporated eastern Italian provinces as Italian citizens. The 1937 census found 57,000 Italian citizens living in Egypt. It was a complex community having a stratified social and professional structure, complete with its own social, educational, and cultural institutions, and economic activity in the agrarian, industrial, and commercial sectors. The list of professions from the 1927 census includes, among others, 2,079 domestic workers (Petricioli, 2007, 6–7). Among them were numerous Slovenian maids and nannies. Leaving for Egypt remained an important part of family economies and the economies of certain communities in the Gorizia region (Milharčič Hladnik, 2015).

In their search for income, particularly in the 1930s, many migrants turned to the inland of the Italian peninsula rather than abroad. Even Italian colonies became destination countries. According to the new economic model and migration politics, Fascism encouraged colonization in Italy and its African colonies in place of other countries. Migration to the colonies became an important social valve because of the slow internal colonization and the entire economy's inability to absorb the surplus workforce. In the eastern provinces, the propaganda encouraging migration to the colonies and support for it also became the vehicle of the *bonifica etnica*. The programs to strengthen deslavicization included separating the *allogeni* from their Slavic roots. They included itemized steps to move them to large cities and the agrarian colonies in Italy and Africa

(Sala, 1974; Grossutti, 2013, 242–243). As many as 2,580 Slovenians and Croatians from the Julian March moved to Ethiopia in the second half of the 1930s. After they had finished their military service there, some stayed on as laborers. Others moved there as a part of the colonization programs or were recruited as workers and professionals to carry out the infrastructural colonial construction projects. Most colonists returned before World War II with the mass Italian return. In 1938, fewer than 600 remained (Vovko, 1992, 88). In 1939, based on an agreement between Mussolini and Hitler, the migration of Italian workers to Germany restarted after a long time. Many Slovenian construction workers from the Julian March were among them (Grossutti, 2013, 245–246).

Traditional Migration Patterns in New Circumstances

Internal Colonization

The Kingdom of Serbs, Croats and Slovenes created a colonization policy as part of its efforts to address the agrarian question. Internal colonization became a vehicle to rationalize the use of agrarian resources in the country and ease the problems of smallholders and rural populations that could not live off the land because of the insufficient and fragmented land property and backward style of farming. At the same time, colonization played a role in the policy of ethnic and religious homogenization, changing the ratios in the ethnic and religious structure of the population in the territories where national minorities lived.

Just like in the decades before the war, colonization in the first Yugoslavia did not generate particular interest among the rural population of Yugoslav Slovenia (Melik, 1938, 157). From the beginning, colonization was mostly linked to the Slovenian and Croatian refugees and migrants from the Julian March. These refugees arrived with little or no property and needed help and work, which further aggravated the already difficult social and economic situation in Slovenian and Croatian territories, created a huge burden for the Yugoslav authorities, and generated dissatisfaction among the local population. The main question was where to settle the newcomers from the Julian March. For rural families, one solution seemed to be colonization. Migrants were interested, and, as early as 1919, they checked the possibilities for resettlement in Macedonia, Serbia, and Kosovo (Štrekelj, 1919). In 1921, 254 families with 1,300 members settled in Prekmurje on the land allocated to them with the agrarian reform (Komac, 2015, 65; Ravnik, 2002, 55).

Slovenians also settled elsewhere in the country, independently of the state, throughout this time. Farming colonists in Slavonia, who took over land as half-tenants or purchased it from the Church, were one example (Kalan, 1936, 84).

The economic crisis made internal settlement increasingly important and the Slovenian migration route southwards intensified. Slovenian migrants in Westphalia and France were also interested in colonization, as many wished to return. The state first addressed the question of internal colonization in 1920 with the Uredba o naseljevanju novih južnih pokrajin (Decree on Colonization of the New Southern Regions) and later with the 1929 Uredba (Decree) and 1931 Zakon o naseljevanju južnih krajev (Law on Colonization of Southern Territories). It defined which land was intended for colonization, the criteria for its allocation, conditions, and obligations for the land recipients. As the main body of the Ministry of Agriculture for Southern Regions, the Supreme Office for Agrarian Reform (Vrhovno poverjeništvo za agrarno reformo) in Skopje, together with different regional offices, was in charge of executing the colonization legislation (F. F., 1931). In the early 1930s—once all the system's weaknesses had become apparent and the land was becoming scarce—the state process of settlements slowed down significantly. Colonization was only possible by purchasing private land and farms or settling rural families or agrarian laborers on large private estates (Drnovšek and Kalc, 2014, 101).

As already mentioned, agrarian reform was also used to deal with national and religious issues. Families from the Julian March were resettled in Prekmurje to reinforce the Slovenian presence in the areas populated by Hungarians. As the resettled families were given preferential treatment during the agrarian reform land allocation process, Slovenian farmers in Prekmurje also opposed the settlement of Primorci (Kovàcs, 2007). In Kosovo, where most of the land was in the hands of Albanians and other non-Slavic landowners, the state settled 50,000 Serbians and many Croatians and Slovenians. Internal colonization became a part of migration policy as an alternative to migration abroad. The Slovenian national emigration program placed it, together with accelerating the modernization of agriculture and industrialization, among the methods of limiting migrations abroad that were "negative for the nation." Nevertheless, the planned colonization remained poorly organized and difficult (Merkun, 1932; Izseljeniška spomenica Jugoslovanske Izseljeniške katoliške akcije v državah zapadne Evrope, 1933, 3; Drnovšek and Kalc, 2014, 101–102).

An illustrative case of colonization in the south of the country is the colony in Bisternica on the Vardar River in Macedonia, established in 1931 at the initiative of the political leadership of the Zveza jugoslovanskih emigrantov iz Julijske krajine (Union of Yugoslav Emigrants from the Julian March). The colony was created with the political and financial support of the state authorities in a region where the Turkish population had lived before migrating to Turkey as a consequence of discrimination against Muslims of non-Slavic ethnicities and Atatürk's incentives to attract Turks from abroad (cf. Pezo, 2009). A cooperative was set up and purchased 400 hectares of land and necessary farming equipment. The colonists—entire families and individuals—owned their farms or worked as wage laborers on the land owned by the cooperative, but they also had other professions. Despite mechanization and the attempts to implement modern forms of farming, the cooperative—the colony's economic backbone—failed in 1936 due to mismanagement and debt. The colony stagnated until the war—when the Bulgarian occupiers expropriated the colonists—then dispersed (Rusić and Novak, 1973). In Bistrenica, like elsewhere in Yugoslavia, the Slovenian colonists found it difficult to mix with other ethnic groups and encountered economic, organizational, and other difficulties. For this reason, the colonization programs had little traction in Yugoslav Slovenia and, for the most part, only attracted migrants from the Julian March (Drnovšek and Kalc, 2014, 102–105).

Seasonal Migration

Between the two wars, seasonal migration remained a widespread pattern of labor mobility. After World War I, the extent of seasonal work increased because of the reduced possibilities and the changed regimes of access to the international labor market. It had an increasingly important share in internal and international migrations. Prekmurje continued to be the center of agrarian and other seasonal work, becoming the region with the highest emigration rate on the Yugoslav, and not just the Slovenian, scale. The reasons for seasonal migration's emergence and systemic role in Prekmurje have already been discussed. After the region was annexed to the Kingdom of Serbs, Croats and Slovenes, the situation did not significantly change, despite the agrarian reform and the partial redistribution of the land from the big landowners. Many smallholders could not pay off the land. Most forests moved from the hands of big Hungarian landowners to big Slovenian landowners, and a part of the land went to the immigrants from the Julian March and other newcomers. Most were part of the agrarian proletariat, which depended on wage work (Pörš, 1940; Miklič, 1939). Seasonal work thus became an even more important part of the regional economy and represented a substantial share of family income for many.

During this time, the destinations of agricultural migrant labor also changed. The traditional and decades-old Hungarian destinations were abandoned because of the border and the increased customs duties on produce, with which seasonal agricultural workers were paid. Seasonal laborers thus turned toward destinations in the Kingdom of Serbs, Croats and Slovenes, and elsewhere in Europe. In 1927, a large number of workers from Prekmurje left for France. In 1929, they left for Germany, which had already attracted some workers in the early 1920s, but it turned out to be an inauspicious employment market because of the inflation (Korpič-Horvat, 1992, 48).

Picture 1.12: Seasonal workers from Prekmurje in Baranja (Croatia) in the 1930s. Author: unknown. Courtesy of: Slovene Ethnographic Museum, Ljubljana.

The bilateral agreements, which provided the basis for seasonal work, treated it separately from other work. During the economic protectionism and rational management of labor migrations, seasonal work was subject to specific agreements and contracts that precisely regulated the relationship between employers and laborers and the workers' rights. Like before the war, Germany paid particular attention to the contingents and was strict when applying the deadlines for arrivals and departures in managing seasonal migrations. The Zagreb office of the French immigration agency Société générale d'immigration managed seasonal migration to France; Deutsche Arbeiterzentrale processed migrants to Germany in Berlin. The intermediary was the Murska Sobota branch of the public employment service in Ljubljana, established in 1920 (Drnovšek, 2012, 129).

Between 1929 and 1939, 22,000 seasonal workers from Prekmurje left for France and 17,000 for Germany. Over 29,000 found employment in other parts of Yugoslavia in the same period (Korpič-Horvat, 1992, 61). Emigration toward Yugoslav destinations gradually decreased in the 1930s, particularly after 1937 when, following a hiatus of several years, the migration to Germany grew exponentially. In 1937, the seasonal income of workers from Prekmurje in Germany was 7 million dinars; in 1938, it was 26 million dinars. In the decade from 1929 to 1938, the total was 56 million dinars. The number of departures to France also decreased and, with that, the remittances. The remittances in 1936 were 17 million dinars, and in 1938, 7 million; in the decade 1929–38, they were 107 million dinars (Olas, 1957, 186). The entire scope of seasonal migrations constantly increased in the second half of the 1930s, so, in 1937, the statistics show 7,000 departures and, in 1939, more than 9,000 departures. This number is likely higher because many made their own arrangements to go and did so without the permission of Yugoslav authorities. At least 6,000 migrants—those who did not come back after previous seasons—must be added to this number (Maister and Uratnik, 1938, 21; Korpič-Horvat, 1992, 61). Seasonal workers—most of them from the hilly part of Goričko—were around 30 percent of the working-age population in Prekmurje (Pörš, 1940).

The agrarian sector employed both genders, but the demand for men or women changed depending on the type of work. From 1929 to 1931, around 70 percent of migrants to Yugoslavia and France were men, while among the migrants to Germany, women made up more than 70 percent. In 1939, the percentage of men among the seasonal migrants from Prekmurje was 60 percent in Yugoslavia and 54 percent in Germany; in France, it increased to 77 percent (Škafar, 1939b, 122–123). The interest in France and Germany was generated primarily by wages, which were three to four times higher than in Yugoslavia in the best years. The earnings further increased with the exchange rate and the devaluation of the Yugoslav currency (Maister and Uratnik, 1938, 21). If, in Yugoslavia, one could earn up to 1,800 dinars in a season,[29] in France, that sum was 5,500 dinars, and in Germany it was 7,000 dinars (Šteiner, 1994). On paper, seasonal migrants also had the same rights as the local workforce, which included insurance and legal protection, with their work counting toward pensionable service. In reality, many of these rights were withheld and subjected to more exploitation

29 The state-owned estate Čokl paid first category workers (adult males) 600 kg of wheat and 1,600–1,800 dinars for six months of work. In comparison, fourth category workers (females aged 15–17) earned 400 kg of wheat and 1,232 dinars (Olas, 1957, 188).

than the local workforce. The states differed in their immigration policies: Germany only accepted agrarian laborers, German bodies selected workers and the contracts were only available for one year. France had no limits on the length of stay, so many remained there for longer and found employment in other economic sectors (Drnovšek, 2012, 130–135).

While some seasonal workers from Prekmurje found year-round employment in Slovenia, mainly in construction (Korpič-Horvat, 1992, 57), many migrations to other parts of Yugoslavia started in the early 1920s. Their main destinations were Podravina, Slavonia, and Vojvodina, where they settled on scattered private and state-owned agricultural estates and forests (Škafar, 1939b). This reorientation was promulgated by the Employment Service (Javna borza dela) in Murska Sobota, which attempted to dissuade the seasonal workers from their traditional destinations in Hungary, where they were subjected to (anti-Yugoslav) political propaganda and (pro-Hungarian) nationalistic proposals (Škafar, 1939a, 35).

Some migrants left for spring works and returned for the fall one, but others remained year-long on the estates. Seasonal migration abroad, particularly to Germany, was better organized, and most of all, workers could count on social security and pension. In Yugoslavia, the legislation only recognized full-time workers; seasonal workers had neither insurance nor pensionable service. They were not organized into chambers of labor or agriculture and therefore had no collective agreements. Working hours and other labor conditions were worse and often only generally defined (Olas, 1957, 189–190). In this system of seasonal agricultural work, the role of a *palir*, a foreman or overseer who selected workers and discussed all aspects of the working relationship with the landowner on their behalf, persisted (Korpič-Horvat, 1992, 45). It was difficult to replace this old system with the one championed by Javna borza dela, which was not fully operational before 1929. The palirs, despite changes and increased supervision, retained their positions and found different ways to exploit workers (Pörš, 1940; Škafar, 1939b, 134–135).

Despite improvements and work contracts with employers, seasonal workers in Yugoslavia worked and lived in significantly worse conditions than in Germany or France. They were earning less and were in a legally worse position.[30] They also lived in provisional, non-hygienic communal living spaces without basic equipment and bathrooms. In case of illness, they were not eligible for benefits; if they left their contract early, they had to reimburse the employer for

30 In the decade 1929–1938, just under 28,000 male and female workers earned 45 million dinars (Olas, 1957, 191).

the travel cost. Nevertheless, observers—considering the general labor conditions and the country's economy and in expectation of the modernization of agriculture and development of other economic sectors that would ease the demographic pressure in Prekmurje—felt that the seasonal migration of Prekmurje agricultural laborers was a satisfactory systemic solution (Škafar, 1939a, 35, 1939b, 128–134). Elsewhere in Slovenia, where there was no tradition of seasonal work, the surplus agricultural workers were left without income, while the more progressive farming areas experienced a shortage of workers. Prekmurje was, in the opinion of some economic experts, a model case of harmonizing the supply and demand of the seasonal workforce. However, an observer watching the phenomenon through the lens of the region's development probably only saw seasonal work as a symptom of the socioeconomic system's backwardness and its inability to turn toward modernization (Maister and Uratnik, 1938, 22–23; Drnovšek, 2012, 137–138).

Migrations and Migrants under Control

A New Migration Regime

The interwar period marked a new chapter in the political attitude toward migration and migrants. During the time of the great migration before World War I, the constitutive nations of the newly created Kingdom of Serbs, Croats and Slovenes were subjected to different migration regimes and policies. Also—depending on their countsries' economic, national, and political interests—their former countries' attitudes toward diasporas and migrants were different. With the unification into one state, the country's founding nations were finally in a position to manage the migration processes and align the different facets of the "migration question" with their economic, social, and national interests.

Initially, all migration-related issues were the responsibility of the government and administrative bodies of the provinces into which the new country was divided. However, the mass returns of prewar migrants from abroad and the renewed emigration caused by economic needs required a centralized approach, demanding coordinated organizational, administrative, legal, social, and political interventions, so the kingdom introduced steps to create and incorporate a migration regime and migration policy into its political agenda. In May 1920, it established the Odsek za izseljevanje in priseljevanje (Emigration Department) as a part of the Ministry of Social Policy in Belgrade (cf. Hranilović, 1987, 329; Đikanović, 2012, 94). In 1921, legislation on emigration and immigration was introduced and replaced the existing acts inherited from former countries.[31] The state took direct control over migrations and actively regulated and managed migration processes. Such an approach was necessary even before the law was adopted, not the least because migration regimes also changed internationally. More concretely, as mentioned above, the United States implemented a quota system, which demanded systems for selecting applicants who fit the requirements (Đikanović, 2012, 171–172).

The legislation unified the migration regime in the entire Kingdom of Serbs, Croats and Slovenes and was based on the liberal principle of the freedom of emigration. The state did reserve its right to limit this freedom and regulate migration according to the concurrent situation and the interests of the country and its citizens. The freedom to emigrate was also legally limited under the citizens' military, penal, and other civic obligations. In formulating its Emigration Act, the Kingdom of Serbs, Croats and Slovenes used the Hungarian law of 1901 as a model, as well as the Italian law of 1901 (amended in 1919), which was the most precise in terms of economic, legal, and health insurance status of emigrants. The Italian law allowed state control over all emigrants and all subjects participating in organizing the migration process. It also allowed coordinating migration with the Italian national interest at home and abroad (Miletić, 2009, 91–92; Kalc, 2021a). The kingdom's legislation, of course, only impacted those Slovenians who were citizens and not those who found themselves as stranded minorities in neighboring countries.

The Emigration Act also created the Emigration Service, a bureaucratic system to manage emigration and immigration and implement migration policy. The system developed gradually and was adjusted and reorganized several times. It included the Emigration Department at the Ministry of Social Policy in Belgrade, the Emigration Commissariat, the port, railway, and shipping commissariats, and emigration commissioners of the Ministry of Social Policy abroad. The center of the Emigration Service was in Zagreb where the Emigration Commissariat was located. In 1923, the Emigration Commissariat lost its autonomy to lead the sector because its existence did not align with the concept of centralized state authority. However, it retained the bureaucratic-operative and practical-organizational competencies (from the organization of transportation, legal

31 The legislation was introduced in three steps: first, in July 1921, with the Emigration Decree, which was followed in March 1922 by the slightly altered Emigration Act, and, in 1923, the Regulations for Implementation of the Emigration Act, which defined the individual articles in more detail and interpreted the ideological guidelines of migration policy.

protection of migrants, and information service to recordkeeping and statistics) and control over the activities of the Emigration Service. At the same time, emigration policy came under the jurisdiction of the Emigration Department at the Ministry of Social Policy. The reduction in competencies of the Commissariat in Zagreb was also the consequence of antagonisms between Belgrade and Zagreb. The latter hoped for a higher level of competencies, referring to the scope of Croatian emigration—the most numerous among the Yugoslav nations, its central transportation role, and its experience in administrating migration legislation gained under the Kingdom of Hungary. To centralize the Emigration Service even further and harmonize it with other state political sectors, the Emigration Department in Belgrade also set up the Izseljeniški medministrski posvetovalni odbor (Interministerial Emigration Consultation Committee) (Hranilović, 1987, 330–331; Đikanović, 2012, 92–107; Bućin, 2017, 39–45).

The Emigration Act also established the Emigration Service abroad—a service the citizens did not know before the war. Migration deputies or commissioners worked at the Yugoslav diplomatic-consular missions to maintain ties between migrants and their homeland and channel Yugoslav national and political ideas to the migration communities around the world (Hranilović, 1987, 331). Since 1928, the private organization Union of Emigration Organizations (Savez organizacija iseljenika, or SORIS), established by returnees, played an important role in this system based on the cooperation of the state administrative structures and private institutions. Similar organizations appeared even earlier, but these endeavors intensified with SORIS, which acted as an umbrella institution. The Emigration Service's main tasks were counseling and informing of the situation in destination countries, protecting and caring for migrants, organizing repatriations, disseminating national propaganda among migrants, organizing money transfers from abroad, organizing migrants' legal affairs, regulating emigration, studying social and labor situations, managing the statistics and evidence of migrants and migrants' organizations (Hranilović, 1987, 332).

It soon turned out that the law suited neither the new times and needs of migrants nor the state itself. It was calibrated to the situation before World War I, which changed radically postwar with the sudden decline of transcontinental migration and the restrictions imposed by the main former receiving countries. With its focus on transatlantic migrations, the law was created to fit the migration to the United States. Most migrants went there before the war and in the first years immediately after, and the greatest economic and national interests were linked to those Yugoslav communities. In addition to protecting migrants and regulating economic business linked to the migration process, the law included provisions to "nationalize" Yugoslav overseas migration. Such provisions were key to successfully controlling the migration process and developing the shipping and maritime economy. However, the basic conditions were not met for its realization: both former migration ports, Trieste and Rijeka/Fiume, which could serve as outlets for migrants from the Yugoslav territory, were now under Italian rule, and none of the home ports had the infrastructure for mass passenger traffic. The government thus renewed concessions to foreign shipping companies for transporting migrants and directed migrants to foreign ports (Bućin, 2017, 42, 47). Continental migrations, which grew as the overseas ones decreased, were left out of the law's protection scheme. Yugoslavia only started regulating these migrations at the end of the 1920s, with a series of bilateral agreements with the countries that received its workforce (Brunnbauer, 2016, 216–220).

The law had other shortcomings and inconsistencies and thus could not become the foundation of the country's migration policy as envisioned. The first attempts to revise it came in the second half of the 1920s (Bartulica, 1929). The revision never happened, partly because Yugoslav politics considered emigration a minor issue, which remained a marginal interest among the country's political leaders (Hranilović, 1987, 333). Emigration conferences, where delegates from Yugoslavia and the migrant communities debated the needs and guidelines for further activity, were key in the development of the Emigration Service and its activities. These efforts gained new impetus in the late 1920s because of the economic crisis and social situation at home and among the migrants. In the first half of the 1930s, the work of the Emigration Commissariat and the Emigration Service expanded. They did more work covering the employment and social security of those leaving to go abroad, helped returnees reintegrate, and implemented the civil interests of the migrants. They also boosted national-cultural and educational activities and communication between the homeland and the emigrant communities (Bućin, 2017, 50–53).

Ambitions and Conundrums of the Migration Policy

In the Yugoslav state between the two wars, emigration was considered a negative phenomenon because it deprived the country and the nation of their vital life forces. In the spirit of the thinking formed during the great migration period before the war, it was thought that the population should be guaranteed a life at home; their demographic and labor potential should be used for the progress of the state and the nation. However, because slow progress and social circumstances did not allow for that, emigration remained an inevitable exit to fulfill material needs and ease social burdens. Therefore, it had to be organized and run rationally: under the state's complete control and following the appropriate migration policy to best serve migrants' needs and the nation's interests.

Creating and executing such a policy was difficult and often controversial, so its plans were only realized in a limited scope. The problem was inadequate legislation, which did not regulate migration as a whole or adequately for its changing modalities. The second source of problems was that any issues related to migration and the Yugoslav diaspora were only dealt with within the narrow government apparatus, private organizations, and enthusiastic individuals despite their resonance with the public. Migration and diaspora issues only made it to the agenda of "big" politics in moments when the internal and foreign political situation required closing the Yugoslav "national ranks" at home and abroad (Brunnbauer, 2016, 211). For this reason, in the first half of the 1930s, the "enslaved brothers" in the Julian March also became interesting for King Alexander I's dictatorship, and their political emigration had financial support in Yugoslavia (Zobec, 2021a). Another problem was that the Emigration Service was plagued by organizational, personnel, and financial problems. The first annual report about its work was only issued for the years 1925 and 1926, and the parliament never discussed the document. The Emigration Service thus called on the National Assembly to show more sensitivity to migrants and pass political directives for further work (*Izveštaj narodnoj skupštini za 1926/27. godinu*, 1928, 3). The dualism of competencies and subsequent frictions between the Emigration Commissariat in Zagreb and the ministry in Belgrade also created problems and prevented the more efficient work of the Emigration Service (D., 1927).

The state administration was not up to speed with either the declared political principles or the concrete needs dictated by the circumstances. It was also late in providing legal and social care for migrants abroad. Bilateral negotiations with European countries only had concrete results with Germany in the form of conventions or agreements about the social protection of workers (Drnovšek, 2006). Because of adjustments to circumstances, the course of migration policy was not entirely determined and consistent. Until the end of the 1920s, the state administration promoted short-term or circular migration so that the migrants would not adapt to foreign countries and assimilate. But once the economic crisis began, it advised emigrants not to return because no help was available at home, and they would only increase unemployment and pressure on the social funds. Emigrants were also advised to take the citizenship of the countries where they resided to get better social protection. Regarding the United States, the Yugoslav authorities hesitated throughout the 1920s between encouraging migrants to return, which would increase the influx of savings short term, and remaining in the United States, which would preserve more long-term sources of income from abroad (Hranilović, 1987, 327–329; Miletić, 2009, 158).

Migration policy had an important purpose that, while only partly implemented, was the most visible in its intentions and endeavors. That purpose was to attract Yugoslav communities worldwide into the nation-building process and connect them to their country of origin by strengthening the sense of Yugoslavdom and establishing an integrated national Yugoslav state with transnational dimensions. Yugoslavia was imagined as a nation of three constitutive "tribes." The state-forming idea and politics were based on the ethnic principle and, as such, nurtured an exclusionist attitude toward the non-Yugoslav nations or communities in its territory while claiming—and trying to exercise—sovereignty over the members of the constitutive nations in the countries of emigration. By implementing the migration policy, the Emigration Service and the diplomatic corps did their part by setting up tight bonds with the migrant communities "inherited" from previous countries and shaping them into loyal and nationally aware Yugoslav diaspora. This extraterritorial sovereignty was based on the notion of emigration as a part of the Yugoslav ethnonational body with 800,000 members, most of them in the United States. They called it "the tenth banovina" (Brunnbauer, 2016; Larson, 2020; Đikanović, 2012).

Picture 1.13: A family picture sent from Little Falls, New York, to relatives in Slovenia in 1934. Author: unknown. Courtesy of: Breda Kolenc, Bočna, Slovenia.

The Kingdom of Serbs, Croats and Slovenes faced the question of Yugoslavs abroad since its establishment when it had no directives for this field and managed the issue with decrees as needed. The question appeared in connection with citizenship and issuing travel documents for returnees. Citizenship policies became a vehicle for migration policy and promoting national interests inside and outside the state territory. At first, they issued a document to return to the new homeland to all applicants but soon tightened the formal procedures and demanded proof of Yugoslav origin. They checked loyalty and refused persons with a negative attitude toward the state. They thus prevented the return of "undesired" migrants, persons with criminal records, and people with mental disabilities, but most of all, persons with radical political views (Đikanović, 2014).

Passports were also an instrument of the migration policy, both in the attitude toward the constitutive Yugoslav nations and national minorities. In the first case, there were restrictive principles, so the Ministry of Social Policy only issued passports after checking the applicants' circumstances and per national interests. On the other hand, the emigration of members of national minorities was such a priority that those emigrants did not even need the basic requirement (completed military service) to obtain travel documents. A passport application, in such cases, was considered an expression of a desire to belong to another country. The Citizenship Act also reflected an attitude toward different national minorities that was, in fact, diverse and favored Slavic communities (cf. Kosnica, 2018). The exclusionist politics were most acutely felt by the German and Hungarian minorities, which represented 3.9 and 3.4 percent of the population, but made up 20 percent of emigration.[32] Specific incentives were given to non-Slavic Muslims wishing to emigrate, representing an important share of migration from the Yugoslav state (Miletić, 2009, 110–111; Pezo, 2009; Brunnbauer, 2012; Đikanović, 2014, 89).

---

32 In 1924 and 1925, this share was almost one-half, but on average, between 1921 and 1928, 47 members per 100,000 inhabitants of these two national communities emigrated, while among the constitutive nations the number of those leaving was less than 10 per 100,000 inhabitants. The state led a diversified migration policy toward national minorities and equated some to the "national" element (Miletić, 2009, 109–110). The members of the German community were, considering specific circumstances, subjected to this diversification policy, while, according to the experts for migration policy, constant emigration of Hungarians was to be favored unconditionally because they, as a consequence of the Hungarian territorial aspirations, represented an anti-national element ("Mišljenje eksperta Iseljeničkog komesarijata Artura Benka Bojničkoga: Imigraciona politika prema narodnosnim manjinama, Zagreb, 17 April 1925").

The citizenship policy was a tool of migration policy, and this was particularly clear when it came to the attitude toward Yugoslavs abroad. The Citizenship Act was quite controversial when it came to this point. It stipulated that any citizen of the countries that united into the Kingdom of Serbs, Croats and Slovenes and with the right of domicile in its territory was a citizen of the kingdom. So were their descendants who obtained citizenship based on the *ius sanguinis*, which meant that most migrants abroad were Yugoslav citizens. Dual citizenship was not permitted, and although there were possibilities of renouncing or returning citizenship, the inconclusiveness regarding many points created serious complications. A male emigrant with US citizenship, for example, could not renounce his Yugoslav citizenship unless he completed military service, so during any visit to the homeland, he risked forced recruitment, but by serving the Yugoslav army, he also risked the loss of US citizenship (Đikanović, 2014, 91). Many problems and instrumental (non-) admissions of citizenship introduced tension into the relations with the host countries. The legal status of migrants in relation to Yugoslav citizenship was better defined only in 1937 (Đikanović, 2014, 93–4).

These problems notwithstanding, the reinforcement of the Yugoslav spirit and congregating migrants around the Yugoslav national idea went hand in hand with the national, cultural, and educational work and propaganda activities. The privately run SORIS, which, as already mentioned, joined the state Emigration Service, informed the public at home about Yugoslav emigration and strengthened cohesiveness between Yugoslavs at home and around the world. With the January 6, 1929 coup d'etat, the state started demanding that migrant associations abroad use Yugoslav in their names (e.g., Yugoslav Association). The state authorities' attempts to construct and strengthen the exterritorial Yugoslavdom, which, in the 1930s, became a priority of the Emigration Service and migration policy, were physically manifested in the move of the Iseljenički muzej (Museum of Emigration) (1933–1941), the Emigration Commissariat, and other migrant organizations into a dedicated building, symbolically located near the Zagreb railway station (Bućin, 2017, 50; Larson, 2020, 282).

Nevertheless, the endeavors to bring Yugoslavs living abroad to their "homeland's embrace" was not altogether successful. The migration policy and its organizationally and financially malnourished apparatus faced many obstacles. The insufficient and inadequately qualified migration officers, working in migration communities, were among them. Additional obstacles were the unresponsiveness of the predominantly Serbian consular staff to serve the primarily Slovenian and Croatian emigration, the inadequate consular network, and often conflictual relationships between diplomatic missions and migrant communities. For example, of all the Yugoslav diplomats, whether ambassadors or envoys,

Slovenians in the United States only accepted the Slovenian Leonid Pitamic as "their man" (Grill, 1979, 461). Likewise, the relationship between the Slovenian community in Argentina and the Yugoslav representatives only improved with the arrival of Slovenian envoy Izidor Cankar in 1936. However, he was not a favorite of the Croatians (Mislej, 1995b). A segment of the Slovenian migrant community also did not support Cankar's politics and did not identify with the Kingdom of Yugoslavia (Zobec, 2021a). Things were similar in France and Germany. The language was another important obstacle, as the staff of the Yugoslav diplomatic service did not speak Slovenian, while the migrants had difficulties communicating in Serbo-Croatian; also, the mission's official written communication was in the Cyrillic script (Kristan, 1927b, 868; Drnovšek, 2010a, 180). The criticism toward the Emigration Service (at home and in the diaspora) was that it could not forge strong emotional bonds between emigrants and the homeland (Hranilović, 1987, 327; Drnovšek, 2010a, 150).

Distrust and uneasiness among the migrants were also spurred by cross-expectations and interests of the Yugoslav state and the receiving countries. Each country saw immigrants or emigrants from its own political point of view and wished to subject them to national and political interests. Foremost, the interests, priorities, and views of the migrants themselves were at play. Having faced many difficulties in constructing their social position in receiving countries, migrant communities were not pleased with the interferences and the broadening of the systems from the "homeland" into their life contexts, as such activities limited them in their practical life and even endangered their hard-gained reputation. The relationships between the homeland and migrants also depended on worldviews and political orientations in the homeland and migrant communities worldwide. Migrants aligned with different political options, from conservative to progressive; considering their mostly working-class origins, leftist ideas were relatively widespread. In the United States, Yugoslav migrants were active in union organizations. Revolutionary ideology was present in Argentina (and France and Belgium), and many participated in the Communist movement (Drnovšek, 1990; Kalc, 2016b; Zobec, 2019). In general, republican ideas resonated strongly among migrants. The Yugoslav unitarist politics opened a series of questions, both from the national point of view and the point of view of the internal Yugoslav political scene. Controversial—and completely unacceptable for many in the diaspora—was the January 6, 1929 Dictatorship of King Alexander.

While some migrants showed loyalty to the Yugoslav centralist and absolutist line or even supported it, others were critical of the regime and its politics, with those on the left accusing it of anti-working class and fascist orientation. In his bestselling book *The Native's Return* (1934), Louis Adamic, the most renowned

American-Slovenian writer, mercilessly decried the social and political circumstances under King Alexander's dictatorship, so the Yugoslav censors banned the book. Likewise, many diaspora newspapers were banned because of their stances on "the old homeland" or their ideological orientations (Grill, 1979). In Argentina, the Yugoslav embassy intervened with Argentinean authorities because of the critical writing of Slovenian newspapers (Zobec, 2021a, 7–9). Communist activists and their organizations and "anti-national" (i.e., anti-Yugoslav) nationalist and separatist groups were particularly subject to surveillance by the Yugoslav authorities, as well as the governments of the receiving countries (Drnovšek, 2010b). Not even the migrants supportive of the regime could avoid disputes. For the Slovenians, for example, the problematic issue was the politics of the Yugoslav dictatorship toward Catholicism. In general, the disputes were influenced by the endeavors of the Yugoslav authorities to treat migrants abroad as if they lived on the Yugoslav territory. The diaspora opposed this and formed an alternative image of Yugoslavdom. It referred to their own lived experience in "new homelands" and demanded the position of independent socio-political subjects (Larson, 2020). Opinions on organizing migrants on a Yugoslav or national basis were also divided. SORIS promoted the principle of a common, integrated Yugoslav organization; others prioritized national (Slovenian, Croatian, and Serbian) organizations that would unite into a Yugoslav association (*Prvi slovenski izseljenski kongres v Ljubljani dne 1. julija 1935*, 1936, 83; Drnovšek, 2010a, 161).

The very starting points of the national and national-homogenization policies remained undefined and contradictory. As the generations born abroad came of age and the numbers of new migrants decreased, the problem of assimilation became more pertinent. During the Great Depression, the Yugoslav Emigration Service supported the line of retaining and reinforcing the national consciousness. In contrast, others even recommended that emigrants assimilate into the receiving social and cultural milieu so that their national pride would not make their life abroad more difficult. Not all countries were seen in the same light: for example, the United States, where naturalization did not necessarily mean the erasure of the original nationality, and countries like Germany and Austria, where prewar immigrants tended to abandon the Slovenian ethnic character and assimilate during the first generation.

The concern for the migrants' needs and the work in Yugoslav communities around the world was certainly far greater in the newly established country than in Austria-Hungary. This is true for the entire interwar period. Through its Emigration Service, the state financially supported educational work among migrants and encouraged their cultural and political organization. Of course, the state encouraged organizations that were favorable to the state or, at least, did not openly oppose it. The state institutions helped maintain the migrant press, supported national homes, enabled education and professional development, organized cultural exchanges, holidays, and trips where young descendants of migrants could meet in the homeland of their parents, involved migrant communities in cultural and other important initiatives, managed legal issues among migrants and returnees, helped maintain close ties between the homeland and the migrants, which was particularly important for retaining the remittances that migrants sent home. These were the areas in which the state and its Emigration Service achieved the best results during this time (Hranilović, 1987, 327–329; Drnovšek, 2010a, 154; Kavrečič and Koderman, 2014).

State and Private Initiatives in Yugoslav Slovenia: The Role and "Power" of the St. Raphael Society

The mass return from abroad and renewed emigration in the first few years after the war posed a real test for Yugoslav Slovenia—itself also a transit territory—and its administrative-political and institutional bodies, which were just taking shape. Initially, migration-related issues fell under the authority of the provincial government authorities (Drnovšek, 2007b). These government and administrative structures could not cope with the organizational, security, and social needs and solve the many known and newly emerging questions in connection with migrations. In Slovenia and elsewhere throughout the country, requests immediately appeared to manage migration issues through appropriate regulations and a centralized institutional governing apparatus. A political and academic debate also started on migration as a social, economic, and national-political question within the newly formed country. At the congress of Yugoslav cooperatives in the summer of 1919, Slovenian politicians took part in a debate on the economic potential of Yugoslav emigration. They participated in creating a plan to organize return migrations. Experts publicly called for measures to protect the economic interests of migrants, the value of their assets on their journey home, resolving inheritance and other civil cases of the deceased migrants abroad, and any other cases that came to the administrative and judicial bodies, as well as legal offices (Kalc, 2021a).

The debate also addressed migration legislation and the state Emigration Service. The career diplomat Ivan Švegel, a very experienced former Austro-Hungarian consul and ambassador to the United States and Canada,[33] promoted

33 He was also the author of a comprehensive study about emigration from the Austro-Hungarian Monarchy to the United States (Schwegel, 1904).

the office of emigration commissioner as a core body that would manage and supervise migration processes, intervene in the international labor markets, and manage all the aspects of the emigration question. He pointed out the possibilities of economic and financial collaboration and the moral and political importance of relationships between migrants and the homeland. Migration legislation, which was primarily the result of expert and political work coordinated between Belgrade and Zagreb, bore the signatures of two Slovenian members of the state government, Vekoslav Kukovec and Gregor Žerjav, who, in 1921 and 1922, were in charge of the Ministry of Social Policy (Kalc, 2021a; cf. Miletić, 2009; Đikanović, 2012; Bućin, 2017).

Once the migration legislation was passed and the management of migration issues institutionalized at the state level, these issues remained a pertinent topic in Slovenia either because of the practical encounters with migrations in the new international circumstances or because of the Emigration Service's difficulties and shortcomings. In the second half of the 1920s, the concern about maintaining contact with migrants also increased. Because of the halt to (transatlantic) migration and ensuing generational shifts, the largest migrant communities required a specific approach, particularly in the United States. Much effort went into averting the threat of assimilation of migrant communities, mostly by promoting national consciousness among the migrants and the emotional bond to the Slovenian or Yugoslav homeland (and thus also helping to retain the remittances and other material support that migrants sent to the "old country"). Some saw particular value in the cultural contribution migrants could bring, based on their experiences and achievements in the receiving countries (Miklavčič, 1927). The first official visits occurred in Slovenia (and Yugoslavia) and the transatlantic and continental migrant communities during this period. Individual and organized tourist visits to the homeland also increased; they were important economically, representing 70 percent of tourists coming from abroad (Kristan, 1927a; Friš, 1996; Kavrečič and Koderman, 2014).

Increased interest in emigration and all its aspects was reflected on the institutional, organizational, and academic levels. In 1928, the Ljubljana Provincial Authority established the Emigration Office, which became an office of the Department of Social Policy and National Health within a year. The office dealt with the social, cultural, and economic needs of migrants, returnees, and refugees from Italy. Within its competencies and always working with the state, the Drava Banovina created its own migration policy in 1929 (Kuhar, 1939, 534–535; Drnovšek, 2010a, 157). In 1928, Slovenian delegates took part in the establishment of SORIS and, as its representatives, set up a sister organization in Ljubljana. The organization did not work out, but the collaboration with SORIS and reports in its bulletin remained. In 1929, the Izseljenska liga (Emigration League) was established in Ljubljana as a socialist, cooperative, and economic organization to support migrants and link them to the homeland (Drnovšek, 2010a, 146–147). In 1938, the Izseljenska zbornica (Emigration Chamber) was established to connect all migrant organizations in Yugoslavia with its migrant organizations worldwide (Zakrajšek, 1938). It became the most important institution of collaboration between the Slovenian migrants and their homeland (Drnovšek and Kalc, 2014, 112). In 1939, when Croatia gained administrative autonomy and transferred the migration issues to the banovina, the Emigration Service in Zagreb lost its centralist character. Slovenia planned to reach similar autonomy in migration policy. In the expectation of transferring competencies to the Drava Banovina and forming the bureaucratic apparatus, the Emigration Department in Belgrade continued to manage all things related to Slovenian migration. The question of the personal and legal protection of migrants was managed in Zagreb. The plans for an independent Slovenian emigration service were cut short by the war (Ureditev slovenske izseljenske službe, 1940).

Migration as a process and Slovenian migrants worldwide became "institutionalized" topics in the 1920s and found a place in official publications and specialized literature dealing with migration (Mal, 1928; *Izseljenski zbornik Izseljenske zbornice v Ljubljani*, 1938; Avsenak, 1939). The establishment of the University of Ljubljana in 1919 and other specialized institutions whose fields included issues related to migrations made the migration question a topic of academic research (Zwitter, 1936; Šuklje, 1936/37; Maister and Uratnik, 1938; Lipoglavšek-Rakovec, 1950) and steps were taken to study migration as a particular multi-disciplinary field was conceived. In 1937, a department for minorities and migration was set up within the Zavod za mednarodno javno pravo (Institute for International Public Law) at the Pravna fakulteta Univerze kralja Aleksandra (Faculty of Law at the King Alexander I University of Ljubljana, now the University of Ljubljana). The university was to become a center of migration research for all of Yugoslavia (Drnovšek, 1993, 21–22).

The representatives of the organizations in Slovenia worked with great enthusiasm for the joint Yugoslav initiatives concerning the question of migrations and the relationships with the Yugoslav communities abroad. They participated in migration conferences from the late 1920s to the mid-1930s—when Ljubljana first hosted a Yugoslav emigration congress. They started debates and presented their points of view regarding the Emigration Service's tasks and the questions of migration policies, but their voices were unheard or marginalized for a long time (Drnovšek, 2010a, 152). The position and the role of Slovenian organizations and representatives in the centralized migration management system and the state Emigration Service was, in fact, less influential than one would expect, considering how numerous the Slovenian diaspora was and how important migration was in the Slovenian ethnic territory. Few Slovenians were members of the diplomatic service, consular personnel, and the Emigration Commissariat (*Prvi slovenski izseljenski kongres v Ljubljani dne 1. julija 1935*, 1936, 40). The Slovenian political representatives were disregarded or excluded from foreign policy, which was the domain of the Serbians (Rahten, 2018, 164). Although the Slovenian positions in the government structures improved in the mid-1930s, Slovenian organizations were increasingly critical of the Emigration Service and its migration policy. They accused the central authorities of wanting to "rule the emigrants" by "helping" them in a dictatorial way and turning them into great patriots overnight instead of treating them as partners and simply supporting them in the activities they had envisioned for themselves (*Prvi slovenski izseljenski kongres v Ljubljani dne 1. julija 1935*, 1936).

Like before World War I, the Catholic Church played a significant role in Yugoslav Slovenia. More precisely, the St. Raphael Society became the ideological and operative motor of issues related to migration and a link in the state's system represented by the Emigration Service. In the first years after the war, the Catholic Church's interest in migration and migrants around the world decreased. Thus, in 1924, the Ljubljana branch of the St. Raphael Society was dissolved. It was re-established in 1927 and, in 1929, it was taken over by Father Kazimir Zakrajšek. Zakrajšek, who before the war was the soul of the St. Raphael Society in New York and gained rich practical and theoretical experience in the United States, became an undisputed leader of the organization and the most propulsive Slovenian personality for migration questions and state policies regulating them. With the support of the Catholic Church, the St. Raphael Society took over the role that the state administrative and political institutions could not play. In the 1930s, in Drava Banovina, this institution was the Emigration Office. Because the office lacked funding and full-time staff, the scope of its work was limited—consequently, so was its influence on the relationship between the Emigration Service and the Slovenian migrants worldwide (Ureditev slovenske izseljenske službe, 1940). Following Zakrajšek's idea, the St. Raphael Society tried to bring attention to the problems related to migration. It encouraged a wide circle of personalities, experts, and institutions to solve them, from governmental bodies to businesspeople, social and cultural workers, and, being a Catholic society, clergy. The St. Raphael Society connected with different migrant communities around the world. It set up branches and published an information bulletin, *Izseljenski vestnik* (Emigrant Newsletter), and other publications. The St. Raphael Society encouraged debate about emigration and Slovenian diaspora in the press and during public meetings and events, such as the first Slovenski izseljenski kongres v Ljubljani (Slovenian Emigration Congress in Ljubljana) in 1935, where delegates from Slovenian communities from all over the world participated.

Zakrajšek and other members of the Catholic circle formulated—congruent with the state—the main points and the directives of Slovenian politics concerning migration and wrote them down as the national migration program in 1931. Like before the war, migration remained a great national problem, a "cancerous wound," which, given the economic and social situation in Slovenia, was incurable. However, the discourse on migration lost its sharpness compared to the prewar period. The negative emphases were mollified, and there was more space for concrete work. At least within its borders, the new state context relieved the Slovenian space of the feeling of national endangerment. There were visible signs of economic and social development, which, together with the international changes, eased the negative social and emotional impact associated with the phenomenon. Moreover, in particular, the new state context allowed politics, the administration, and private organizations to develop independent and targeted approaches toward migration. The program encouraged taking advantage of the benefits migration provided. It could benefit individuals, their families, and the entire national community. However, it had to be "properly organized," managed, and controlled (Merkun, 1932; Zakrajšek, 1938, 11–13). The first Slovenian Emigration Congress, organized in 1935 at the Eucharistic Congress in Ljubljana, coordinated the work based on the national migration program and improved the development of the national Emigration Service. The establishment of the Emigration Chamber in 1938 allowed the Church and the Drava Banovina to work together and enabled the strengthening of the Slovenian migration policy (Zakrajšek, 1938, 14; Drnovšek, 2010a, 180–185, 188). In addition to regulating migration processes and the concern for the labor conditions and legal and social status of migrants, Slovenian politics also placed moral and religious education first, followed by retaining the national consciousness. Since the St. Raphael Society, in contrast to other, more centrally oriented parts of the Emigration Service, supported the principle of ethnically based emigrant organizations first, and only later their unification in the Yugoslav Alliance, it advocated the establishment of a World League of Slovenian Emigrants.

The comprehensive activity of the St. Raphael Society reflects the great zeal of the Slovenian Catholic Church concerning the "emigrant issue." This zeal must be linked to the difficult and, in many ways, insufficient engagement of the state bodies. Zakrajšek emphasized that the state's role in the implementation of an efficient and useful migration policy was of key importance, as it had an administrative apparatus and financial means available. However, the system needed the support of private institutions and the Church (Zakrajšek, 1938, 8–9). The St. Raphael Society tried to support the migrants when the legislation and the state with its services abandoned them without real assistance or even entered into conflicts with migrants regarding controversial questions (Kalc and Zobec, 2021).

The St. Raphael Society and the Slovenian Catholic Church based their commitment in response to the migration question—which they understood as a social, moral, and national mission—on Catholicism. In their vision and work programs, there was a desire for a close connection between migrants and the homeland under the Church's and clergy's patronage. These programs, presented at congresses and other occasions—and the activities themselves—show that the Church worked even harder than the state to control migration and create a transnational Yugoslavia and, particularly, a transnational Slovenia. The far-reaching (albeit not easily realized) thinking about establishing a cross-border integrated national space reached into all areas of the migrant communities and their comprehensive institutionalization into the homeland's organizational, economic, and national-cultural frames. To supplement the state register of migrant organizations compiled by the Drava Banovina, they also planned to create parish lists of migrants, publish names of people living abroad as annexes to the *Mohorjev koledar* (St. Hermagoras Society Almanac), collect and preserve migrant correspondence, and send Slovenian clergy on missions among migrants. They also wanted to teach their Slovenian parishioners about migration-related issues and the life of their fellow villagers abroad and include migration as a phenomenon and the history of the Slovenian and Yugoslav diaspora into school curricula as a part of civic and patriotic education. They investigated how to provide schooling for migrant children and academics in Slovenia to carry the Slovenian and Yugoslav national spirit worldwide. They planned to establish a migrant educational association that would set up Catholic societies among migrants, connect them and guide their activities, and make sure that the diaspora was organized and connected migrants on different continents, as well as connect the diaspora with the homeland (Mrkun, 1932; Drnovšek, 2010a, 168).

The society planned to construct a complex near the Ljubljana railway station, similar to the one built in Zagreb for the central Emigration Service. The complex, which was never realized, was to include a Migrants' Home with a hostel for migrants and returnees and a central migration library. The home would also house state, church, and private migration services whose work could be reciprocal and centralized. A Migrants' Club (a Slovenian version of SORIS) that would connect returnees and an Emigration Museum to gather material about emigration and educate the public and Slovenians around the world about this issue were also planned. Although there was no shortage of emphasis on the social, material, and legal problems of migrants, the Church and the St. Raphael Society put the care for their national, moral, and religious integrity at the forefront. This emphasis was dictated by growing secularization, the spread of liberalism—particularly socialism and communism—in migrant communities, and from there, to people at home (Zakrajšek, 1938, 14).

The St. Raphael Society's activities were also, at the same time, a reaction to the efforts of the state authorities to reinforce the Yugoslav spirit and the fact that Slovenian emigrants, particularly those in Europe, were joining the Yugoslav organizations (Drnovšek, 2010a, 170–179). Zakrajšek had aimed to set up a Slovenian missionary in every larger Slovenian settlement abroad because each dedicated priest—he maintained—would compensate for the work of many paid consular officers. He also supported the establishment of Slovenian schools, newspapers, libraries, and a series of other incentives to prevent the migrants' assimilating and limit the foreign environment's impact on their hearts and minds. To protect family life, he called on wives to follow their husbands with their children and believed young men should be assisted in traveling home to find Slovenian girls. Migration had to be pragmatically used because it would enrich the homeland with foreign money and spread the influence of "the state and the nation" outside of it (Drnovšek, 2010a, 187).

A special task of the Slovenian migration and national politics was caring for Slovenian migrants in other Yugoslav provinces. This migration had become a new, burning national problem with the new state and new migration routes now leading to the Yugoslav space. Such movements still had all the characteristics of the international ones but raised even more questions about retaining the national identity and the Catholic faith. Public institutions and the St. Raphael Society paid much attention to the "Slovenian diaspora," as they called the Slovenians in the non-Slovenian Yugoslav regions. There, they supported the creation of Slovenian societies and other organizations. However, many organizations already active in the area were set up by the émigrés from the Julian March and were, at their core, politically charged, particularly when it came to the territorial question of the Julian March and the activism against Fascist Italy; their members mostly did not find themselves in religious organizations. The St. Raphael Society worked to aid international and internal migrant communities and, of course, help them retain their national consciousness and Catholicism. The work on national and religious issues in the Yugoslav territory was pertinent not only because of the pressure of what they considered to be the Great Serbia policy but also because Slovenians in other parts of Yugoslavia were subjected to other forms of cultural assimilation—particularly their descendants, who had no access to Slovenian schools and cultural and social ties to their Slovenian places of origin. Despite cultural differences, assimilating into brotherly nations raised fewer social objections and problems and was thus easier than assimilation abroad. In the eyes of the St. Raphael Society, this possibility of assimilation made any support for the colonization and general internal migration policy contestable. After all, both national and religious assimilation were seen as threats (Drnovšek and Kalc, 2014, 112–114).

2 Social and Cultural Organization of Slovenian Immigrant Communities

The Organization and Scope of Activities

Like migrants of other nationalities, Slovenians in receiving countries organized into smaller or larger local and regional communities, creating different mutual aid and solidarity structures. The need to socialize with their fellow country people in an environment that was new in language and culture coincided with the need for greater socioeconomic security, particularly in extreme cases such as physical injuries, illness, disability, and death. The first associations were thus primarily geared toward creating fraternal benefit societies. Following the example of other migrants, Slovenians created the initial formal structures of self-organization in cultural, religious, and political life. With that, a feeling of belonging to Slovenehood emerged, something new in the life of many Slovenians in the first large cohorts of migrants. Before emigrating, their identity was often local or regional, mainly tied to their local place of origin rather than their language or nationality. Abroad, however, the language functioned as the most efficient unifier, separating them from other immigrants.

Most Slovenian organizations and societies in receiving countries served several purposes, from economic (insurance and other forms of mutual aid) and political to social (organizing events, entertainment, sports) and cultural (publishing, literature, theatre, choirs, traditional folk and modern folk-pop music, and dance activities) to religious (church and propaganda) and educational (schools, language courses, lectures). The first things migrants built in their communities were churches. In some places, the first Slovenian church communities emerged already in the late 19th century. All the activities typical for migrant parishes and other religious centers soon followed: constructing churches and parish houses and establishing schools and religious publications (Kolar, 1991, 394). After a church, followed a Narodni dom (Slovenian National Home), which became the hub for various meetings, events, and performances. It also housed immigrant community libraries and extensive archives of meeting records, programs, and other organizational documentation. Migrants maintained contact with their places of origin, language, and culture through ethnic organizations and used different cultural activities to introduce their traditions and values to the host society.

The larger Slovenian organizations, often created by merging smaller, local units, functioned primarily as the organizational link between the communities' separate, more distant parts. The migrant press helped unite the dispersed communities even better than ethnic organizations. Before the publication of the first newspaper, the communities in different parts of the new homeland could only foster contact through correspondence and infrequent meetings. However, once the first Slovenian newspaper, *Amerikanski Slovenec* (The American Slovene), appeared in Chicago in 1891, author and editor Ivan Molek wrote that, with it, the Slovenian migrants "stepped into their public cultural life" (quoted in Friš, 1991, 405). The diversity and abundance of periodical publications significantly influenced a rich literary production, encouraging and chronicling debates on many cultural issues. The periodicals were primarily published in places where the concentration of Slovenians was the highest. Their role in the diaspora was invaluable because they "chronicled what was happening among them, but also encouraged political and economic activities among the Slovenian migrants" (Klemenčič, 1991, 300). The press thus involved the Slovenians in the diaspora in a dialog as much as it contributed to their ideological and political polarization.

Like the population in the Slovenian ethnic territory since the mid-19th century, the Slovenian migrants were characteristically split into two opposing ideological and political camps: the Catholic conservative and the progressive liberal. Once the common Yugoslav state was created in 1918, the unitarian pro-Yugoslav camp was also formed in the interwar period. Ever since the first migrants left their homes, Slovenians abroad were clearly divided into those who put the Church and religious faith at the center of their lives and those who based their worldviews on liberal and progressive values. Although in both cases these were working-class people, and their interests were meant to align for this reason alone, in organized life and all other forms of ethnic association they were entrenched in their fundamental ideological and political divisions. More radical groups existed within the two main ideological and political camps, which often lent a non-compromising tone to the media altercations between the two poles. This division was reflected throughout the organizations, from their press and publications to their insurance, cultural, artistic, and humanitarian activities. Both camps experienced further fragmentation into organizational units, groups, and media with different political preferences, which, even within the same group, could change more or less visibly over the decades, depending on the organization's leadership or editorial board.

When Yugoslavia was occupied in 1941, and Slovenia was divided between the occupying forces of Italy, Germany, and Hungary, the Slovenians in the diaspora stepped together in their concern for the "old country." But as soon as it turned out that the two political sides from the homeland—the official Yugoslav government in London with the Slovenian People's Party on the one side and the Partisan Army and the Liberation Front on the other— sought their support, the transient unity first loosened and then mostly evaporated. The circumstances of a fateful schism between the supporters and the opponents of the occupying forces in the ideologically and politically divided Slovenia (Luthar et al., 2013, 416–440) caused even stronger polarization in certain parts of the Slovenian diaspora than ever before.

Many Catholic and socialist bulletins of the diaspora organizations, generally intended for the broadest circle of potential readers, i.e., the working class, announced from the start that they would stand up for workers' issues, stating so in their mastheads. The declarative objectives for which both camps stood thus had to coincide in many aspects, although their paths differed significantly. This differentiation was perhaps why the border between the left and the right was, despite all the conflicts and reproaches, less pronounced before World War II than toward its end. At that time, the eager expectation of the outcome of the military combat and the Yugoslav revolution demanded a non-compromising political determination among Slovenian migrants.

The United States of America

Among the Slovenians who emigrated before World War I, and even in the interwar period, those who settled in the United States reached the highest level of organization. They settled primarily in larger industrial cities where they created relatively contiguous Slovenian neighborhoods: around 800 were formed over time (Velikonja, 1996, 384). A Slovenian settlement included a Slovenian fraternal benefit society, a national home, the headquarters of the ethnic press, and (at least partly) a Slovenian Catholic or Protestant parish (Klemenčič, 2001c). Settlement members also founded Slovenian enterprises, banks, shops, cultural institutions, publishing houses, museums, language schools, taverns, and travel agents in these neighborhoods. Immigrant communities' organizational structures helped immigrants to survive materially and culturally. On the one hand, they represented a bridge between the immigrant community and the host country. On the other, they provided the transnational and translocal preservation of social, cultural, and political ties with the homeland and the place of origin.

The key to the structural organization of the Slovenian immigrants was connecting to other, particularly Slavic, nations. They were linked through the similarity of languages and, in the case of Austro-Hungary, also the historical background, including the idea of Pan-Slavism. The representatives of Slovenian, Croatian, Russian, Czech, Polish, and Slovak immigrants established a Slavic American League in 1895. "Such events were part of a strategy to present themselves to the English-speaking public, but they also depended on minor linguistic differences between the Slavic languages, combined with similar sending regions (in Austria-Hungary), and the still-lingering Pan-Slavist and Austro-Slavist ideas" (Steidl et al., 2017, 151).

In the last two decades of the 19th century, the concentration of Slovenians in urban centers of the United States was relatively low compared to the numbers in the following decades. As the Czechs had moved to the United States earlier, Slovenian immigrants in certain places joined Czech churches (for example, in Cleveland) and Czech organizations and later established their own. One of the societies that later helped to establish the Kranjsko-slovenska katoliška jednota (Grand Carniolian Slovenian Catholic Union), or KSKJ, first joined the Czech Catholics. Likewise, the Slovenian club from Chicago joined the liberal Czech-Slavic fraternal support union established in 1884 in Cleveland because, in the first decade of its operation in the United States, there was no liberal Slovenian umbrella fraternal organization. Adopting and adapting the Czech associations' organizational strategy and regulations, Slovenian migrants used the Czech experience in forming the first Slovenian benefit associations.

Jakob Zupančič, the first editor of *Mladinski list —Juvenile*, the most popular paper for Slovenian migrant children and young adults, modeled his editorial practices after similar papers of other immigrant nations, particularly the monthly journal *Svobodná škola* (Freethinking School) of the Czech liberal educational institution that was first published in 1896. The successful preservation of the mother tongue among the children and grandchildren of the Czech and German immigrants—also helped by youth magazines rich in content and design—made a strong impression on Slovenian editors. Thus, they emulated their rich experience (Hočevar, 1977, 235–239; Klemenčič and Javernik, 1996, 345; Friš, 1994, 41, 48; Evanish, 1996, 59; Milanič, 2001, 21).

Slovenian and Croatian migrants frequently joined each other's associations. When Slovensko-hrvaška zveza (Slovenian-Croatian Union) from Calumet, Michigan, merged with the Hrvatska bratska zajednica (Croatian Fraternal Union) in 1925, the latter had already been accepting Slovenian members for over two decades and by then counted around one thousand of them. Slovenians also collaborated intensely with other South Slavic nations in the political arena. In 1905, the Slovenian Socialist Association of America was created in Chicago; in 1909, the Croatian Socialist Union; and in 1910, the delegates from different South Slavonic Clubs in Chicago founded the Yugoslav Socialist Federation, which between 1909 and 1917 and again between 1922 and 1940 was a member of the Socialist Party of America. It was divided into three sections: the Slovenian one initially had 18 local units, the Croatian 13, and the Serbian three (Klemenčič, 1987, 82). In the first decade, the federation was primarily a political organization working to spread socialist ideas and enrich the cultural life of its members. From 1940 until its cessation in 1952, the members were primarily Slovenians. In 1906, the Slovenian (later Yugoslav) Socialist Federation began publishing the Slovenian newspaper *Proletarec* (Proletarian), in 1912, *Ameriški družinski koledar* (American Family Almanac), and in 1917, the monthly *Prvi maj* (The First of May; since 1933, *Majski glas/The May Herald*). Additionally, the Yugoslav Socialist Federation published *Radnička straža* (Workingman's Guard, 1907–1918) (Hranilović, 1988).

Slovenians also built national homes with Croatians and Serbians, such as the American Yugoslav Center in Euclid, Ohio, and the Slovenian-Croatian Club in Escanaba, Michigan, which often rented out their premises to other national associations. "Tight resources facilitated trans-ethnic South Slav cooperation, maybe even more pervasively than any ideology" (Steidl et al., 2017, 162). During World War I and immediately afterward, as a part of the Jugoslovansko republičansko združenje (Yugoslav Republican Alliance, or JRZ), the Slovenians in the United States worked for Yugoslavia to become a republic and secure "just borders" in the postwar border disputes. They also raised funds, providing financial aid for their fellow Slovenians back home. During World War II, they carried out these tasks with members from other Yugoslav nations as part of the Jugoslovanski pomožni odbor (Yugoslav Auxiliary Committee) and the Združeni odbor Jugoslovanskih Američanov (United Committee of South-Slavic Americans, or ZOJSA).

The first centralized fraternal benefit organizations were most often established by merging the existing local associations that Slovenians had started setting up soon after arriving in the United States. At the end of the 19th and the beginning of the 20th centuries, the United States had no social security in place, nor were employers obliged to provide it. Likewise, occupational health and safety were not legally regulated, which made workplace accidents frequent and often fatal. Those workers who were not members of a fraternal benefit society were left on their own in cases of unemployment, injury, permanent disability, or illness. For this reason, the first Slovenian benefit society, St. Joseph's Society, was established in Calumet, Michigan; in 1884, it developed into the Slovenian-Croatian Union (Friš, 1994, 38). Around the turn of the century, similar societies were established elsewhere in the United States, wherever the work circumstances and the concentration of Slovenian immigrants dictated it. Because the primary activity of these societies was insurance, local associations started to merge into centrally managed umbrella organizations, all to minimize the business risk and, in times of crisis, prevent their financial collapse. So, before the end of World War I, nine such fraternal benefit societies were operating around the United States; they were also instrumental for the Slovenians in terms of cultural, religious, and/or political association. Each one of them published their official bulletin—they are discussed in the segment about the migrant press. Broadly speaking, the organizations were divided into Catholic on the one side and liberal or socialist on the other. Such divisions, however, were often neither completely unambiguous nor permanent.

Over the years, the percentage of Slovenians in the United States who were members of organizations changed. Once the United States entered World War I, the US government required the fraternal organizations of nations hailing from or associated with the Central Powers countries to exclude all those members who spent time in their countries of origin during the war (Burjak, 1996, 373). The financial crisis in the early 1930s also caused a significant drop in membership for most organizations, but the situation normalized in a couple of years (Friš et al., 1997, 267, 270; Vovko, 1990a; Klemenčič, 2013, 2001c, 1996; Klemenčič and Maver, 2017, 90; Friš, 1994, 38–41). The membership totals in Table 2.1 show that in the last few years before World War II, Slovenian organizations and societies had nearly 145,000 members. The table does not include the members of all Slovenian societies in the United States. However, one could be a member of several organizations simultaneously, and some Slovenian fraternal organizations had members from other Slavic nations, particularly Croatians. We can certainly speak of mass membership of Slovenians and their descendants in the Slovenian societies and organizations in the United States.[34]

34 At the end of the millennium, the SNPJ, with its headquarters now in Imperial, Pennsylvania, had more than 54,000 members. The KSKJ, which since 1962 has been operating as Ameriško-slovenska katoliška jednota (American Slovenian Catholic Union), had over 32,600 members, the Ameriška bratska zveza (American Fraternal Union, or ABZ, former JSKS) had around 16,100, and the Ameriška dobrodelna zveza (American Benefit Association) around 15,400, while the Zapadna slovanska zveza (Western Slavonic Association) had around 6,400 members (Klemenčič, 2001c).

Table 2.1: Slovenian umbrella organizations in the United States before World War II. Sources: Friš et al., 1997; Klemenčič, 2013; 2001c; 1996; Friš, 1994.

Slovenian name of organization	English name of organization	Year of establishment	City of establishment	Number of members around 1911	Number of members around 1937
Kranjsko-slovenska katoliška jednota – KSKJ	Grand Carniolian Slovenian Catholic Union or KSKJ	1894	Joliet, Illinois	11,500	35,258
Jugoslovanska katoliška jednota – JSKJ (from 1940 Ameriška bratska zveza – ABZ)	South Slavonic Catholic Union	1898	Ely, Minnesota		23,097
Podporno društvo sv. Barbare (1917 joined SNPJ)	Slovene Benefit Society Saint Barbara	1900	Forest City, Pennsylvania	8,000	
Slovenska narodna podporna jednota – SNPJ	Slovene National Benefit Society	1904	Chicago, Illinois	7,100	49,478
Zapadno slovanska zveza – ZSZ	Western Slavonic Association	1908	Denver, Colorado	725	3,760
Slovanska delavsko podporna zveza – SDPZ	Slavonic Workingmen's Benefit Union	1908	East Conemaugh, Pennsylvania	2,000	

Slovenian name of organization	English name of organization	Year of establishment	City of establishment	Number of members around 1911	Number of members around 1937
Slovenska svobodomiselna podporna zveza – SSPZ (1941 joined SNPJ)	Slovene Free-thinking Benefit Association	1908	Chicago, Illinois	2,000	9,100
Slovenska dobrodelna zveza – SDZ	Slovene Mutual Life Association	1910	Cleveland, Ohio		11,240
Prvo south bethlehemszko szlovenszko betezsne pomagajocse drüstvo vu Ameriki	The First Windish Fraternal Benefit Society of America	1912	Bethlehem, Pennsylvania		1,470
Jugoslovanska podporna zveza Sloga – JPZ Sloga	South Slavic Benevolent Union Sloga	1915	Milwaukee, Wisconsin		2,500
Slovenska ženska zveza – SŽZ	Slovenian Women's Union of America	1926	Chicago, Illinois		8,934

The Main Mutual Aid Organizations

The two most important mutual aid organizations were the Slovenska narodna podporna jednota (Slovene National Benefit Society), or SNPJ, and the Kranjsko-slovenska katoliška jednota (Grand Carniolian Slovenian Catholic Union), or KSKJ. The Jugoslovanska katoliška jednota (South Slavonic Catholic Union, or JSKJ) and the Slovenska dobrodelna zveza (Slovene Mutual Life Association, or SDZ) also played a role.

The KSKJ was established in Joliet, Illinois, in 1894 and only accepted active Catholics as members. In the early years, only men could join. Later, it also allowed women. From 1902 on, women could establish their own lodges within the KSKJ but did not have the right to have a female delegate at the union convention. Like all benefit organizations established later, the organization rewarded lodges and individual members for bringing new members; they published ads in their own newspaper and elsewhere and organized competitions to increase membership. The administrators of the KSKJ and its local lodges regularly checked if the members lived by the tenets of the Catholic faith and attended weekly service and confession. The KSKJ's strict rules and overbearing control over its members' lives created resentment in some of its lodges and caused members to switch to other local organizations. In 1914, the editorial desk of *Amerikanski Slovenec* (The American Slovene), the KSKJ's official paper, distanced itself from the organization; the organization thus established the *Glasilo K. S. K. jednote* (KSKJ Bulletin). In 1916, it set up a Youth Department. Between 1920 and 1923, it published *Angelček* (Little Angel), a monthly for young readers. Sports clubs established by the local lodges were also crucial for youth recruitment. During World War II, the KSKJ materially supported its drafted members, particularly those who were wounded, and their families; it informed the American public about the situation and collected funds for occupied Slovenia (Friš et al., 1997, 27–28, 267–273; Kolar, 1999, 69; Friš, 1999, 286–287; Klemenčič, 1987, 79).

In 1898, four years after establishing the KSKJ, two lodges seceded from it and set up the Jugoslovanska katoliška Jednota (South Slavonic Catholic Union, or JSKJ) in Ely, Minnesota. The following year, another lodge shifted from KSKJ to JSKJ. This shift triggered insults and slander in the papers of both unions—the Chicago-based *Amerikanski Slovenec* and the New York-based *Glas naroda* (Voice of the People), which would become the official paper of the JSKJ. The board members and administrators of the new organization were "the most influential members of the Slovenian community in the United States, bankers, journalists, innkeepers, merchants [...] who had broad authority" (Friš, 1994, 42). When established, the new organization was no less Catholic than the KSKJ, and its lodges were named after Catholic saints. The spiritual leader also managed the religious affairs; the main and the daily assemblies began and ended with a mass. Additionally, the members had to take the annual "community confession and communion."

Like the KSKJ, the JSKJ only allowed male members for its first 15 years, admitting women lodges only from 1913. From 1916 onward, any "healthy male or female of a 'south Slavonic language' between the ages of sixteen and fifty residing in the United States" could become a member (Friš, 1994,

43). There were also quite a few Croatians among the members. In 1913, the organization set up a youth division that welcomed only healthy children from one to 16 years of age. Under the influence of its growing progressive faction, the JSKJ officially adopted the policy of complete religious and political freedom in 1920. In 1924, it stopped collaborating with the *Glas naroda* and established its own weekly *Nova doba* (New Era) (Bajec, 1980, 75–76). Over the next couple of years, the socialist camp within the JSKJ grew to the point that, in 1940, the organization was renamed the Ameriška bratska zveza (American Fraternal Union, or ABZ), which, a year later, following the attack on Yugoslavia, initiated a joint campaign for material aid.

The first liberal fraternal organization, SNPJ, was established in Chicago in 1904 and has functioned as the liberal and socialist counterweight to the conservative organizations of Slovenians in the United States ever since. The SNPJ recognized its members' religious and political freedoms. Women got the right to establish their own lodges in 1908. A year later, their rights within the SNPJ were equal to men's (the KSKJ, for example, only achieved that in 1920). In 1908, the society started publishing the *Glasilo SNPJ* (Bulletin SNPJ), which in 1916 changed its name to *Prosveta* (Enlightenment). The paper soon clearly profiled its socialist orientation and became a staunch advocate for workers' rights. Local SNPJ lodges were known for their lively political activity: they organized political lectures, meetings, and other events. On several occasions, they raised money for striking workers who were not SNPJ members (Blagotinšek Turk, 2001, 101–109; Klemenčič, 1991, 305, 2001c, 2013, 618; Evanish, 1996, 59–63).

In its first decade, the SNPJ proved itself to be the core unifying force of the liberal pole of the Slovenian communities for both fraternal benefit and cultural-political activities. The SNPJ lodges organized social and entertainment events, such as balls and picnics, sporting meetings, anniversary celebrations, and other cultural events, particularly plays and recitals. In 1925, the youth lodge SNPJ no. 559 Pioneer was established in Chicago, the first English-speaking SNPJ lodge. Others soon followed. A year later, *Prosveta* introduced English pages, thus securing a wider readership and even new members among the descendants of immigrants. In 1935, the SNPJ had 82 active lodges (Milanič, 2001, 19–26). In the 1920s and 1930s, the SNPJ positioned itself as the biggest organization of Slovenian immigrants in the

United States. After World War II, its membership increased, and many of its lodges remain active today.[35]

At the end of January 1914, an attempt was made to merge all the Slovenian benefit organizations, but the initiative garnered too little support at the unification convention (Blagotinšek Turk, 2001, 107). Disagreements between the conservatively oriented Catholic associations on the one side and the liberal and socialist ones on the other first manifested as early as 1905 when the predominantly Catholic associations in Chicago organized a joint picnic, and, out of protest, the socialist side organized an excursion on the same day. From then, two distinct political camps formed among the Slovenian Americans. Their clashes took place in the arena of political and social events, on the pages of their newspapers, and within a significant part of their cultural and artistic production.

Two Central Women's Organizations

In addition to numerous women's lodges within the central Slovenian fraternal organizations, Slovenian women in the United States also established their own organizations. After World War I, the Zveza jugoslovanskih žena in deklet (Union of Yugoslav Women) from Chicago actively collected material aid. Its Slovenian section held fundraisers and variety shows to help compatriots in Yugoslavia and organized a social event honoring Slovenian American soldiers who returned from the battlefields as American heroes (Chicaški Slovenci na noge, 1919, 5, 7). The organization had a Catholic orientation; it collaborated with the Red Cross, the Slovenska narodna zveza (Slovenian National Alliance), and the American Foundation for Aid to Yugoslavia but had concerns regarding the left-oriented Slovensko republičansko združenje (Slovenian Republican Alliance) (Chicaški Slovenci na noge, 1919, 5). While local women's organizations existed in all Slovenian settlements, two central organizations of Slovenian migrant women branched out across the United States.

On the initiative of Marie Prisland, the Slovenian Catholic women established the Slovenian Women's Union of America (SWUA) in Chicago in 1926. Its goal was "on the one hand preserving Slovenian culture and heritage, and on

35 US legislation clearly separates ethnic fraternal benefit insurance organizations from the for-profit private insurance organizations—the former, for example, have certain tax benefits in comparison to the latter (Susel, 1996, 101). Perhaps this was an additional reason for the unexpected longevity of certain Slovenian ethnic organizations, which in addition to insurance, also keep Slovenian traditions and provide a lively social and cultural life for their members.

the other helping members to integrate into the American society more easily" (Valentinčič, 2015). The SWUA first had a column in the *Amerikanski Slovenec* but established its own monthly in 1929: *Zarja—The Dawn*. Marie Prisland was the SWUA president for 20 years and firmly believed that, through organized activity, Slovenian women could reach positions of authority and respect in the Slovenian community in their new country, as well as the country of origin. The SWUA had 83 lodges in 1937, with almost 9,000 members (Friš, 1994, 40; Prisland, 1964). In the 1920s and 1930s, the organization published several books (a collection of articles and essays, a Slovenian songbook, and a Slovenian cookbook) and set up a scholarship fund and a youth section; since 1938, it also organized trips to Europe and Slovenia. It wanted to help Slovenian women scattered throughout the United States gain financial security (through insurance services), education, active citizenship, and mutual moral support. Men and even many women were initially skeptical and uneasy about the new organization. Yet, once the SWUA expanded and developed beyond all expectations, it gained much respect in the Slovenian community in the United States. Its local branches were established in 14 states; its membership at the end of World War II reached 14,000 and, in 1957, at its peak, 16,000 members (Milharčič Hladnik, 2007). The union sought to give Slovenian women immigrants and their descendants an equal voice in all matters—within the Slovenian communities and the larger American society.

The second organization of Slovenian women was the liberally oriented Progresivne Slovenke Amerike (Progressive Slovene Women of America, or PSWA). It was established in 1934 at the home of its first president, Anne Grill, in Lyndhurst near Cleveland, Ohio, to provide organizational activities and initiatives in culture, education, and humanitarian aid. From its foundation, the PSWA supported the political position of the Slovenian American writer Louis Adamic who, in his books, *The Native's Return* (1934) and *Struggle* (1935), criticized King Alexander I's dictatorship in Yugoslavia. The conservative circles of the Slovenian Americans, supported by the consular representatives of the Kingdom of Yugoslavia, had relentlessly attacked Adamic and his liberal ideas, which was another trigger for establishing the PSWA. The organization's statute guaranteed its members religious, philosophical, ethical, and political freedom. To improve the social and economic situation of Slovenian women in the United States, the organization worked to disseminate democratic beliefs and general knowledge and to strengthen social awareness among them. It worked with other liberal cultural and humanitarian organizations and promoted the cultural values of both homelands (since 1939, for example, with an annual exhibition of Slovenian embroidery and other handicrafts). Since 1939, the organization had its supplement in the Cleveland-based newspaper *Enakopravnost* (Equality). During World War II, the organization joined forces with other organizations to aid occupied Slovenia and Yugoslavia and was among the most active in fundraising. By the war's end, they raised and sent 38,000 dollars to the Slovenian Red Cross. Their charity work helped many health, cultural, scientific, and educational institutions in Slovenia, and they also participated in American humanitarian campaigns (Žitnik Serafin, 1995b, 370; Slokan, 1983; Ivanush, 1964; Cafnik and Cingerli, 1996).

Organizations of Immigrants from Prekmurje

Small numbers of Slovenians from Prekmurje started arriving in the United States at the end of the 19th century.[36] Most, however, arrived between 1906 and 1913 and settled chiefly around Bethlehem, Pennsylvania; Bridgeport, Connecticut; Cleveland; and Chicago. Either as part of their Catholic or Protestant churches or not, they immediately started establishing fraternal benefit, religious, charity, cultural, and political associations and clubs. As citizens of the Kingdom of Hungary, they participated in Hungarian, Slovak, Croatian, and Styrian migrant organizations before founding their own organizations, associations, and parishes.

In 1912, they formed, in Bethlehem, the Prvo south bethlehemszko szlovenszko betezsne pomagajocse drüstvo vu Ameriki (The First Windish Fraternal Benefit Society of America), the first general association of Slovenians from Prekmurje in the United States, which in the years 1913–1922 developed six branches in different cities in Pennsylvania, New Jersey, and Connecticut (Kuzmič, 1996, 357). The parent organization built a Slovenian Home in Bethlehem (in 1916), where it staged plays, recitals, cabarets, and social events. The association was not linked to the Church, and it was open to all Slovenians regardless of religion or worldview and, later, also to members of other ethnic groups. Women

36 Slovenians from Prekmurje were, in many aspects, a curiosity among Slovenian migrants. Their regional identity was more prominent, beginning with the fact that they used the Prekmurje Slovenian language in their benefit and cultural associations, newspapers, and literary works. The question of their Slovenehood was often a question of political differences within the community. The Slovenian ethnic belonging of many members of the community after World War I clashed with the political pressure of the Hungarian movement in the United States to preserve Hungarian national and civic identity among the emigrants from the Kingdom of Hungary. These pressures were linked to the attempts to "return" Prekmurje (which became a part of the Kingdom of Serbs, Croats and Slovenes with the Treaty of Trianon) to the new Hungarian state (Kuzmič, 2001, 100–139).

could join as of 1917, and the new regulations in 1925 provided the framework for youth participation. The association's primary activity was insurance, but it also played an important role in humanitarian activities. The membership grew steadily until World War II: from 300 members in 1913 to almost 1,500 in 1941.

The community published the majority of its newspapers in its most important settlement in Bethlehem, Pennsylvania, in the Prekmurje Slovenian, a version of the Slovenian language with its own written standard and literary tradition. The weekly *Szlobodna reics* (The Free Word, Bethlehem, 1916–1928)—replaced in 1928 by *Slovenske novine* (Slovenian Newspaper)—and *Zvona glasz* (The Voice of a Bell; Bethlehem, 1921) were for the Protestant readers. The Catholic Prekmurje immigrants in Bethlehem and elsewhere read the weekly *Vogrszki Szlovenecz* (Slovenians from Hungary; Bethlehem, New York, 1916–1917). Only one of the most notable papers was published in Chicago: the Catholic monthly *Zvejzda vogrski Szlovencov* (Star of Hungarian Slovenes; 1916–1922), whose editors also published the pro-Hungarian *Kalendar za vogrske Slovence vu Ameriki na navadno leto 1922* (A Calendar for Slovenians from Hungary for the Common Year 1922). These papers published articles on a range of political and social topics, informative pieces, and polemics, in addition to religious content. The disagreements between different fractions gave way to the need for greater connection and collaboration, which brought about the establishment of a joint newspaper for all the Prekmurje people in Bethlehem, the weekly *Amerikanszki Szlovencov glasz* (Voice of American Slovenians; Bethlehem and Pittsburgh in Pennsylvania; Bridgeport in Connecticut, 1921–1954).

Slovenian Parishes

Most Slovenian immigrants belonged to the Roman Catholic Church, and one of their first needs was to have a Slovenian priest. The Catholic Church supported establishing ethnic parishes, which were vital organizations among Slovenian and other immigrants in the United States. The Church (despite its numerous warnings, mainly about the "moral traps" of life in emigration) throughout the period defended the right of people to move and to have access to the liturgy in their own language. American government data shows that, in 1916, there were 5,600 ethnic Catholic parishes in the United States; most of them were Spanish, followed by Polish, German, French, and then Italian. An ethnic church community, if it was large and contiguous enough, often included different activities that helped preserve its mother tongue and culture. Between 1871 and 1924, in addition to the Franciscan monastery in Lemont (Illinois), 32 Slovenian parishes were established in the United States and a further seven that Slovenians established with members of some other nations, particularly Croatians, Czechs, and Slovaks (Friš, 1999, 7–8; Kolar, 2001, 126, 133–134; Klemenčič, 2001c). Slovenians built more than 50 churches between 1871 and 1924 (Klemenčič, 2001a, 170). From the turn of the 19th to the 20th century, around 100 Slovenian priests worked in 46 Slovenian or mixed parishes in the United States (Klemenčič, 1987, 80). Some parishes also ran parish schools, which also taught the Slovenian language.[37] Priests organized choral and other music associations and theatre groups; parish social activities included all sorts of social events, and they even organized trips to the Slovenia (Klemenčič, 2001c, 2001b, 182–183, 2013; Susel, s. d.).

Slovenians in the United States built churches following examples of other immigrant nations. But while some others—for example, the Norwegians—actually owned those buildings, the Slovenian Catholics built them for the bishop, who was officially the owner. Slovenian parishes were mostly poor, but the parishioners diligently contributed to purchasing the land for the church and its premises (buying an existing building or constructing a new one), church equipment, parish houses, etc. In addition to their parish, they co-financed the diocese, the Vatican, Catholic universities, missions, and Eucharistic meetings. The most important Slovenian friars in the United States were the Franciscans; some were Benedictines. Five Slovenian priests were appointed bishops, mostly for their missionary work among the indigenous and immigrant populations (Plut Pregelj, 1992, 391; Friš, 1999, 9–13).[38] The Society of St. Raphael for Slovenian immigrants, established by Rev. Father Kazimir Zakrajšek in New York in 1908, provided for the Catholic Slovenians who lived in settlements without a permanent

37 As early as 1894, the Slovenians in Cleveland built the St. Vitus Church and added a Slovenian public school to it in 1912. At its peak, in 1922, the school had 1,680 students. The Saturday Slovenian School at St. Vitus Church in the first Slovenian parish in Cleveland still operates, the lessons are twice monthly, and the parish continues to publish its bilingual *Bilten* (Bulletin).

38 These bishops were Friderik Irenej Baraga (1797–1868) as the most famous among them, Ignacij Mrak (1810–1901), Jakob Trobec (1838–1921), Janez Vrtin (1844–1899), and Ivan Stariha (1845–1915). To compare: Polish Americans made up one-eighth of all the Catholics in the United States and had one bishop in 1908 (Friš, 1999, 12–13). Friderik Irenej Baraga was a grammarian of Native American languages. He published *Otawa Anamie-Misinaigan,* the first book written in the Ottawa language, and a dictionary and grammar of the Ojibwe language. He was nicknamed "the Snowshoe Priest" because he would travel hundreds of miles each year on snowshoes during the harsh winters.

Slovenian priest. The society had a representative on Ellis Island to help Slovenian immigrants upon their first contact with the new country (Podbersič, 2018; Smolik, 2013).

National Homes

The hub for all sorts of activities within the Slovenian community in the United States was the Slovenian National Home. Initially, national homes were built by the Slovenian benefit societies in the first decade of the 20th century. The first one opened in 1901 in Pueblo, Colorado (Valenčič, 2002, 166). In 1905, the national homes in Chicago, Illinois, and Johnstown, Pennsylvania, were built and, in 1908, one was built in Herminie, Pennsylvania. By 1913, there were national homes in Frontenac, Kansas, Ely, Minnesota, and Rock Springs, Wyoming. Several national homes in Pennsylvania were built by the branches of the Podporno društvo sv. Barbare (Slovene Benefit Society Saint Barbara). Even during World War I, when Slovenian Americans united in their concern for the "old country," several national homes were built, including one built in Pennsylvania by the Slovenians from Prekmurje. The money for construction came from selling shares and organizing social events (Valenčič, 2001, 187–188). Most national homes were built in the 1920s and 1930s. In the wider Cleveland area, the Slovenian immigrants and their descendants opened eight national homes in the interwar period (Susel, 1996, 104).[39]

Just as the KSKJ often participated in constructing Slovenian Church centers, the SNPJ was the main initiator of the construction of national homes. Preparations for constructing the celebrated central Slovenian National Home on St. Clair Avenue in Cleveland started before World War I. The main hall had 1,324 seats; other halls could host parallel events (Valenčič, 2001; Klemenčič and Emeršič, 1990; Stanonik, 1997, 28–29).[40] Today, the SNPJ still has national homes

39 Slovenians in Cleveland and their descendants contributed several important people to the politics of the city and the state of Ohio. In the 1930s, the community had four members in the city council (Klemenčič, 1992, 2001c). The best-known Slovenian American from Cleveland was Frank J. Lausche (1895–1990), the mayor of Cleveland (1942–1944), the governor of Ohio (1945–1947 and 1947–1957), and a member of the US Senate (1957–1969).

40 The building had eleven offices, seven shops, a (Sokol) gymnasium, a reading room, etc. Along with the Slovenian school, the premises were rented out to different choirs, a drama school led by Avgusta Danilova, a travel agency, a photo studio, a modern art school led by Harvey Gregory Prusheck, the Slovenian National Museum, the

in many places throughout the United States. Most of them—around 50—are between Cleveland and Pittsburgh (Stanonik, 1997, 28). The large number of Slovenian National Homes built in the United States in the interwar period is unparalleled elsewhere among Slovenians worldwide, either before World War II or after. These national homes, most of which still operate today, testify most vividly to the emphasized need for social, economic, and cultural associations of the migrant communities.

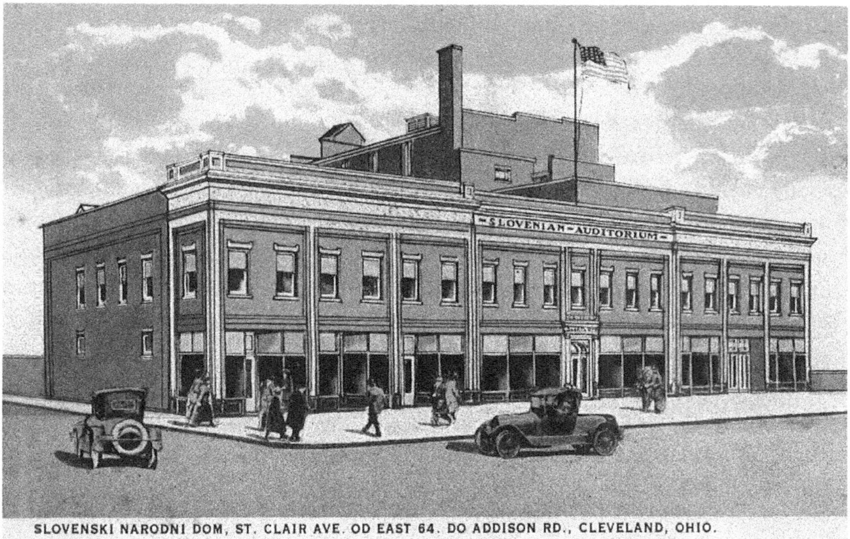

Picture 2.1: Postcard of the Slovenian National Home, Cleveland, Ohio. Courtesy of: Joseph Valenčič.

Cultural and Art Activities

Slovenians in the United States developed a rich publishing practice, with many publishing houses and printers, including the book publishers Prosvetna matica (Cultural and Educational Center) and SNPJ in Chicago and Ameriška domovina (American Home) in Cleveland. Several authors also self-published (e.g., the popular poet Ivan Zorman). Slovenian American literature flourished between

headquarters of the Slovenian Benefit Society, and almost 100 associations and organizations (Valenčič, 2001, 189).

the two wars, and the works were more likely published in periodicals than they were as book editions. This practice is also true for the literary output in other Slovenian communities abroad at that time. However, in neither scope nor quality can it be compared to that in the United States. Yet this diversity (of literary works and different periodicals) was crucial for preserving Slovenian identity in every receiving country.

In communities, choirs and theatre groups were very popular. Among the first to be established were church choirs, which, from 1890 onward, transformed into choral societies, and others followed. The Slovenian community in Cleveland alone had 19 choirs and 10 drama groups between the two wars. The Cleveland Slovenians and their descendants formed scores of choirs, the majority of which—like their fraternal benefit organizations, newspapers, and almost all other ethnic organizations and activities—showed the apparent split into the two ideological camps, the conservative Catholic on the one hand and the liberal and socialist on the other. We can see this split in the choirs Lira, established in 1912, mainly linked to the Slovenian Catholics from St. Clair Avenue, and Zarja, established in 1916 under the auspices of the Socialist Club and considered at the time the leading Slovenian American choir. Zarja further split into the independent Zarja and the socialist Zarja; both continued to produce concerts and operas. Slovenian migrants regularly put on opera productions with their performing troupes from the late 1920s to the 1950s. In 1940, the independent Zarja reorganized as Glasbena matica. In addition to these two, one of the longest-active choral societies in Cleveland was Jadran, established in 1920 (Valenčič, 2002; Valenčič and Milharčič Hladnik, 2011; Klemenčič, 2013, 619–620).

Instrumental or vocal-instrumental groups playing traditional or pop-folk music were very popular. Some migrants brought instruments—accordions, zithers, tamburicas—and so Slovenians started to form different bands, which later often developed into orchestras. They started brass bands, and by 1920 the first big bands were established; their concerts also attracted members from other ethnic groups. For many, dances in Slovenian ballrooms were the only accessible entertainment. Slovenian polkas and folk songs were modernized using banjos and were Americanized with dance and jazz arrangements. The first records were made during World War I. Matt Hoyer and his trio in Cleveland recorded more than 100 songs between 1919 and 1930. One of the most famous descendants of the Slovenian immigrants in Cleveland was the popular author and performer of pop-folk music Frankie Yankovic (1915–1998). As a teenager, he was already a leader of a music group in the city. After World War II, he became famous around the nation as "the American King of Polka."

Picture 2.2: Slovenian Radio Program, Cleveland, Ohio. Author: unknown. Courtesy of: Joe Valencic.

Choirs and youth drama groups developed into regular community drama groups. Jože Zavertnik (1925) lists 17 associations dedicated to theatre activity before 1910, some only for a short while. In the interwar period, several drama associations and sections were created in and around Cleveland that lasted longer: Ivan Cankar (1919–1955);[41] Lilija (established in 1919), which was still active in the 1990s, as new immigrants joined it after World War II and, later, their descendants; and the Anton Verovšek Society (1924–1955).[42] A series of

41 Ivan Cankar (1876–1918) was a writer, playwright, essayist, poet, and political activist. Together with Oton Župančič, Dragotin Kette, and Josip Murn, he is considered as one of the pioneers of modernism in Slovenian literature. He is regarded as the greatest Slovenian writer. Many cultural clubs and societies in Slovenia and in the Slovenian diaspora are named after him.

42 Anton Verovšek (1866–1914) was one of the first Slovenian professional actors.

other drama clubs were active there. Most were ideologically less profiled than other associations; the exception was the drama sections of socialist clubs that often staged politically engaged plays. The Ivan Cankar Dramatic Society was the most ambitious: it staged the works by several Slovenian classical authors (Ivan Cankar, Fran Levstik[43]) and numerous American and European authors (Wilde, Tolstoy, Nušić) as well as original works by Slovenian migrants (Andrej Kobal, Etbin Kristan). The society also staged plays in English, including Ivan Cankar's *Scandal in the Valley of St. Florian* in the English translation by Anthony J. Klančar. In the first five seasons, the society staged 50 different productions.

Picture 2.3: The gramophone record of the song "Moja Micka" issued on the Columbia label (1923). Courtesy of: Joseph Valenčič.

In painting, sculpture, and architecture, the following, in particular, became well-known: painter Harvey Gregory Prusheck (1887–1940), who was briefly the president of the Chicago Society of Artists and who established the Yugoslav

43 Fran Levstik (1831–1887) was a Slovenian writer and playwright, as well as an influential editor, literary critic, and political activist.

School of Modern Art in Cleveland in 1931; architect Ivan (John) Jager (1871–1959), best known for his contribution to the urban planning of Minneapolis in 1906; ceramic artist Paul Bogatay (1905–1972), whose works are exhibited in American museums; and Michael Lah (1912–1995), a leading animator and cartoon director at MGM studios, who rose to fame in the 1940s and 1950s (Stelè, 2013; Pajsar, 2005; Mesesnel, 2013; Valenčič and Milharčič Hladnik, 2011; Klemenčič, 2013, 620).

The material Slovenian cultural heritage in the United States consists predominantly of churches and national home buildings that Slovenian immigrants built in all larger Slovenian settlements. Church paintings of the Stations of the Cross with Slovenian titles are also important. In Leadville, Colorado, priest Jurij Trunk of St. Joseph's Church painted them in 1927–1928. As for the Slovenian National Homes, whose construction peaked between the two wars, the largest and most characteristic is the one on St. Clair Avenue in Cleveland that opened in 1924. Slovenian material cultural heritage in the United States also includes several monuments and sculptures, like those in the Slovenian (then Yugoslav) cultural garden in the Collinwood neighborhood in Cleveland. Some monuments linked to the Slovenians in the United States predate World War I. For example, in 1911, a monument was erected at the cemetery in Trinidad, Colorado, in honor of 35 Slovenian and Croatian victims of a mining accident (Valenčič, 2002; Klemenčič, 2001a). The first Slovenian museum opened in Cleveland in the 1940s. Documentaries made in the Slovenian community in the interwar period comprise another segment of the material cultural heritage. Anton Grdina, a Slovenian businessman from Cleveland, and Božidar Jakac, a Slovenian painter who visited the United States in the 1920s and 1930s, are the authors of the most documentary film footage on the cultural, religious, social, and economic life of Slovenian immigrants to the United States (Tršan, 2001, 181–182).

Political and Humanitarian Transnational Activity

After the American entry into the world wars, Slovenian Americans exhibited a high level of American patriotism. Like fraternal organizations of other nations, Slovenian benefit societies allocated a large part of their funds to purchasing war bonds and rallied among their members to buy them as well. Still, most of their organized activity during both wars was geared toward care for their country of origin. In the history of their organized activities, for Slovenian immigrants and their descendants in the United States, the two world wars represent the peak of their political and humanitarian engagement and simultaneous connection to other Yugoslav/Southern Slavic nations.

During World War I, the Slovenian—later Yugoslav—Republican Alliance was established, and during World War II, the Slovenski ameriški narodni svet, or SANS (Slovenian American National Council, or SANC) and Združeni odbor južnoslovanskih Američanov (United Committee of South-Slavic Americans, or UCSSA). All three organizations were very successful in collecting material aid. Additionally, with its broad scope of activities, the Yugoslav Socialist Alliance played a significant role in spreading the Yugoslav idea in the United States in both periods.

Early on, the Slovenians in the United States joined the movement to establish a Yugoslav state. On March 10 and 11, 1915, at the 1st Yugoslav National Convention in Chicago, 468 delegates of different South Slavic organizations from the United States and Canada gathered, adopted a resolution on South Slavs uniting in a single state, and elected the Yugoslav National Committee. At the next convention in Pittsburgh, the Yugoslav National Council was elected (Čizmić, 1974). The pro-Austrian Slovenians and Croatians were also active, but they did not form an organization. A part of the Slovenian and Croatian clergy opposed the disintegration of Austria-Hungary and rejected the political programs for a single state with other South Slavic nations. Yet, political and humanitarian organizations mostly supported the united Yugoslav state and the postwar self-determination of Yugoslav nations regarding the organization of their state. This was the position of the Slovenska liga (Slovenian League), established in early 1915 (led by Frank Sakser of New York), and its successor, the Slovenian National Alliance, established on November 11, 1917, in Cleveland. Both worked with the Yugoslav National Committee, whose representatives came to the United States in 1916 to, among other things, encourage gathering signatures for a petition against the Italian demands for parts of the Slovenian ethnic territory. Both organizations held rallies, published statements and petitions, and raised material aid for people in the "old country." Influenced by the democratic ideals of the US federal government system, most Slovenian Americans preferred a federal to a unitary Yugoslav state, many even favoring a federal republic (Klemenčič, 1987, 85–139; Čizmić, 1974).

The Yugoslav Republican Alliance was created by merging the Slovenian Republican Alliance, established in Chicago in 1917, and the Croatian Republican Alliance, established in 1918. With much grassroots support, the Slovenian Republican Alliance had chapters in all Slovenian settlements around the United States. Among its chief goals was to spread the idea of a Yugoslav federal republic among the Slovenian Americans and the greater American society. The alliance fundraised for political action and other material aid for the people in the "old country." The most urgent issue at the establishment of the new Yugoslav state was the disputed borders, that is, the contested territories on the western part of Croatia and the eastern, northern, and western parts of the Slovenian ethnic territory. A favorable border was established only in the east, where the present-day Prekmurje became part of Yugoslav Slovenia. In the west, Slovenian Istria, the Trieste hinterland, the entire Goriška, Croatian Istria, and the Kvarner Islands were already promised to Italy in 1915 by the British and the French. Italy occupied these spaces immediately after the war ended. The border dispute was resolved by a bilateral agreement between the Kingdom of Serbs, Croats and Slovenes and the Kingdom of Italy, the Treaty of Rapallo, in November 1920. Around 340,000 Slovenians in Primorska (former regions of Goriška, Istria, parts of Inner Carniola, and the city of Trieste, altogether forming a part of the Julian March) and 160,000 Croatians remained on the Italian side of the border. "The loss of the entire western part of what would later be Slovenia was catastrophic for the Slovenes, who had also permanently lost Carinthia" in the north (Luthar et al., 2013, 378). Carinthia, where more than 80 percent of the population was Slovenian, was lost to Austria due to the plebiscite held in October 1920.

From the fall of 1918 on, the Yugoslav Republican Alliance concentrated on solving the Yugoslav border question—and stated so in their memorandum regarding this question sent to US President Woodrow Wilson when Austria-Hungary capitulated. The resolution, signed by 250,000 Yugoslav Americans and presented to the Senate committee by their delegation, had a similar goal. Inventor Mihajlo Pupin, the president of the Serb Federation "Sloga," played an important role in appeals to the US administration for a favorable solution to the Yugoslav border disputes. In the fall of 1919, the representatives of 11 migrant organizations co-signed the *Appeal to the President and the people of the United States of American in behalf of Jugoslav nation* in which they also emphasized the dangerous consequences of annexing Primorska (or the "disputed territory," as they called it in the *Appeal*) to Italy (Klemenčič, 1987, 130–134). The Yugoslav Republican Alliance functioned until 1924; during World War II, its funds were allocated to the Slovenian American National Council, which in many aspects continued its mission (Klemenčič, 1987, 85–139; Švent, 2009).

Between 1918 and 1920, while endeavoring for a favorable solution regarding the disputed territories in Carinthia and Primorska and their borders, the collaboration between Slovenian and Yugoslav Americans continued to strengthen. The predominantly Catholic Slovenian National Alliance did reproach the Yugoslav Republican Alliance in many articles about bolshevism but still worked closely with it in the activities against Italy's annexation of Yugoslav regions to Italy (cf. Slovensko Republičansko Združenje in Slovenska N. Zveza za skupni shod, 1918) and in sending aid. Priests, individual benefit societies, and other organizations critical of the JRZ's activities also organized such fundraising.

After the war, Slovenian Americans' interest in the events in Yugoslavia had not waned. In 1934, Louis Adamic's bestselling book *The Native's Return* (1934) strengthened it. The book attacked King Alexander I's dictatorship and predicted his forthcoming death, which came true with the assassination in Marseille on October 9, 1934. In his book, Adamic also criticized the political repression in the Kingdom of Yugoslavia. *Struggle* (1935b), Adamic's translation of Edvard Kardelj's report on the torture of political prisoners in the jails of the Kingdom of Yugoslavia had a similar impact.[44] Both books were banned in Yugoslavia, and King Alexander's regime introduced heavy prison sentences for selling and even owning either (Grill, 1979, 191). Despite the ban, there was a great interest in both books in Yugoslavia. *The Native's Return* was reprinted during World War II and distributed to American soldiers leaving for the European battlefields.

The first response of Slovenian immigrants and their descendants to the attack on Yugoslavia on April 6, 1941, was humanitarian activity. As early as April 19, 1941, the representatives of Slovenian fraternal benefit organizations—both liberal and Catholic—set up the Jugoslovanski pomožni odbor - Slovenska sekcija (Slovene Section of the Yugoslav Auxiliary Committee), or JPO-SS, on the initiative of the American Fraternal Union in Chicago, to coordinate collecting material aid for Yugoslavia. By the end of the war, the JPO-SS raised 80,000 dollars. On May 10, 1941, the representatives of Slovenian, Croatian, and Serbian benefit organizations tried to coordinate the fundraising for the people in the "old country" with the formal establishment of the Yugoslav Auxiliary Committee, consisting of the Slovene, Croatian, and Serbian sections. Still, inner antagonisms rendered this attempt at a coordinated action inefficient. The sections raised funds separately, and the connection between them was weak. At the same time, there was a campaign for collecting aid under the umbrella of Zveza slovenskih župnij (Union of Slovenian Parishes), which was led by Rev. Father Kazimir Zakrajšek and Father Bernard Ambrožič. The campaign reflected the Slovenian Catholic conservative leadership's political endorsement of the Yugoslav Government in Exile in London. The union raised over 40,000 dollars and sent it to the bishop of Ljubljana, Gregorij Rožman, who actively supported the Italian Fascist occupation of the Slovenian ethnic territory. This action caused

44 Edvard Kardelj (1910–1979) was an important figure in the prewar Yugoslav Communist movement and, after World War II, became one of the leading creators of the Yugoslav socialist self-management system.

dissatisfaction among the Slovenian Americans, not only among the socialists and the liberals but also among some Catholics who were convinced that, in the time of occupation, such an act constituted indirect aid to Hitler and his allies (Klemenčič, 1987, 166–171; Čizmić, 1998, 19).

Together with the Slovenians, other supporters of Yugoslavia also collected aid for Yugoslavia during the war, mainly as a part of the United Yugoslav Relief Fund of America (UYRF). With the mediation of Konstantin Fotić, the then ambassador of the Kingdom of Yugoslavia to Washington, DC, it was established in the summer of 1942 and operated under the auspices of the American Friends of Yugoslavia (established in April 1941) with a seat in New York. The fund organized campaigns with the approval of the special supervisory President's War Relief Control Board, which the US government established to supervise humanitarian activities. It permitted organizations to collect material aid and encouraged them to merge into larger, joint humanitarian organizations. In 1943–1944 alone, the UYRF sent almost two million dollars' worth of various kinds of aid (Lees, 2007, 173; Čizmić, 1998, 19).

In December 1942, the representatives of all the political affiliations at the Slovenian National Congress established the SANC, with its seat in Chicago. It worked on several issues: recognizing the leading role of the Yugoslav Partisan Army in the fight against the occupier; creating an autonomous and unified Slovenia that, after the war, would thrive in a federal, democratic republic of Yugoslavia; and raising aid. Etbin Kristan was elected the first president, Louis Adamic the honorary president, Marie Prisland and Janko Rogelj vice presidents, and Rev. Father Kazimir Zakrajšek the general secretary. SANC soon took a stand against the Yugoslav government in London, whose interests in 1941–1943 were represented by the Yugoslav Information Center in New York (Klemenčič, 1987, 205–213). In its first year, SANC set up 90 branches throughout the Slovenian settlements in the United States. The number reached 117 in total. They raised money, collected food and clothing, and organized cultural and social fundraising events to help Slovenia. Although there were disagreements within the council, SANC was the most unified political organization of Slovenian immigrants in the United States (Žitnik Serafin, 1998, 56).

On December 19, 1942, Louis Adamic published an article in *The Saturday Evening Post*, which charted the fight of Yugoslav partisans against the occupiers and the Chetniks led by Draža Mihajlović. Once this information became known to the US public, the myth of national hero Draža Mihajlović, as the resistance leader in occupied Yugoslavia, began to crumble. At the same time, this shift triggered further polarization of the Slovenian community in the United States, along with its organizations and press, as one side increasingly leaned toward supporting the Liberation Front and the National Liberation Army under Josip Broz Tito's command, while the other warned against the Communist takeover in the partisan movement. SANS firmly supported Adamic in his disclosure of Mihajlović's links to the occupiers and voiced support for the partisan resistance, which made the general secretary Rev. Father Kazimir Zakrajšek, who was of a different opinion, resign.

In June 1943, the representatives of Slovenian, Croatian, and Serbian Americans met in Pittsburgh and established a coordination committee of Yugoslav organizations in the United States, the United Committee of South-Slavic Americans, or UCSSA, and appointed Louis Adamic as president. The committee worked for the unity of the South Slavic Americans in the fight for democracy and against fascism, for moral and material support to the partisan army in Yugoslavia, and for the disclosure of the intent of the prevalent group in the Yugoslav government in London to re-establish a military dictatorship in Yugoslavia. When, in December 1943, the organizations in the United States found out about the decisions of the second AVNOJ assembly, the UCSSA supported them immediately. The UCSSA distributed the publications of the Yugoslav partisan press to the Yugoslav papers in the United States and Canada, and the partisan press reported on the UCSSA's activities (Lees, 2007, 190; Klemenčič, 1987, 252–255). In March 1944, Adamic resigned from the committee for health reasons and was succeeded by the renowned Croatian-American violinist Zlatko Baloković.

In December 1944, on Adamic's initiative, the War Relief Fund of Americans of South Slavic Descent was established, a charity fund which, by the time of its cessation in 1949, had raised around 3.26 million dollars in currency and other material aid for Yugoslavia (Đikanović, 2022; Čizmić, 1992, 224, 1998, 19). In the fall of 1945, the fund was taken over by the American Committee for Yugoslav Relief, or ACYR, led by Adamic and Baloković, which also included Americans who were not of Slavic origin. The honorary president was Eleanor Roosevelt, and the sponsors included Thomas Mann, Eugene O'Neill, Elmer Rice, Yehudi Menuhin, Orson Welles, Edward G. Robinson, and others (Žitnik Serafin, 1992, 32). By 1949, over 20 million dollars' worth of aid was sent to Yugoslavia with the help of the fund (Klemenčič, 1987, 272). Adamic's action also contributed to the eligibility of Yugoslavia for United Nations Relief and Rehabilitation Administration, UNRRA, aid despite the Cold War.

Never were different segments of the Slovenian community more united and more successful than when they were collecting material aid for Yugoslavia and supporting its people in their endeavors to win the border disputes than during both world wars and the first years following them.

Canada

Slovenians in Canada first organized local associations, which were lodges of the main fraternal benefit organizations based in the United States, particularly the SNPJ and the KSKJ. Zavertnik claims that the SNPJ had four lodges in Canada in 1923: two in Alberta and two in British Columbia, totaling 186 members. Other organizations with lodges in Canada included the Slovenian-Croatian Union, with its association in Sault Sainte Marie in Ontario (Zavertnik, 1925, 548), and the US-based Zapadna slovanska zveza (Western Slavonic Association, today WSA Fraternal Life, Inc.) (Trebše Štolfa, 2001, 211). The most regular contact between the Slovenians from both countries occurred within the common fraternal organizations or benefit societies and during church activities. From the mid-1930s, the Slovenian Franciscans from Lemont, near Chicago, cared for Slovenians in Canada. In 1933, the members of the Slovenian club Pipa (The Pipe) in the mining town of Kirkland Lake, Ontario, established a Canadian-Slovenian benefit society, the first Slovenian fraternal society with a seat in Canada. At its establishment, it already had over 90 members. In 1940, it transformed into the Bled Mutual Benefit Society. During World War II, 14 chapters of the still-functioning Bled Mutual Benefit Society were established in Ontario, Quebec, British Columbia, and Alberta. In 1944, it had over 500 members (with slightly over 800 at its peak in 1960). The first, and the only unchangeable, article of its rules guarantees its members individual freedom of religious and political beliefs. For this reason, there was no room for religious or political activities inside the society. During World War II, Slovenians in Canada set up organizations, societies, and committees to help those in occupied Slovenia. After the war, their support went toward reconstruction (Trebše Štolfa, 1996, 243–256, 2001, 211–214; Jurak, 1999, 316–317; Genorio, 1992, 343–345).

Several cultural associations, including independent theatre groups and choirs, emerged in Canada between the two wars. In 1927, the Lira Cultural Association was established in Windsor, Ontario: it had a choir, a drama section, and the first Slovenian library in Canada. Their productions were also attended by members of other Slavic nations (Kocjančič, 1998, 150). A year later, the Slovensko kulturno društvo Zvon (Slovenian Cultural Society Zvon) was established here and in 1943 the Slovensko pevsko društvo Zvon (Slovenian Choral Society Zvon). Several Slovenian societies and associations operated in Kirkland Lake, Ontario, in the early 1930s: a social club, a drama group, and the Pevsko društvo Triglav (Triglav Choral Society). The society operated successfully for three decades; in 1935, a drama group was added and, in 1938, a Slovenian library. Before World War II, the third town with a Slovenian cultural society was Sudbury, Ontario;

from 1936, the choral group Vigred (Spring) was active here. After World War II, Slovenian migrants and their descendants developed more organized Slovenian parishes, educational societies, Slovenian language and regular schools, and sports activities (Genorio, 1989; Trebše Štolfa, 2001, 214, 1996, 246).

Argentina

Slovenians in Argentina started setting up societies after World War I. However, to support the idea of a common state of South Slavs, the Jugoslovanska narodna odbrana (Yugoslav People's Defense, or JNO, 1919–1925) was established already during the war. Its seat was in Chile. It also had the Center of the Southern Atlantic in Buenos Aires, where it coordinated members' work in Argentina, Brazil, and Uruguay and published the multilingual newspaper *Jadran* (Adriatic). At the same time, there were already groups of Croatians and Montenegrins in South America, primarily seamen and merchants, who already had several joint societies and organizations. At the end of the war, a group of patriotic Slovenian intellectuals, who had moved to Argentina just before the war, set up the Slovenia section within the JNO, primarily to protest the annexation of Primorska to Italy. The core members of this group were Ciril Jekovec, Gabrijel Brinšek, France Krašovec, Bogomil Žnidaršič, and Janez Benigar, later a well-known anthropologist (Brecelj, 1992, 167; Mislej and Jurak, 1995, 16).[45]

In Buenos Aires, three Slovenian political camps formed in the interwar period. The largest was led by the left-oriented group that in 1925 established the Delavsko kulturno društvo Ljudski oder/Asociación Cultural Eslovena Escenario Popular (Workers Cultural Society Popular Stage); the second, center-left, grouped around the Slovenian Cultural Association 1; and the third, organized since 1930 as Sokol La Paternal, supported the Yugoslav idea and King Alexander I. Sokol, renamed the Izseljensko društvo Tabor (Tabor Emigrant Association) four years later, was established by a group of Yugoslav nationalists that had previously separated from the Slovenian Cultural Association 1. The two united in 1938 as the society Slovenski dom (Slovenian Home) partially due to the efforts of Izidor Cankar, the envoy of the Kingdom of Yugoslavia to Argentina, who

45 Janez (Juan, Ivan) Benigar (1883–1950) moved to Argentina in 1908 and lived among the Mapuche or the Araucaunian native people whose life, culture, and language he studied. He tried to improve their living circumstances and fought for their rights. He received several high accolades for his work and was a member of the National Academy for History (Lipovec Čebron, 2017; Žigon, 1999, 93–101; Jevnikar and Mislej, 1999, 11; Horvat, 1997, 37–38).

was working toward greater unity. The society bought land in Villa Devoto and opened a Slovenian Home by the end of 1938. Each ideological camp expressed political opinions in their press, and the split was also visible in the cultural and educational work (Mislej, 2001, 237–238; Brecelj, 1992, 169–172; Jevnikar and Mislej, 1999, 11–14).

Picture 2.4: Slovensko prosvetno društvo I (Slovenian Cultural Society I), Buenos Aires. Author: unknown. Courtesy of: Marjan Drnovšek.

The first Slovenian male choir in Argentina was established in 1924. Primorje, another choral society, was active there between 1927–1928.[46] In the Villa Devoto quarter in Buenos Aires, the Mutual Benefit Society Naš dom (Nuestro Hogar/Our Home), involved in education, was established in 1935. The Ivan Cankar Benefit Society was also established in Buenos Aires a year later. Slovenians united within the Yugoslav benefit and supporting organizations, such as the Yugoslav Immigrant Protection Foundation, led for several years by architect Viktor Sulčič (1895–1973).[47] The Slovenians from Prekmurje settled primarily in the Avellaneda quarter of Buenos Aires and, in 1940, established the society Slovenska krajina (Slovene March). Since 1930, there had been unsuccessful

46 Primorje, or the Littoral, a version of the toponym Primorska.

47 Viktor Sulčič is a co-author of two famous buildings in Buenos Aires: the fruit and vegetable market Mercado de Abasto, completed in 1934, and the Boca Juniors' stadium La Bombonera, completed in 1940.

initiatives to unite the Protestant and Catholic Slovenians from Prekmurje into one organization. Due to internal disagreements, the Slovenian Benefit Society Bernal was established only in 1943. In Córdoba, several societies formed: the Yugoslav Reading Society operated there between 1928 and 1932 and, since 1933, the third chapter of the Slovenian Cultural Society from Buenos Aires and the Slovenian Benefit Society Edinost (Unity) since 1940. In Rosario, where about 1,000 Slovenians (mostly from the Primorska region) lived, the Slovenian Workers' Society Triglav, established in 1930, produced theatre plays and held meetings with religious masses, dancing, and bowling. Its members also built a national home (1937) and established a well-regarded Slovenian school. Choral and drama societies were instrumental in retaining the Slovenian cultural tradition in South America in general before World War II. In fine arts and architecture, two names appear in the interwar period: architect and painter Viktor Sulčič, one of the prominent personalities in the cultural life of Slovenians in Buenos Aires; and painter and graphic artist Sergej Hočevar (1896–1973), also known as Sergio Sergi, who worked in Santa Fe and Mendoza. Posthumously, an award for graphic arts and a graphic arts salon in Mendoza were named after him (Mislej, 2001, 238–243, 1990, 223, 225; Žigon, 1998, 41–48; Banko and Mouzakis, 2003, 156; Toplak, 2008, 79–81).

During World War II, the community revived the work of the JNO. Because of the news about the actual role of Draža Mihajlović, collaborating with the occupiers on the one side and the partisan movement on the other, the JNO eventually experienced a schism. The Slovenians from Primorska within the JNO established a committee to unify Primorska with Slovenia or Yugoslavia after Italy's capitulation. The committee soon became independent and continued its work as the Committee for the Yugoslav Primorje. It joined the organized help for the compatriots in the destroyed homeland and gathered 10,000 signatures supporting the Yugoslav demands at the Paris Peace Conference in 1946 (Žigon, 1998, 41–42; Mislej, 2001, 240).

Brazil and Uruguay

In São Paulo, where Slovenian immigrants, mainly from the Primorska region, were most numerous, the leftist benefit, cultural, and educational society ORNUS, Ovekovečiti Rojstvo Novega Udruženja Slovenov v Braziliji (Eternalizing the Birth of the New United Slavs in Brazil), was established in 1928. The society had a library, a drama group, a benefits section, a tamburica group, and a chess club. They published their articles in the press of Argentinean Slovenians because *Izseljenec* (Emigrant), then the only Slovenian paper in Brazil, was published only once (1930). The society organized Portuguese language courses for newcomers and, in the second half of 1930, a three-year Slovenian language course for young people. In 1941, the society started building its own home. Those members of the Slovenian community in São Paulo who disagreed with the ORNUS's political orientation expanded their activities within the existing choir Primorje in 1929. In addition to the choir, Primorje had a Slovenian library and a women's gymnastics section. It shared premises with a Czech society and worked closely with it. In 1931, the immigrants from the Kingdom of Yugoslavia formed the Jugoslovansko podporno udruženje (Yugoslav Benefit Society), with Slovenians making up a quarter of the members. Due to internal friction, it only worked for five years. During World War II, Slovenians in São Paulo, working with counterpart organizations in Argentina, collected aid through the Jugoslovensko podporno udruženje (Yugoslav Support Association to aid victims of the war (Mislej, 2001, 244; Žigon, 1998, 85–87).[48]

In addition to around 500 Slovenians from Prekmurje, some Slovenians from Primorska also moved to Uruguay in the interwar period. The former set up the Prvo slovensko prekmursko društvo/Primera Sociedad Eslovena Transmurana de Montevideo (First Society of Slovenians from Prekmurje) in 1935, the oldest still operating Slovenian society in South America. From the early days of its operation, the society boasted a youth orchestra, a mixed choir, a male choir, and a radio program. In Montevideo, Slovenians also participated in the more right-wing choral society Slavček (Nightingale) and the leftist association Ivan Cankar, which merged into Slovenski krožek (Slovenian Circle) in 1937. This organization included a drama section, a choir, and a Slovenian language course. Priest, diplomat, and the initiator and co-founder of the Slovenian Circle, David Fortunat Doktorič, founded the choir and led the Slovenian lessons. He regularly contributed to the Slovenian monthly *Duhovno življenje/La vida spiritual*, published in Buenos Aires (Mislej, 2001, 244–245, 1999, 32–34; Žigon, 1998, 80–83; Prešeren, 2001, 120).

Before World War II, the Slovenians in South America were scattered all over the vast continent, yet they still constantly connected on several levels, both within each country and internationally. During and after World War I, as a part of the JNO in Chile and its Slovenian section in Argentina, they protested the annexation of parts of the Slovenian ethnic territory to Italy and supported

48 Jože Prešeren (2001, 120) also mentions the Slovenian benefit society Beritoga (named after a coastal town near São Paulo), which worked in the interwar period and all the way to the 1970s.

the establishment of the Yugoslav state. During World War II and immediately afterward—again as a part of the revived JNO and its Committee for Yugoslav Primorje—they were united in collecting aid and in the endeavor to unite Primorska with Yugoslavia.

Slovenian communities in Argentina, Brazil, and Uruguay differed significantly in numbers but shared numerous traits. Each community had benefit, humanitarian, cultural, and educational societies and organizations, priests, diverse educational activities, libraries, and newspapers. The ideological-political antagonisms between the Slovenians in South America were visible in the community's division into three camps: each established its own societies and published its own papers. The communities were most united in 1946 when all 16 organizations and societies of Slovenians in Argentina, Uruguay, and Brazil joined together in the Committee of Yugoslav Primorje and—in their shared concern for a fair solution to the border disputes—organized rallies and collected signatures in support of the political goals of their compatriots in the "old country."

Europe

Similar to transatlantic migrant communities, Slovenians' organizational and societal activities in European countries had a support, humanitarian, educational, cultural, and patriotic mission. Such self-organization, based on ethnic or linguistic identity, had a significantly longer history in those parts of Austria-Hungary, Italy, and the territories that formed the Kingdom of Serbs, Croats and Slovenes to which Slovenians traditionally had migrated in large numbers in previous centuries. Until the founding of the University of Ljubljana in 1919, almost all Slovenian intellectuals studied at the then "home" universities in Vienna, Graz, Prague, etc., and many continued to do so even after 1919. Only a minority studied at other European universities. The largest number of Slovenian students and associations was in Vienna at the end of the 19th century and the beginning of the 20th century.

Slovenian workers tended to congregate primarily within liberal or Catholic workers' societies, which, in some places, they created and operated in collaboration with Slovenian intellectuals and students. In Vienna, Slovenian migrants participated in the Croatian Workers' Association Sloga (Unity), established in 1891. In Graz, they established a social-democratic Yugoslav Workers' Association Danica/Südslawischer Arbeiter-Bildungsverein Morgenstern in 1896, which had a library, a choir, a tamburica group, and provided educational activities; the association's statute was written in Slovenian. Three years later, the workers' association Naprej was established in Graz; Catholic students helped to set it up, but the leadership soon passed to the liberal students' society Triglav, established in 1875. Slovenian Catholics thus started the workers' educational society Domovina in 1902 (it had its own nursery school from 1907), which was also taken over by the liberals in 1909, so the Catholics gathered in the educational society Kres (Bonfire) (Drnovšek, 1999a, 36).

Relationships forged upon similar ideological foundations also existed among Slovenian associations outside the monarchy, particularly in Germany. Besides compulsory schooling for children in German schools, the increasingly common intercultural marriages and the high mobility of labor migrants between German mining regions hindered a more vital development of the social and cultural life of migrants there. Nevertheless—in Germany before World War I and elsewhere in Western Europe in the interwar period—a network of both Catholic and leftist workers' benefit societies evolved that tried to provide social security, worked on maintaining the use of the mother tongue, and supported the social and cultural life of their members.

Migrant laborers, who often suffered work accidents and work-related injuries, depended on bilateral social agreements between their country of origin and host country if they existed. Thus, the support from their ethnic associations was even more crucial. The associations soon added other humanitarian, cultural, or educational activities to their mutual aid and support. The communities set up choirs, brass and tamburica orchestras, theatre groups, and chess and other sports clubs. Many associations had libraries and organized cultural and social events on religious and state holidays and other occasions, wine harvests, trips to the home country, etc. (Drnovšek, 1999a, 44–48).

There were three worldview-based types of Slovenian associations: Catholic, pro-Yugoslav nationalist (monarchist), and socialist or communist; in Germany, there were also pro-Austrian ones. Each group had its own newspaper—although there were far fewer periodicals and other Slovenian media in the European countries than in the United States. At this time, Slovenians in Europe also did not build national homes or churches. It was also mostly not possible to establish ethnic parishes in European countries. Instead, Slovenians and other migrant nations set up their own pastoral centers within their local church communities. Slovenians had several such centers in the interwar period, particularly in Germany. Slovenian migrants in Western Europe who did not take Yugoslav citizenship joined Austrian associations. Certain leftist Slovenian migrants joined leftist organizations and parties, such as the French Communist Party. Slovenian societies mostly collaborated with the other nearby Slavic associations, as evident from the reports about shared social events. The ideological pressures of Fascism and Nazism on the one side, Communism on the other, and—third—Yugoslav centralism, significantly influenced the operations of the associations in Europe. For this reason, they often did not reach an equally extensive organization and material base as the Slovenian organizations in the United States. Once Germany occupied the Netherlands, Belgium, Luxemburg, and France, the Slovenian associations in those countries mostly ceased their operations (Kolar, 2001, 135–138; Drnovšek, 1999a, 45–48, 2002a, 153–157).

Germany

Slovenians in Westphalia were less active in organized associations and social life than their compatriots in the United States. Still, in the years before World War I, they had nevertheless been gathering in mining, choral, and women's societies (Wörsdörfer, 2017). Around 1898 in Essen-Borbeck, they established a Slovenian choral society, Ilirija, another choral society of the same name in Bergeborbeck, a Slovenian choral and benefit society, Bratovska ljubezen (Fraternal Love), in Homberg-Hochheide, and a music society in Osterfeld. In 1904, the Slovenian colony in Zell im Wiesental set up the Slovensko delavsko podporno društvo (Slovenian Workers Benefit Society), which operated for more than three decades without any contact with other Slovenian or German associations. From 1902 to World War I, several Slovenian Catholic mining societies of Saint Barbara appeared in various places in the Rhine-Ruhr region (or North Rhine-Westphalia). At the same time, Slovenian Rosary Societies for women and Austrian associations operated parallel to the miners' and women's associations, which Slovenians also joined (Drnovšek, 1999a, 37–41, 2012, 161–164; Werner, 1985).

After 1918, approximately 40,000 Slovenians lived in Germany (Drnovšek, 1999a, 42). Twenty-five Saint Barbara miners' societies were operating there in 1926 with 750 members, and nine women's Rosary societies with a further 820 members. In 1922, the priest Janez Evangelist Kalan established the Zveza slovenskih katoliških društev (Union of Slovenian Catholic Associations). All but two of the mentioned associations became members of it in the following few years. Although the Nazi Regime in Germany in 1933 banned Catholic societies, all these Slovenian Catholic societies and those established later continued working uninterrupted. The numbers from 1937 show that 23 Saint Barbara societies and 16 women's Rosary societies were active in Germany. The Union of Yugoslav Catholic Associations had 2,243 members in 34 societies in 1938. It was linked to the erstwhile Slovenska ljudska stranka (Slovene People's Party), which in the second half of the 1930s was operating in Drava Banovina as a part of Jugoslovanska radikalna zajednica (Yugoslav Radical Union). It was linked to the erstwhile Slovenska ljudska stranka (Slovene People's Party), which in the second half of the 1930s was operating in Drava Banovina as a part of Jugoslovanska radikalna zajednica (Yugoslav Radical Union). The Zveza jugoslovanskih podpornih društev (Union of Yugoslav Benefit Societies) had 1,690 members in 36 associations; with a Slovenian leadership, two choirs, a theatre section, and a benefit and funeral fund, it was connected to the Yugoslav Democratic Party. The parallel women's societies had a further 218 female members. In Germany, there were several Slovenian choral societies. Slovenians also set up gymnastics and sports associations, but most societies had a benefit, educational, humanitarian, or social character (Drnovšek, 1996, 34–36, 2012, 161–167; Žumer, 1996, 397–399).

With the proclamation of the January 6 Dictatorship in 1929, many Slovenian associations were renamed as Yugoslav. The changed regime in Yugoslavia propagated the unitarist politics of the dictatorship through consular and other Yugoslav legations. There were ideological disagreements between the Zveza jugoslovanskih narodnih katoliških društev (Union of Yugoslav National Catholic Societies) and the Zveza jugoslovanskih delavskih in podpornih društev (Union of Yugoslav Workers' and Benefit Societies), which the two unions tried to mitigate by establishing the Jugoslovanski izseljeniški odbor (Yugoslav Emigrant Committee) in 1928 in Karnap near Essen. But because the two unions did not reach the expected agreement and collaboration, the Centralna zveza jugoslovanskih delavskih društev v Nemčiji (Central Union of Yugoslav Workers' Societies in Germany) was established in 1933 (Drnovšek, 2012, 167–169, 2008, 255–256; Wörsdörfer, 2017).

France

Around 16,000 Slovenians lived in France in the early 1930s. By the end of the decade, there were more than twice as many, around 34,000 (Drnovšek, 2000). The Slovenian miners in northern and northeastern France were the most numerous and best organized. With migrants from other Yugoslav nations, they developed strong mutual support activities. Again, their associations fell into three groups: Catholic, pro-Yugoslav, and socialist or communist. Like elsewhere in Western Europe, the Catholic Slovenian migrants in France organized within the mining benefit society of Saint Barbara (for men) and Rosary societies (for women). Most were established in the 1920s. In the department of Pas-de-Calais, there were three Rosary and five Saint Barbara societies. Two smaller Saint Barbara societies merged with the Slovensko društvo šahistov (Slovenian Chess Players' Society) and the Slovensko prijateljsko društvo (Slovenian

Friendship Society) into Zveza slovenskih delavskih društev Liévin (Union of Slovenian Workers' Associations Liévin) (Bogovič and Canjko, 1983, 106). All the Catholic mining associations were members of the Yugoslav Mining Association of Northern France in Liévin. In the nearby Sallaumines, there was also the Christian Yugoslav Section of the Free Union of Miners of Northern France. In the 1920s, Slovenian benefit societies of Saint Barbara and other Slovenian or Yugoslav workers' benefit societies (Edinost, etc.) were established elsewhere in the country as well, most of them in northeastern France, particularly in the departments of Moselle, Meurthe-et-Moselle, and Meuse in the "iron" Lorraine. At the Slovenian Eucharistic Congress in Aumetz in 1931, the Yugoslav Emigrant Catholic Action was established to spread the teachings of the Catholic Church among the migrants. In many places, particularly in the departments of the north and northeast, there were chapters of the pro-Yugoslav Zveza jugoslovanskih delavcev v Franciji (Union of Yugoslav Workers in France), established in 1933 in Metz. In 1936, it had 37 chapters in France and Luxemburg and expanded its influence into the Saarland (Drnovšek, 1996, 38–44).

The French Communist Party, as well as the Slovenian and Yugoslav communists, influenced the Yugoslav Workers Benefit Society (est. 1928). It had its seat in Lens and eight chapters (mainly in Pas-de-Calais) with nearly 1,400 members—Slovenians, mostly, with Serbians and Croatians. By the second half of the 1930s, Slovenians in Pas-de-Calais also had a girls' ballet ensemble as part of the Društvo prijateljev Sovjetske zveze (Association of Friends of the Soviet Union), the Ivan Cankar Choral Society, a library, and the Benefit and Educational Society Vesna (all in Lens). Towns in other departments had a tamburica section, a cycling section, and a chess and sports club called Delavec (Worker). At this time, the union movement flourished in France—it was particularly strong in Pas-de-Calais, which was seen as the hub of the Slovenian migrant left. Between 1936 and 1938, during Blum's Popular Front—the Front Populaire led by French socialist prime minister Léon Blum, during which workers fought for and won some fundamental rights—Slovenians there made unsuccessful attempts to unite leftist societies with the Catholic ones. The Catholic migrant press fought against the influence of the liberal and socialist periodicals among the migrants and attempted to prevent the "merging of the Christian and communist Yugoslav workers in France into one block" (Drnovšek, 1999a, 58).

In 1936, the Zveza jugoslovanskih rudarjev v severni Franciji (Union of Yugoslav Miners in Northern France) was established in Lens, and the Jugoslovanska sindikalna zveza (Association of Yugoslav Unions), which had chapters in several places around Pas-de-Calais and other departments. In Pas-de-Calais, the Slovenian section of the Association of the Friends of the Soviet Union organized cultural events and Slovenian and French lectures. In several towns around Pas-de-Calais, women established anti-war and anti-fascist associations and collected money and other aid for the Yugoslav volunteers in the international brigades in the Spanish Civil War. The Catholic Yugoslav miners' society Saint Barbara La Machine (dep. Nièvre) in central France expressed support for Republican Spain in an open letter in 1937 (Drnovšek, 1996, 44), proving that Catholic societies also defended the anti-fascist stance.

In 1925, the Yugoslav section of the French Communist Party was established in Paris as well as the Yugoslav inter-union committee of the Confédération générale du travail unitaire (United General Confederation of Labor, or CGTU). The Slovenian Club had met in the French capital since 1928. The seat of the Yugoslav colony established in 1928 was also in Paris. The community had sections all over France and, for a short while, even published the magazine *Jugoslavija u Francuskoj* (Yugoslavia in France), which also published Slovenian poetry and other contributions. In the second half of the 1930s, the Yugoslav Emigrant Committee, the Jugoslovenska radnička zajednica (Yugoslav Workers' Union, est. 1936), and the Yugoslav Sokol Society were active in Paris. In the city, where, by 1930, the Združenje jugoslovanskih študentov (Association of Yugoslav Students) was also formed, several other Slovenian or Yugoslav associations, organizations, and committees supported political migrants. Financial aid was collected for victims of political persecution in the Kingdom of Yugoslavia by at least two groups: the Yugoslav section of the International Red Aid starting as early as 1933 and the Emigrant Aid Committee in 1938. From 1937, the Yugoslav National Committee operated in Paris to help the Spanish nation. Since the beginning of the Spanish Civil War, Yugoslav migrants had been joining the Spanish Republican side, with 420 joining from France and 191 from Belgium. Upon returning to Paris after the defeat, they established the Odbor nekdanjih španskih dobrovoljcev iz Jugoslavije (Committee of Former Spanish Volunteers from Yugoslavia) (Drnovšek, 1996, 45–48).

Belgium and the Netherlands

Between 1923 and 1940, Slovenians in Belgium set up 20 societies, independently or in collaboration with other Yugoslavs. In 1934, the Association of Yugoslav Societies was established in Eijsden, where the Central Committee of the Communist Party of Yugoslavia played a significant role. The activity of the member societies was like that in other countries of Western Europe: material self-help, cultural and social events, and preserving the Slovenian language. In the Catholic clubs, there was more emphasis on Christian values. They rejected the anti-religious press and banned political participation. In Belgium, the Catholic Church was strong, but so was the Communist movement as, for a while, Brussels was the center of the Zveza revolucionarnih delavcev in kmetov iz Jugoslavije (Union of the Revolutionary Workers and Farmers from Yugoslavia), which was a central subsection of the Communist Party of Yugoslavia. In 1940, most societies closed their doors; only a handful tried to operate illegally (Drnovšek, 2012, 298–309, 1996, 36–37, 1997b, 357–389, 1990; Lipoglavšek-Rakovec, 1950, 49).

Between 1926 and 1931, eight Saint Barbara societies were set up in the Netherlands. They were all members of the Union of Saint Barbara Societies of Yugoslavia, established in 1928 in Heerlen. All the societies had a choir and a library. Some also had a youth club and drama, tamburica, and cycling sections. Return migrations during the economic crisis resulted in fewer societies before the beginning of World War II. Their activities decreased also because the descendants of the immigrants did not show enough interest in the associations' activities. The exception was the Slovenian choir Zvon (Bell) from the society in Heerlen, which also had a children's choir (Drnovšek, 1996, 37–38, 2012, 268–274).

The Yugoslav Territory

Before World War I, most Slovenians living in the territory of today's Croatia lived in Zagreb. Their share in the entire city population was, according to the 1890 census, more than 15 percent. In 1908, the Lipa Association for education and entertainment was established; it operated until 1920. Only Slovenians could be members, and the official language was Slovenian. In 1906, the Istarsko planinsko društvo (Istria Mountaineering Association) was established in Pazin as a chapter of the Slovenian Mountaineering Association. Two years later, Slovenian reading societies opened in Pazin and Pula. These reading societies often had a tamburica orchestra, a choir, and a drama section that staged comedies and only used their mother tongue to communicate. Between 1911 and 1913, the association Skup Slovencev (Gathering of Slovenians) operated in Pula. It organized parties and had a choir and, like most other associations, gave the profit from the entry ticket sales to the Society of Saints Cyril and Methodius. From 1913 onward, there was a Slovensko pevsko zabavno in podporno društvo (Slovenian Choral Recreation and Benefit Society) in Pula, most likely the descendant of Gathering of Slovenians. Its working language was Slovenian, and its activities included singing and acting, organizing excursions, lectures, and mutual aid, and running a library.

After the Austrian occupation of Bosnia and Herzegovina in 1878, more than 3,000 persons moved there from the Slovenian ethnic territory. Among them, in 1910, there were 400 government officials. Around 1,000 Slovenians lived in Sarajevo by the end of the 19th century and, during World War I, many Slovenians from Primorska moved there. The first Slovenian association in the city was established in 1897 as the Slovensko omizje (Slovenian Table), which operated from 1910 as the Slovenski klub (Slovenian Club). The goals of both associations were alike: operating a reading room, a library (established in 1911), and a choir, and organizing events, parties, lectures, and excursions. Both maintained a neutral political stance (Koblar, 2008, 335–339; Kržišnik-Bukić, 2007, 20, 69; Žitnik Serafin, 2014, 155–158).

Despite the ban on Slovenian societies during World War I, members of specific associations continued to meet and attempted to revive them after the war. However, this was not easy. The proclamation of the dictatorship in 1929 meant that many associations were dissolved. Despite all the prohibitions, the interwar period was the time of the liveliest social life of Slovenians in other parts of Yugoslavia. It was the time when Slovenians established the most associations in Zagreb. In 1920, after the prewar Lipa stopped working, the Academic Association Triglav was established as a chapter of the Triglav student network, created in 1875 in Graz. In 1923, a chapter of the Yugoslav Catholic Academic Association Danica was established. A year later, the Slovensko prosvetno društvo Zagreb (Slovenian Cultural Society Zagreb) started, which organized events and lectures and opened a library. Most important for the further development of the social life of the Slovenians in Zagreb was the establishment of the National Library and Reading Room in 1930. It had over 1,000 members in the first year and was the center of Slovenian cultural and educational activities in Zagreb. The library functioned as a society because it had the organizational, propaganda, and educational sections, a section for excursions, a chess, dance, and choral section, which formed three choirs, and a youth section that organized its own drama activities, lectures, youth excursions, and tea parties. The educational section organized weekly lectures, while the propaganda section collected around 15,000 addresses of Slovenians in Zagreb. In 1936, the association changed its name to Narodni dom (National Home). During World War II, the National Home was banned and dissolved and started working again after the war under the new name Slovenski dom (Slovenian Home) (Strašek, 2006, 97–104; Jerman and Todorovski, 1999, 5, 24; Kržišnik-Bukić, 1995, 147–148, 152–154; Založnik, 2009, 58–60).

The second significant association of Slovenians in Zagreb was the Catholic Anton Martin Slomšek Society (1933–1941), intended for cultural, educational, and supporting activities. It had a children's and a girls' choir, and it organized church concerts and theatre productions. By the beginning of World War II, its membership was over 800 Slovenians. In 1937, Slovenian students established the academic club Slovenski jug (Slovenian South). Two years later, in the Kustošija suburb, the Slovenian Cultural Society Kustošija was established. It organized educational lectures, excursions, and parties, helped its members in need, and had its own library (Strašek, 2006, 97–104; Kolar, 1995, 119; Kržišnik-Bukić, 1995, 148–150).

After the end of World War I, the Slovenian Club in Sarajevo could once again operate legally. It peaked in the mid-1920s when it had more than 200 members. It often used members' contributions to help other Slovenians but had little understanding of the vital issues affecting the Slovenian community in Sarajevo. Once the leftist Workers' Cultural Association Ivan Cankar was established in 1934, disagreements emerged between it and the Slovenian Club, as well as within the club itself. The new association soon boasted a library, a choir, a drama group, and an orchestra, and organized socials, lectures, performances of their groups or groups touring from Slovenia, and excursions. It published its own newspaper from 1935. In 1938 and 1939, the association set up the sports and chess section and a mutual aid fund, but it stopped functioning in early 1941 (Koblar, 2008, 339–358).

As the need for mutual material aid increased, the Sarajevo Slovenians established the Slovenska kreditna zadruga (Slovenian Cooperative Bank) in 1935 (Koblar, 2008, 339–358). In the interwar period, the Združenje emigrantov Gortan-Bazovica (Emigrant Union Gortan-Bazovica) also operated in Sarajevo, established by the Sarajevo Slovenians and refugees from Goriška, Friuli, and Istria escaping Fascism. The authorities tolerated the fraternal supportive, humanitarian, and nationalist activity of this association—and other associations from the Julian March around the country—and, at times, in the periods of chilly relations with Italy, even used them for political propaganda. On such occasions, they printed brochures about the Yugoslav rights to Slovenian and Croatian ethnic territories annexed by Italy after the war and the suffering of the Yugoslav population there (Jergović, 2015). Slovenian families from Istria and Primorska also found refuge around Banja Luka in early 1920. In 1923, they set up the Slovenian choral and educational society Triglav in Banja Luka, with at least 150 members.

Around 6,000 Slovenians lived in Belgrade in 1931. By the second half of the 1930s, they had five associations (Ivan Cankar, Edinost, Prosvetno društvo, Triglav, and Istra–Trst–Gorica) and a Slovenian mixed-choir, Ljubljanski zvon. In 1935, the Društvo Slovencev v Beogradu (Society of Slovenians in Belgrade) was established as a descendant of three smaller Slovenian associations there; it had choral, drama, and education sections, a library, and a reading room (Stanič, 1987, 243; Kržišnik-Bukić, 2003, 126, 2004, 546; Žitnik Serafin, 2014, 163–171). In the interwar period, mostly Slovenians from Primorska moved to the territory of the Vardar Banovina (present-day North Macedonia), from which the Turkish population had emigrated and thus enabled settlers to live there. In 1931, Slovenians from Primorska started settling in the village of Bistrenica and established the first Slovenian association on the Macedonian territory (Rusić and Novak, 1973; Kržišnik-Bukić, 2003, 128; Jovanović, 2001, 320).

When World War II began, thousands of Slovenians were deported from Slovenia to the territories of present-day Croatia, Bosnia and Herzegovina, and Serbia. Around 5,000 Slovenians were resettled in 22 places in Bosnia and Herzegovina, while over 7,000 were expelled to Serbia. Slovenian refugees developed a cultural life, formed choirs and drama groups, organized celebrations, and had several small libraries. In 1944, the Slovensko kulturno-prosvetno društvo Franc Rozman – Stane (Slovenian Cultural and Educational Society Franc Rozman – Stane) was established in Belgrade (Dolinšek Divčič, 2001, 302; Suhadolnik, 2006, 46–48; Roš, 1967, 164; Ferenc, 1993, 80; Bartulović, 2001, 20–23).

Publishing and Media

Even today, the Slovenian press remains one of the most important sources for the general and cultural-historic research on Slovenian migration. More explicitly than all other historical sources, it mirrors the dynamics that governed the Slovenian diaspora's ideological, political, and cultural relations and moods and their diverse manifestations and effects on the relations in different communities worldwide. It also mirrors Slovenians' markedly complex, diverse, and often changing attitudes to the events in their old and new homelands. The migrant press has some specific characteristics that distinguish its content from the periodicals in Slovenia at the same time. Before World War II, it most often contained a combination of news from the old and the new country on the one side and society news in migrant communities on the other. We can also find—often in specialized sections—all sorts of other contributions, from practical advice, literary texts, polemic commentary, entertainment, puzzles, and ads, to special sections for women readers and "children's corners."

Migrant communities had a great need for practical advice on finding employment, acquiring citizenship, and other aspects of adjusting to their new environment. The editors took this need into account because practical content brought them the most readers. The press had enough space for letters and other contributions from ordinary workers who would never have even considered publishing in a paper "at home" in Slovenia. Editors, too, particularly in the early years, were often self-taught (Sulič, 1996, 47–49). Specialized, literary, or cultural journals were fewer than general ones. As a rule, they all differed from each other based on the worldview they reflected. Regarding the target audience, we can divide periodicals into internal association bulletins aimed at the members and publications aimed at a wider circle of Slovenian readership in the new (and also the old) homeland. The publishers of the latter were usually migrant organizations with members and sympathizers throughout the host country.

We can also categorize the migrant newspapers and magazines into the more or less "neutral" ones, the ideologically clearly profiled ones, and the emphatically political periodicals. Young Slovenians often read special pages or supplements in the periodicals for adults because there were barely any specialized papers for children and young adults. Some bulletins were distinctly religious. Even more, they encouraged—alongside their journalistic content—Christian values and championed adherence to the Church. The radically left papers were relatively few. Many more of them could be labeled as moderately left, and most of them were "workers' newspapers"—specifically addressing either the "Slovenian Catholic Workers" in the diaspora or their "liberal" or "progressive" counterparts.

The content of the newspapers and magazines reflected their political orientation. Many bulletins tried using different subtitles to clearly indicate their orientation. These subtitles frequently changed, reflecting the occasional (and often temporary) corrections of the political position of the paper. The politically profiled papers were most common in Western Europe, but very few were published entirely in the Slovenian language. Among the explicitly political papers in the United States, there were, for example, the Cleveland monthly paper *Socialistična zarja* (1920–1923), published by the Yugoslav section of the SLP (Socialist Labor Party), the little-known mimeographed *Iskra*, which was, also in Cleveland, published by the Slovenian communists in the mid-1930s, presumably for three years (Bajec, 1980, 101; Klemenčič, 1991, 305), and the communist weekly *Naprej: Tednik slovenskih proti-fašističnih delavcev v Ameriki in Kanadi* (Forward: A Weekly of Anti-Fascist Workers in America and Canada, 1935–1941, Pittsburgh, from 1940, Cleveland) (Munda, 1969, 68–69). The migrant press was instrumental in developing literary creativity before World War II (Habjan and Koron, 2014), most visible in the Americas. Cleveland and Chicago stood out in the number of relevant literary and cultural immigrant periodicals, followed by Buenos Aires. In the centers of Slovenian migration around Western Europe, the number of such publications was significantly lower.

The gap in the absolute number of titles may be deceiving if we simply compare the number of titles of Slovenian newspapers published simultaneously in the two countries with the most active Slovenian publishing communities, the United States and Argentina. But as soon as we consider the relatively greater difference in the number of all Slovenian immigrants in each country when these papers were published, we can see that Slovenians in Argentina were at least as active at publishing periodicals as the Slovenians in the United States. In other parts of Europe, Slovenians published around 12 Slovenian (or partly Slovenian) newspapers in the entire period before World War II, which mostly had a short life span.

The United States of America

The first migrant paper in Slovenian, the Catholic weekly *Amerikanski Slovenec*, was first published in Chicago in 1891. Between 1899–1914, this herald of the Slovenian migrant periodicals was, as its masthead explained, the paper of the Kranjska slovenska katoliška jednota (KSKJ) and, later, also of other Slovenian Catholic organizations in the United States. Since 1924, it was the paper of "Slovenian Catholic Workers in America" (Bajec, 1980, 11–14). *Amerikanski Slovenec* appeared at the start of the most intense migration of Slovenians to the United States (1890–1914). The newcomers to the community needed, on the one hand, practical information, and on the other—as the publishers were convinced—the appropriate "spiritual" support (moral or religious) and, above all, ideological or political orientation. Therefore—partly in response to the Catholic orientation of the first paper—others quickly followed. Already in 1893, there appeared the New York-based weekly *Glas naroda* (The Voice of the Nation), subtitled *List slovenskih delavcev v Ameriki* (The Paper of Slovenian Workers in America) (Bajec, 1980, 11–16). *Glas naroda*, a centrist paper for workers, had one of the largest print runs among Slovenian dailies in the United States, 14,000 copies. The Catholic share of Slovenians in Cleveland responded six years later with the monthly *Narodna beseda* (The National Word, 1899–1908), which after six years also became a daily. The latter was also the first Slovenian paper with an English subtitle:

The Only Slovenic Newspaper in Cleveland. Between 1906 and 1952, one of the most significant leftist-workers' newspapers, *Proletarec*—a bulletin of the Slovenian section of the Yugoslav Socialist Federation/Alliance in America (a member of the Socialist Party of America)—was published in Chicago. Later, it also served as the bulletin of Prosvetna matica (Bajec, 1980, 27–28). Throughout its run, the paper championed socialist values and significantly contributed to Slovenian fiction writing. It played a particularly valuable role during World War II when it informed the Slovenians in the United States about the war events in Slovenia.

When *Glasilo SNPJ,* from 1916 onward *Prosveta*—one of the most influential newspapers—started in 1906, at least twelve other Slovenian newspapers were being published in the United States (Bajec, 1980, 32–36). It was first a monthly, from 1910 a weekly, and from 1916, when it changed its name to *Prosveta*, a daily. During World War I, in addition to the usual society news and news from the old and the new country, it was primarily dedicated to the news from European battlefields and the consequences of war. The newspaper vividly reflected the minority position of Slovenians in American society and the changes in their life. In 1908, Rev. Father Kazimir Zakrajšek co-founded the St. Raphael Society in New York and, in 1909, started to publish and edit the oldest Slovenian religious paper as a part of it, the monthly *Ave Maria*, which after several moves settled in Lemont, Illinois. Between 1917 and 1922, there were up to 22 or 23 Slovenian newspapers published at a time, the first of the two peaks of the Slovenian press in the United States.

In the interwar period, and particularly toward the end of World War II, Cleveland was the setting of a political, ideological, and market-share battle between the newspapers *Ameriška domovina* and *Enakopravnost*. The printing houses for both dailies were only a couple hundred meters away from each other on St. Clair Avenue. The last editor of *Ameriška domovina*, Rudolph N. Susel (1992, 242–243), writes that in its four-decade-long battle with *Enakopravnost*, *Ameriška domovina* had the advantage of support from traders and other businesspeople, who regularly contributed to the paper.

AMERIŠKA DOMOVINA

AMERICAN IN SPIRIT FOREIGN IN LANGUAGE ONLY

AMERICAN HOME

SLOVENIAN MORNING DAILY NEWSPAPER

NO. 208 — CLEVELAND, OHIO, FRIDAY MORNING, SEPTEMBER 5th, 1930 — LETO XXXII.—VOL. XXXII.

Zanimive slovenske vesti iz širnih naselbin Zedinjenih držav

Dva meseca je Andree stražil v ledu pri svojih mrtvih tovariših

Silne so žrtve, ki jih zahtevajo avtomobili v Ameriki

Ogromne časti, odlikovanja in denar za francoske zrakoplovce

Osemsto mrtvih v silnem orkanu, ki je zadel dominikansko republiko

Nekaj drobnih novic iz lorainske naselbine

Lahi sodijo Slovence kot teroriste

Organizirano delavstvo v Clevelandu na delu, da porazi Zmunta

Preiskava radi nerednosti pri primarnih volitvah v Euclid vasi

Hudo je, kadar zdravnik toži zdravnika, da mu pobira proč bolnike

Društvo, kjer so nepoljubljena dekleta

Zopet ena žrtva suhaških agentov

Francozi v Dallasu

Tudi poštnim uslužbencem grozi brezposelnost

Karl Mayeva žena pride na obisk k Indijancem

Picture 2.5: The front page of the newspaper *Ameriška domovina*. Courtesy of: Digital Library of Slovenia – dLib.si

Jutri je maskeradna veselica SRZ ženskega odseka v Collinwoodu

ENAKOPRAVNOST

SLOVENIAN-EQUALITY-DAILY

NEODVISEN DNEVNIK ZASTOPAJOČ INTERESE SLOVENSKEGA DELAVSTVA

ENAKI PRED LJUDMI IN POSTAVO

LARGEST SLOVENIAN DAILY BETWEEN NEW YORK AND CHICAGO

EQUAL BEFORE MEN AND LAW

ADVERTISING MEDIUM

LETNIK II. VOLUME.

RUSKA VLADA SE SESTANE Z ALIIRCI

BOLJŠEVIKI SO SPREJELI PONUDBO ZA RAZGOVOR.

HOČEJO PRIZNATI RUSKE DOLGOVE ENTENTI IN GOTOVE TRGOVSKE KONCESIJE.

Izvirno poročilo o razmerah na Slovenskem.

Dne 24. oktobra narodni praznik jugoslovanske neodvisnosti. - Slava Wilsonu, Kreku in Korošcu.-Umikanje Nemcev in Madžarov.-Meščanje so beračili na kmetih.-Strašna draginja.-Trgovci in kmetje so delali dobro kupčijo.-V Ljubljani so srbski vojaki.-Gorica popolnoma razbita.

VELIKA PRORAČUN. PREDLOGA

SPREJETJE PRIČAKUJEJO V HIŠI V SOBOTO ZVEČER.

Fany Demojzes in njena sestra Albina v narodni noši na slavju slovenske neodvisnosti.

Fany Demojzes in njeni sorodniki.

Picture 2.6: The front page of the newspaper *Enakopravnost*. Courtesy of: Digital Library of Slovenia – dLib.si.

The newspapers most clearly expressed Slovenians' ideological and political polarization. Their disagreements—reflected in verbal disputes between the representatives of the circles close to the Catholic Church on the one side and the "progressive," leftist-oriented portion of Slovenian Americans on the other—increased significantly during World War II. *Ameriška domovina* was—like all other papers of the Catholic pole of Slovenian Americans—extremely critical toward Louis Adamic because he supported Tito and the partisan struggle. On the other hand, *Enakopravnost* and other progressive Slovenian press in the United States enthusiastically supported both Adamic and the national liberation struggle. These battles in print continued after World War II. At the same time, these publications revealed internal disagreements in both camps that occasionally threatened the stability of the alliances on both sides, outside the otherwise united decision of Slovenian Americans to help the wartime and postwar Slovenia.

South America

In 1920–1921, the first Slovenian texts published in South America came out in *Jadran*, the Buenos Aires-based paper of the Center Jugoslovanske narodne odbrane (JNO) for the Southern Atlantic, which was primarily not published in Slovenian. The texts were the contributions of a group of Slovenian intellectuals who moved to Argentina before World War I and established the Slovenian section of the JNO. The leftist Ljudski oder in Buenos Aires started publishing the occasional bulletin *Delavski list*. The paper was published for only two years because of different problems and bans during the military regime in 1930–1932. It was replaced in 1932 by *Delavski glas* (Jevnikar and Mislej, 1999, 14). The goal of both papers was political and ideological propaganda rather than cultural issues. Thus, the Ljudski oder established the workers' cultural magazine *Njiva* in 1937. A group of Slovenian intellectuals published in *Slovenski tednik* (1929–1936), which liberals established along with the Slovensko prosvetno društvo I (Slovenian Cultural Society I). In 1936, the monthly *Duhovno življenje* started coming out and, a year later, *Slovenski list*. The unitarian Yugoslav nationalists wrote for *Slovenski dom*, which only published three issues in 1932, and later for the weekly *Jugoslavija* (Buenos Aires, 1932), which had a Slovenian segment in the following years. At first, *Slovenski dom* was also the paper of choice for the members of Sokol, who later moved to Viktor Kjuder's *Novi list*. Architect Viktor Sulčič led the consortium that published *Novi list*.

During World War II, the Slovenian press in South America, just like in other countries of the Slovenian diaspora, reflected the plurality of their political standpoints and views. *Njiva*, in addition to news, brought socialist and communist-oriented content, which is why the authorities banned it in 1943. It published the first reports about the partisan movement under Tito's leadership in mid-1942. In August of that same year, it published the *Klic osvobodilne fronte civiliziranemu svetu* (The Appeal of the Liberation Front to the Civilized World). *Slovenski list* also commented on the war in Yugoslavia and the contested role of Draža Mihajlović. *Slovenski list* inevitably withered and, in the fall of 1945, Slovenians from Primorska started publishing *Pravica* (Justice), to which former writers for *Njiva* and other workers' periodicals contributed. One-off publications and leaflets appealed for help under the auspices of the newly established organization Svobodna Jugoslavija (Free Yugoslavia). At the same time, leftist activists led efforts to unify all Slovenian societies in Argentina (Mislej, 1992, 186). The contributors of the *Naš dom*, the bulletin of the Gospodarsko podporno društvo (Economic Benefit Society) that was published sporadically between 1938 and 1943, had been promoting such an idea some years earlier (Brecelj, 1992, 174). Slovenians in Argentina published—for a shorter or longer time—around 16 newspapers before 1945.

Europe

The press in Europe usually targeted Slovenian readers in all the countries of Western Europe where there were workers' settlements. They also received certain migrant newspapers from both North and South America, so the need among Slovenians there for their own newspaper activity was slightly lower. Even after the establishment of the University of Ljubljana in 1919, many Slovenians studied or taught in Paris and other university cities of Western Europe. Neither the students nor the educators were in contact with the Slovenian migrant miners and other workers in these countries. As a rule, the activity of Slovenian intellectuals in the territories where there was a concentration of Slovenian migrants was linked to political activity. Only migrant priests and teachers—who, in addition to other goals (religious or ideological propaganda), also worked to preserve the Slovenian language and culture among their compatriots in that part of the world—were the exception. Despite the lack of motivation, Slovenian printing houses (local printing houses had problems with Slovenian diacritics), and appropriate cadres, several migrant newspapers were published in Western Europe in the decade before World War II that reflected the ideological and political standpoints of the publishers and editors. Like the immigrant societies in this area, the press was divided into three groups: the pro-Yugoslav papers; the Catholic papers; and the workers' or communist periodicals. In the 1920s and particularly in the 1930s, about 10 pro-Yugoslav newspapers were published and often co-financed by the Yugoslav state. Additionally, the international funds of the Comintern financially supported 19 of the communist or otherwise leftist-profiled papers. Many of these papers had Slovenian editors and Slovenian supplements. The majority of these newspapers, both pro-Yugoslav and communist or socialist, were published in France, far fewer in Belgium, with a sprinkling of titles in Switzerland and the Netherlands. The pro-Yugoslav papers advocated—particularly after the proclamation of the January 6 Dictatorship—unitarism and the "one Yugoslav nation." Among them were *Jugoslavija u Francuskoj* (Yugoslavia in France, Paris, 1928–1929), *Jugoslovenski glasnik u Francuskoj* (Freyming, 1930), and *Jugoslovenska sloga* (Merlebach, 1930), which was published in Slovenian and dedicated many articles to the migrants from Primorska and the dangers of fascism. Slovenians also contributed to the newspapers *Pariske novine* (1930–1931) or *Jugoslovenske pariske novine* (Paris, 1931–1939) and *Glas jugoslovenskih iseljenika* (The Voice of Yugoslav Emigrants, Amnéville, 1933–1939) (Drnovšek, 2012, 1992, 273–282). The leftist migrant press linked the illegal activity of the Communist Party of Yugoslavia with Slovenian and other Yugoslav migrants, for example, in campaigns to help the volunteers of the international brigades in Spain.

The most widely distributed leftist paper among the Slovenians in France was the Parisian *Glas izseljencev* (Voice of Emigrants, 1936–1939), which had as many as 3,200 subscribers in 1938. In Zurich, in the early 1930s, the Slovenian *Delavski list* was available, a mimeographed bi-weekly mostly read by Slovenian seasonal mason workers in Switzerland. The following papers in the 1930s were also partly published in Slovenian: *Naša pravica: List delavcev in kmetov iz Jugoslavije v Belgiji in Franciji* (Our Right: The Paper of Yugoslav Workers and Farmers in Belgium and France), which was declaratively against the dictatorship and fascism in Yugoslavia and advocated for the release of political prisoners; *Glas delavcev in kmetov—radnika i seljaka iz Jugoslavije u Francuskoj i Belgiji* (The Voice of Workers and Farmers from Yugoslavia in France and Belgium, 1926–1928), published by the Yugoslav workers' section in the socialist union CGT (Confédération générale du travail); *Sloga radnika i seljaka* (The Unity of Workers and Farmers, 1929–1936), a communist paper also aimed at the Yugoslav migrants in Belgium and France, and *Borbeni radnik / Borbeni delavec / L'Ouvrier militant* (1928–1930), the body of the Yugoslav inter-union committee CGTU (Confédération générale du travail unitaire).

The third group consisted of Catholic papers. Of the four Slovenian Catholic bulletins in Western Europe, two were only published once, both in France: *Vestnik Jugoslovanske katoliške misije v Merlebachu* (The Newsletter of the Yugoslav

Catholic Mission in Merleback), edited and managed by Slovenians, came out in 1930, while the *Jugoslovenski rudar* (The Yugoslav Miner), the bulletin of the Yugoslav section of the Christian organization Free Union of the Miners of Northern France in Sallaumines, came out a year later (Bajec, 1980, 90–91). One of the more interesting papers from the literary point of view is *Naš zvon* (Our Bell, 1925–1927), published in Bottrop-Boy, one of the centers of Slovenians in Westphalia (Drnovšek, 2012, 151). In Heerlen, the Netherlands, the Slovenian priests published the monthly *Rafael* (1931–1935), the primarily informative bulletin with reports about the current activities of Slovenian societies in Belgium, the Netherlands, France, and Germany. Because of financial constraints, the paper merged in 1935 with *Izseljeniški/Izseljenski vestnik: Glasilo slovenskih izseljencev celega sveta* (Emigrant Newsletter: The Bulletin for Slovenian Emigrants Around the World), published by the Society of St. Raphael in Ljubljana from 1932, into a joint bulletin *Izseljenski vestnik Rafael* (1935–1940) and, from 1938, again as *Izseljenski vestnik*. The bulletin continued to be published by the Society of St. Raphael (Žitnik Serafin, 1999, 89–97).

Like the migrant press in North and South America, the periodicals in Western Europe also contained harsh polemics that reflected the mostly unbridgeable political and ideological differences among the three camps of Slovenian migrants in Western Europe, particularly when it came to the events in the country of origin.

The Yugoslav Territory

In Zagreb, Slovenians published the papers *Novi izseljenik*, *Zagrebški Slovenec* (1932), and *Nova ljudska pravica* (1939). The monthly *Naša misel* (1935), the first Slovenian paper in Sarajevo and the first bulletin of the newly established Delavsko društvo Ivan Cankar was replaced by *Zora: Časopis za družbeno kulturo in prosveto* (Dawn: The Newspaper for Social Culture and Education, 1937–1940). At first a quarterly and then a monthly paper, it published leftist articles, cultural news, and literary texts. In the interwar period, Belgrade Slovenians published *Naša radost* (Our Joy, 1925–1940), *Preporod* (Renaissance, 1928–1940), *Naše ognjišče* (1932), *Slovenski glas Beograda* (1933), the monthlies *Beograjski Slovenec* (1936–1938) and *Slovenski odmevi* (1938–1939), and *Izseljenički svet* and *Bilten*, both around 1939 (Drnovšek, 1992, 273). The short-lived weekly *Slovenski glas Beograda* was replaced by *Slovenski beograjski tednik: Neodvisen list za kulturno, gospodarsko in socijalno reportažo* (1933–1934). Another bulletin worth mentioning is *Istra*, published by the Zveza jugoslovanskih emigrantov iz Julijske krajine (Union of the Yugoslav Emigrants from the Julian March) from

1931 to 1940. It led a severe anti-Italian and anti-fascist campaign on behalf of the Yugoslav minorities in annexed territories in Italy.

Educational Activities

The preservation of the Slovenian language and cultural heritage and their transmission to the descendants of the migrants were among the most emphasized tasks of the youth sections, drama and recitation clubs, children's choirs, etc., organized within the Slovenian migrant organizations and societies, both secular and Church-affiliated. They were also the primary interest of the ethnic press dedicated to youth, particularly in the early phases. However, these activities could not have successfully withstood the assimilation pressure from the receiving societies had they not been supported by supplementary Slovenian schools and Slovenian language and history courses. In organizing these, migrants most often needed help from the country of origin.

Before World War I, the care for the migrants and the upbringing and education of their children in the spirit and values of the "old country" were mainly carried out by the Slovenian Church. If it wanted to realize its religious mission among the migrants, it had to stand for the national values that were often the cornerstone of religious life, particularly in an environment where the majority followed different religious rites. The Austrian government did not take any systematic measures in this area, although some notable individuals, including Janez Evangelist Krek, suggested initiatives and several proposals to help and support migrants.

After World War I, the newly established administrative bodies of the Yugoslav state took over the care for migrants. From 1921 onward, the (General) Emigration Commissariat operated in Zagreb and was tasked with nurturing contacts between migrants around the world, supporting their cultural and other organizations, and promoting the Yugoslav idea. It was renamed Banovina Emigration Bureau in 1929. The Catholic Church was still active in producing informative and educational press for Slovenian migrants in the interwar period. Besides the Škofijski izseljenski sklad (Episcopal Emigration Foundation) of the Ljubljana diocese, there were two societies actively publishing books, including fundraising for their publishing and distribution: Mohorjeva družba (St. Hermagoras Society) and the Society of St. Raphael. The latter organized courses, events, lectures, and different campaigns in the homeland for emigrants and, in 1935, started the initiative for the Izseljenska zbornica (Chamber of Emigration) (Kolar, 2001, 130–132; Vovko, 1990b, 234). From 1929 onward, the series Izseljeniška/Izseljenska knjižnica, a collection of educational books for all those who were either moving abroad or coming back to visit the homeland, was published in Ljubljana.

To preserve the mother tongue in diaspora, the Drava Banovina sent other Slovenian books to migrants, either directly or via the General Emigration Commissariat in Zagreb. Several thousand different books from Slovenian publishers were sent to compatriots around Western Europe and the Americas, so that the Slovenians there could maintain contact with the standard Slovenian language more easily. In the 1930s, enthusiastic Slovenian individuals, organizations, and societies in Western Europe, Canada, and South America thus started setting up migrant libraries. In contrast, this activity had started even earlier in the United States. In the cities with a larger concentration of Slovenians, Yugoslav consulates were established (Drnovšek, 1999a, 53, 2002a, 158, 2012, 69–108, 114–117, 2015, 133–134; Kolar, 2001, 128; Vovko, 1990b, 233–234).

Participants in the two Catholic Slovenian emigration congresses in Ljubljana (1935 and 1937) emphasized the need for permanent priests and teachers to preserve the mother tongue in the diaspora and reinforce national awareness. A series of representatives from many migrant communities participated in the first congress and adopted the guidelines for the "national education of migrant youth." They stipulated that the Yugoslav state should finance and organize Slovenian language instruction for emigrant children and youth. The state should, among other things, publish core textbooks for the diaspora in bilingual versions, that is, in Slovenian and the languages that Slovenian children spoke in the new environments. The state's care for its emigrants should be modeled after some other countries, for example, Italy, Germany, and Poland. The majority of the emigrants' needs were already stated in the Izseljeniška spomenica Jugoslovanske Izseljeniške Katoliške akcije v državah zapadne Evrope (Memorandum of the Yugoslav Emigrants' Catholic Campaign in the Countries of Western Europe), published in 1933 (Izseljeniška spomenica Jugoslovanske Izseljeniške katoliške akcije v državah zapadne Evrope, 1933; Drnovšek, 1998a).

In 1935, the Slovenska izseljenska zbornica (Slovenian Chamber of Emigration) was established. Besides the educational division, it consisted of divisions for statistics, religion, culture, propaganda, social affairs, and economy, the divisions for an emigration archive, and a division for seasonal workers and the diaspora (which in this case meant Slovenians who lived in other parts of Yugoslavia). Based on Drava Banovina's records from 1937, the priests and teachers were, at the time, operating with adequate financial support in all major Slovenian settlements in Europe and South America. One of their principal tasks was strengthening the ties between the emigrants and their homeland and connecting them with representatives from the host countries and Yugoslav diplomatic legations. Occasional visits from the Slovenian bishops contributed to closer ties with the country of origin. In Germany, some German priests learned Slovenian so that they could work among the Slovenian immigrants more efficiently (Drnovšek, 2012, 69–108, 2002a, 158; Pislar Fernandez, 2004, 105–116; Kolar, 2001).

The data from the field was less positive. In France, three priests and two teachers took care of several hundred children, the majority of whom did not have access to learning the Slovenian language. In the north of France, a teacher who taught in seven "schools" scattered around the region had to take extra work to survive. Three priests helped him organize an afternoon school, so that around three hundred children had access to Slovenian instruction. One of the teachers estimated that only a tenth of the parents taught children to read and write in Slovenian. In Alsace and Lorraine, they spoke a mix of French and German; their mother tongue was primarily alien to them. A much better experience awaited a teacher on the German-Belgian border, where 75 percent of the Slovenian children responded to his invitation. At the beginning of the 1930s, only about a third of the Slovenian children in Germany knew the Slovenian language. The first Slovenian "school" was established in 1926 in Gladbeck at the home of a priest who worked among the Slovenians there. The school also had a children's choir that gave performances at home and in other towns. The Ruhr-Saarland region had two teachers from the homeland employed. In Belgium, makeshift Slovenian schools existed in three places. In the Netherlands, Slovenian children did not have an emigrant teacher. The main problem was that the instruction took place on inadequate premises, where they just barely managed to set up provisional classrooms, and not in schools, which is what the Czech and the Polish migrants managed to achieve through bilateral agreements (Drnovšek, 1999a, 48–58, 2001b, 286, 2015, 135–144; Pislar Fernandez, 2004, 105–116).

These conditions reveal why the migrant societies' activity was so important for children—they could participate in different Slovenian celebrations, from Saint Nicholas Day, Mother's Day, Christmas, and Easter to celebrating the association's special days, and state holidays. They performed fairy tales and plays, recited poetry, and played and sang music. On "Emigration Sunday" in Merlebach in 1933, the children's choir that performed a concert of Slovenian songs had 80 children. Immersion vacations for migrant children in the homeland were also important. Drava Banovina, as well as its different religious, educational, charity, and women's organizations and emigrant priests and teachers, helped to organize them; in reality, children residing in Western European countries benefited most because they lived close. In the 1920s and 1930s, several holidays were organized for a few hundred children, primarily from Germany and the

Netherlands. Holidays in Yugoslavia usually lasted for about five weeks. Migrant children were encouraged to correspond with their peers back in the homeland and to share holidays, including exchange stays between migrant and Yugoslav children at home and abroad. The inclusion of migration topics in the textbooks for Slovenian children in Yugoslavia, study programs in the homeland, etc., was also encouraged. In 1939, the increasingly inventive plans for strengthening ties between the old place and the migrant children reached their peak, but the war prevented their realization (Drnovšek, 2012, 93–108, 2015, 139–140).

In the United States, the endeavors to retain the Slovenian language started in the late 1890s. The Slovenian Sokol encouraged the establishment of reading societies. Different organizations had collections of Slovenian books, and they organized cultural and educational events: Cleveland's Slovenian Reading Society, established in 1906, and the Slovensko izobraževalno društvo Vihar (Slovenian Educational Society Vihar) in Krayn, Pennsylvania, established in 1912, are two examples. The fraternal benefit societies organized various types of educational activities. Slovenian parishes in large centers had parish primary schools where the language of instruction was predominantly English, as the state school legislature demanded English instruction in compulsory primary schools. But Slovenian priests and other members of the community—like in parish schools of other immigrant nations—were sometimes allowed to teach religious education, church songs, and recitation in the mother tongue (Valenčič, 2002, 164–166; Plut Pregelj, 1992, 399, 402).

Slovenian nuns started participating in the educational, social, and health activities of Slovenian migrants, particularly the School Sisters of St. Francis of Christ the King from Maribor, who, by running schools and teaching migrant children in Slovenian, significantly contributed to preserving the mother tongue in the diaspora. In 1909, they set up the first community in Kansas City. They were invited to all the Slovenian and Croatian parishes. The number of their institutions increased enough for them to establish their own North American province and a Mother House in Chicago in 1920, which also became the headquarters of the novitiate and the province (which both moved to Lemont in 1962). From 1922 onward, the province had stations in Illinois (Chicago, Kansas City, and Joliet) and Indiana (Cleveland and Gary. At the end of the 1920s, they took care of five Slovenian and eight Croatian parish schools. As early as 1924, 14 Slovenian parishes in the United States had their own school; in addition to the School Sisters from Maribor, Slovenian American women who were members of other American orders taught there (Friš, 1992, 1999, 15–16; Kolar, 2001, 140).

Since 1908, the Maribor School Sisters had their own institution in Alexandria, Egypt, where Slovenian Franciscans helped them in their work. They taught the Slovenian language to the children of migrants and opened a nursery school. They also opened a hostel for young women, the Asylum of Franz Josef (later St. Francis Asylum), which hosted social events and activities. They ran a similar institution in Cairo (Kolar, 2001, 140–141).

As was generally the case in all countries of immigration, immigrant children in Argentina also had problems with their mother tongue, which mostly emerged once the children entered primary school. A child with a poor knowledge of Spanish was banned from speaking Slovenian at home, which reduced the use of the mother tongue in private life to the language of conversation between adults. When trying to pass the knowledge of Slovenian to children and youth, the parents, associations, and the Yugoslav legation in Argentina were again aided by the Slovenian School Sisters from Maribor, who worked there from the early 1930s. Between 1931 and 1937, 41 sisters arrived from Slovenia, but their work was very difficult because the Slovenian community in Argentina was so dispersed. Organizing school activities was further hindered by the ideological differences and conflicts among the migrants, so for the most part, all they managed was organizing occasional courses.

Slovenian schools could only be organized where the concentration of Slovenians was the highest. After the mass immigration in the interwar period, Slovenians in Buenos Aires first lived in modest circumstances in the La Paternal quarter. Only later, once they improved their economic status, did they start buying plots in the Villa Devoto quarter. The initiative for establishing the first Slovenian primary school course in Buenos Aires came from the parents. They rented a room and equipped it for instruction and acquired books from Slovenska matica and Drava Banovina. The Sokol La Paternal helped with the course and set up a school committee for this purpose. In June 1934, the Tabor Emigrant Association was founded, and it organized primary school courses in two-quarters of the Argentinean capital. During these courses, the need for professional teachers soon arose, so the Yugoslav envoy in Buenos Aires, Izidor Cankar, encouraged the establishment of a Slovenian school on the premises of the Tabor Emigrant Association. It operated from the spring of 1937 onward. Upon its establishment, it enrolled 24 children; the School Sisters managed it. The school lost its Slovenian character in 1942 through a government decree that stipulated all schools had to be accessible to children of all nations, so in 1943, it opened its doors to other Argentinean children as well.

In Villa Devoto, the Gospodarsko podporno društvo Naš dom/Asociación Mutual Nuestro Hogar (1935) opened its premises to the Slovenian primary school course, which was successfully organized there until the Argentinean government banned all the societies of prewar immigrants in 1949. In 1938–1939, the School Sisters also ran a Slovenian school in the St. Rafael parish church in Villa Devoto. Slovenian school courses also took place in Rosario in the Argentinean province of Santa Fe (from May 1938), where the Slovensko delavsko društvo Triglav (Slovenian Workers' Association Triglav, established 1930) organized the well-attended instruction of the mother tongue. Another Slovenian school operated in a suburb of Saladillo between 1938 and 1942 (Kolar, 2001, 140–141; Mislej, 1990, 187–189, 2001, 238–239; Jevnikar and Mislej, 1999, 12–13; Horvat, 1997, 39, 1998, 137–139).

Slovenians in other parts of South America were fewer in numbers, and their attempts to organize Slovenian language courses were comparatively rarer. In São Paulo, Brazil, the ORNUS society organized a Slovenian language course for 25 to 30 participants, which lasted for three years. In Montevideo, Uruguay, the priest and cultural worker David Doktorič led a Slovenian language course. Doktorič, who arrived in Argentina in 1937 as a Yugoslav emigration correspondent, was directed to Uruguay by the Yugoslav envoy, Izidor Cankar, so he could animate the cultural life of Slovenians there and become the chronicler of their social life (Jevnikar and Mislej, 1999, 18–20; Mislej, 2001, 244–245).

However, the dominant society and culture impacted the children of migrants in all the host countries. The struggle between the ancestral cultural tradition and the need to belong to the country where they were growing up created a particular intergenerational tension between parents and children that was characteristic for diaspora families. In the families who lived in more contiguous ethnic communities, the parents' wish to pass the Slovenian identity onto their children was aided by the immigrant organizations and institutions from the "old country." In many places, they were partly successful, but the conditions allowing cultural diversity differed greatly in different host countries and different migration periods. Some Slovenian periodicals, societies, organizations, schools, and religious and cultural institutions have been operating continuously since the interwar period or even longer—a testimony to the persistence and/or the adaptability of the original ethnic identity.

Literary Creativity

Slovenian migrants have produced literary works that reflect the group belonging to the new hybrid culture, still largely Slovenian but noticeably expanded and supplemented with elements of the new environment. At the same time, they reveal an individual's intimate experience of the present and often idealized view of the past. A large segment of the literature that Slovenian migrants produced helped reinforce the stereotypes about the idyllic childhood in the "old place" and the stepmotherly foreign countries. Still, many migrant literary texts contributed to destroying such images. Liberation from the more or less "prescribed" themes and traditional literary approaches is more characteristic for the literary production of the descendants of the Slovenian migrants. Nevertheless, we have to stress the diversity of topics, themes, and styles in the literature of Slovenian migrants and their descendants. Today, this literature can present to the reader how these authors saw the migrant experience or migration background, their complex identity questions, group trauma, distress, dilemmas, and the successfully overcome obstacles. These are characteristic of living in a minority immigrant community in an environment that is sometimes more, sometimes also less, accepting of immigrants.

The largest part of the literary production of Slovenian migrants in the periods studied was created and published within the organizational structures of the Slovenian diaspora. At the end of the 19th century and in the first half of the 20th century, Slovenian migrant writers far more often published their works in Slovenian migrant press and annual publications than they did in books, so their literary production mostly consists of individual poems and short fiction. Despite that, a surprising number of writers also tackled drama. Those who were more productive also published poetry collections, longer narrative works, or plays, either in the diaspora or occasionally in the country of origin.

Considering the characteristics typical for migration literature, we can categorize the literary works of Slovenian migrant authors into works with emigration and immigration topics. For the most part, they deal with the reasons for emigration, the process of migration, the early adjustment periods in the new environment, and life in the new homeland in the later stages of adaptation and acculturation. We also see literary renditions of life in the old place—often idealized and with a strong note of nostalgia. The ideological, religious, or politically engaged treatments of the social and intimate human reality in emigration, with an emphasis on the contemporary social and moral challenges for the individual and the society, also appear. In addition, there are topics from the old or the new country not linked to the theme of migration and other literary themes.

The United States of America

The earliest Slovenian migrant literature takes the form of missionary letters in Latin and German. Since 1833, they were also published in Slovenian and, from 1841, most often in *Zgodnja Danica*.[49] Until the end of the 19th century, missionary texts were written and published in Latin, major European languages, Slovenian, and the indigenous languages of the peoples among whom the authors were doing missionary work. Soon, other literary genres appeared, such as travel journals, autobiographies, biographies, polemical texts, religious-educational and popular science pieces, and poetry. Travelogues in Spanish and poetry in Latin were published by Marko Anton Kappus and Ferdinand Košak. Friderik Irenej Baraga was the author of religious-educational, popular science, and linguistic texts (for example, the Ojibwe grammar and the Ojibwe-English dictionary). Baraga's classmate Andrej Bernard Smolnikar published four comprehensive books of memories and polemics in German, the last one also came out in English translation. Popular science was also written, in addition to Baraga, by Franc Šušteršič and Franc Pirc, while poetry in Slovenian was published by Janez Čebulj, Oton Škola, and Franc Pirc, among others. Biographies and Slovenian poetry were also written by Jakob Trobec (Stanonik, 1999, 15–88).

After establishing *Amerikanski Slovenec* the first Slovenian paper, in Chicago in 1891, which was quickly followed by other papers throughout the United States, Slovenians there soon developed literary activity in their mother tongue. This activity branched out into many directions and genres in the 1920s and peaked in the 1930s and 1940s. In the early years of the Slovenian mass migration, original migrant fiction was rare, and we find a substantial amount of anonymously published short fiction. Published works were overwhelmingly reprints of Slovenian authors and translations of world classics (Pogačar, 2014). The initiators of the original literary creation in Slovenian were mainly newspaper editors, teachers, community organizers, and actors and directors from the drama groups that Slovenians in the United States started establishing as early as 1900. The literary production, diverse in genre and theme, created several distinctive literary currents with specific common characteristics that distinguished migrant literature from contemporary literary production in Slovenia. Many migrant literary authors were also translators from English into Slovenian and vice versa. They contributed to intercultural understanding in the new homeland, thus expanding the literary horizons of their compatriots in the United States and the members of the broader American society.

The overviews of the literature by Slovenians in the United States (e.g., Petrič, 1999; Gobetz, 1979) and the introduction studies and biographical data about

49 Among the authors of these letters are, for example, Franc Pirc, Friderik Irenej Baraga, Ignacij Mrak, Oton Škola, Lovrenc Lavtižar, Jakob Trobec, Jožef Buh, and Peter Jeram.

the presented writers in the anthologies of their works (e.g., Petrič, 1982b, 403–406; Gobetz and Donchenko, 1977, ix–xii, 219–236) list around 90 Slovenian American authors who, between 1891 and 1945, at least occasionally published fiction. Their works were closely linked to the ideological orientations of newspapers and magazines that published them. Thus, a relatively large segment of Slovenian literature in the United States was inevitably tendentious, thinly veiled socialist or Catholic conservative propaganda, often with a clear message, moralizing, and didactic. In general, the aesthetic aspects of the literary works are secondary to the communicative role of literature; the newspapers that published such works depended on subscribers and their literary tastes. Of course, some writers also worked independently. Literary historian Jerneja Petrič categorizes the studied authors into Christians, who primarily published in *Amerikanski Slovenec, Baragova pratika, Novi svet*, and *Ave Marija koledar*, proletarian writers and the moderate left, who published in many papers and annual publications, the most important being *Prosveta, Proletarec, Majski glas, Mladinski list, Cankarjev glasnik, Ameriški družinski koledar*, and *Napredek*, and ideologically neutral writers and satirists and other renowned creators (Petrič, 1999).

According to Petrič (1999, 93–97), the prose works of Slovenian immigrants and their descendants in the United States in the first half of the 20th century can be divided into the following groups: (1) realistic socio-critical prose that likes to use the old place as a theme, with a mixture of nostalgia and criticism toward the exploitative social system that forced migrants to go abroad (Anna Praček Krasna, Katka Zupančič, Ivan Molek, Etbin Kristan, Jože Zavertnik, Ivan Jontez, Andrej Kobal, et al.); (2) psychological realism: the peak in quality is a novel in English by Frank Mlakar *He, the Father* (1950) and his short stories (the short story "My Uncle Polde" was included in the anthology *The Best American Short Stories of 1942*). Of the authors who mainly wrote in Slovenian, the most important representatives are Janko Rogelj and Andrej Kobal; (3) socialist-realist or proletarian prose, written by members of the socialist clubs of the Yugoslav Socialist Alliance in all major Slovenian centers in the United States, with the propaganda far too often overpowering literary elements (Ivan Molek, Frank Tauchar, occasionally also Jože Zavertnik, Ivan Jontez, Anton Slabe, Katka Zupančič, and Anna Praček Krasna, and partly Jože Ambrožič and Milan Medvešek); (4) documentary prose (the leading representative is Louis Adamic); (5) autobiographical-memoir prose (Louis Adamic, Vatro Grill, Jurij Trunk, Andrej Kobal, Anna Praček Krasna, Janko Rogelj, Fred Bahovec, etc.); (6) travelogues (Jože Grdina, Jurij Trunk, Anton Terbovec, etc.); (7) young adult prose (Katka Zupančič, Anna Praček Krasna, Andrej Kobal, Mary Jugg Molek, etc.);

(8) humorist prose—from entertaining anecdotes to social satire (represented mostly by Frank Kerže and Ivan Jontez).

Of the many authors who mainly wrote in Slovenian, only a few published one or more works as a book. The most important ones were Ivan Molek, Andrej Kobal, and Anna Praček Krasna. All three—just like the highly prolific Katka Zupančič—were primarily published in the progressive Slovenian American periodicals.

The stylistic scope of Slovenian American poetry ranges from simple quatrains (most poets) to sonnets and wreaths of sonnets (Zvonko A. Novak, Mirko G. Kuhel) and free verse (mostly in the English language poetry of Mary Jugg Molek). Among the poetry styles, the predominant is socio-critical "proletarian" poetry, written by most of the authors mentioned above (most prosaists mentioned under 1 and 3). Parallel to that, there is expressive and love lyricism, stylistically ranging from simple and descriptive to associatively richer metaphors (the later phases of Anna Praček Krasna's work). Many Slovenian American poets wrote patriotic poems and laments. Ivan Bukovinski, who also published humorous poems, and Mary Jugg Molek wrote reflexive lyricism. The primary authors of socio-critical (including satirical) poetry—most published in the 1930s—are Ivan Molek, Katka Zupančič, and Anna Praček Krasna. Didactic poetry was published by all those authors of young adult prose who were also poets—they were also mostly the authors of humorous poetry for children. Katka Zupančič, as the leading representative of young adult prose, poetry, and drama, published extensively in all the listed periodicals. Aleksander Urankar and Ivan Zupan published their poems primarily in the Slovenian American Catholic press of the time (Petrič, 1999).

Popular Slovenian poets in the United States were Ivan Zorman and Jack Tomšič; the latter only published his poems in Slovenian American papers between 1923 and 1945. Zorman, who self-published five collections of poetry between 1919 and 1938 that were almost "folk" simple in form, yet contained content particularly attractive to immigrants, was undoubtedly the most popular poet. Most of his poems were set to music by renowned Slovenian composers in the diaspora and in the home country. Likewise, many other poets' poems were set to music, as singing Slovenian songs was the most popular way to preserve the Slovenian language throughout the diaspora.

We also see parallels with the socio-critical and "proletarian" prose in poetry and drama. The primary representatives of this current in drama were Etbin Kristan and Ivan Molek. For both, this part of their literary opus is represented by topical drama emphasizing the conflict between capitalists and proletarians. Still, their characterization is black-and-white, and the plays have several other structural weaknesses. Molek and Kristan are also the leading representatives of comedy, satire, and farce. Katka Zupančič was the main author of young adult drama, while Frank Česen and Andrej Kobal wrote popular "folk" comedies. The latter is among the foremost Slovenian playwrights in the United States. His plays were staged in New York, Chicago, and elsewhere, most of them dealing with topics of migration. The main representative of historical drama is Joško Oven, while the plays by Rev. Father Kazimir Zakrajšek emphasize Christian ethics (Petrič, 1999).

South America

In South America, particularly in Argentina, the literary creation of the Slovenian migrant authors began quite a bit later than in the United States. Most literary contributions can be found in the periodicals *Slovenski tednik*, *Duhovno življenje*, *Novi list*, and *Njiva*—all were established in Argentina either at the end of the 1920s or the 1930s. In addition to the Slovenian authors from Argentina, Slovenians from other countries in South America published in these periodicals. The same periodicals also published the three most important authors from this time: David Doktorič from Uruguay and two artists from Argentina—Gvidon Jug, the author of the first Slovenian book published in South America, *Izseljenec: Vtisi mojega potovanja v Argentino* (An Emigrant: Impressions from My Journey to Argentina, 1931), and poet, writer, editor, educator, amateur actor, and director Ivan (Jan) Kacin. Poet and writer Bogomil Trampuž Bratina from Ecuador, who left most of his opus in manuscript form, published a couple of poems in *Duhovno življenje* in addition to a book in Spanish. Other authors, including many women writers, from South America published most poems and short fiction in the mentioned Slovenian publications in Argentina and occasionally in the *Izseljenski vestnik Rafael*. Among the Slovenian women poets from Argentina, whose works were published in migrant publications on both sides of the Atlantic, there are, for example, Justina and Anica Jeram and Pepca Furlan. Although their works were popular among readers and remain interesting today as historical documents, none of the poets showed serious literary ambition (Mislej, 1999; Tavčar, 1999).

Europe

Slovenian migrant authors in Europe followed similar paths. They published their writings in the diaspora press, which displayed three distinct ideological orientations: pro-Yugoslav papers; socialist or communist periodicals; and Catholic periodicals. Of all three groups, the best preserved are Catholic papers. From the point of view of literary contributions, the most interesting is the monthly *Naš zvon* from Bottrop-Boy in Westphalia, edited by the priest, editor, and writer Janez Evangelist Kalan. In the first two years, we find anonymously published original short fiction, mostly fictional contributions by the editor himself, with characteristic moralistic emphases. Kalan also published his editorials and *Letters from Westphalia* in *Naš zvon*; in both cases, his pieces are an essayist combination of fictional and journalistic elements, written in the author's characteristic style. The thematic is diverse, from religious, political, ethical, and social to different cultural and other current themes, such as the concern for the preservation of the Slovenian language in diaspora. Poetry, mostly anonymous, was rare, and some of it was also reprinted from the *Glasilo K. S. K. jednote* from the United States. The Catholic monthly *Rafael*, published in the Netherlands from 1931 to 1935 exclusively in Slovenian and aimed at the compatriots in Western Europe, also brought many fiction pieces. Most of them are original migrant pieces. Some are reprints from the homeland and the Catholic Slovenian magazines in the United States and Argentina, such as *Ave Maria* and *Duhovno življenje*. The same holds true for all the years of this monthly after it merged with the Ljubljana-based *Izseljenski vestnik* (Žitnik Serafin, 1999, 77–97).

Throughout the European diaspora, the texts that were of a better literary and aesthetic quality were mainly written by authors who made a name for themselves internationally in other fields and not in original fiction writing: historian and classical archeologist Vojeslav Molè, literary historian and critic Janko Lavrin, and painter, graphic artist, and photographer Veno Pilon. They established themselves in their fields before 1945, but their literary memoirs were published as books after World War II (Žitnik Serafin, 1999, 98–116). All three also significantly contributed to the Slovenian breakthrough into the European cultural consciousness and vice versa as translators and literary and cultural mediators.

3 Migration Stories

Migrants' own words and interpretations show the diversity of migration trajectories and processes by placing them into social, economic, and political circumstances. Through their subjective perceptions and individual interpretations, the participants of the migration process reveal a kaleidoscope of individual, dynamic, and complex decisions and everyday negotiations. Their narratives speak of the interpersonal connections and emotional dynamics of departure and return while existing in transnational and translocal emotional worlds. The selected stories defined by gender, class, religious, ethnic, and local differences show that migrant experiences are never homogenous. In fact, these stories highlight the differences among the Slovenian migrants regarding the localities from which they migrated and their circumstances in the destination countries. The subjective experiences of those who left and those who stayed indicate the establishment of translocal bonds and transnational families and childhoods. Likewise, the subjective views of those involved in the migration process show the continuities and discontinuities in retaining and changing the national, ethnic, and hybrid identities, traditions, heritage, and ideological and political structures of migrant communities. The purpose of presenting stories is to acknowledge the multi-dimensional circumstances of moves and experiences of individuals. Each story, in its own way, uniquely and unrepeatably paints a picture of a subjective experience of the migration process.

The following life stories illustrate the different periods of Slovenian migrations, the directions of the migration routes, migrants' motives, levels of social participation in the receiving countries, and whether migrations were voluntary or forced. They demonstrate the importance of using narrative sources that disclose the personal, intimate, and emotional experiences of the actors in migration processes. These life stories are based on biographical and autobiographical texts published in newspapers, magazines, and books or saved in archives; testimonies, narratives, and personal memories or memories of descendants; and personal papers and documents, such as photos, diaries, letters, and postcards. The narratives introduce those who leave—always simultaneously emigrants and immigrants—and those who stay as agents and interpreters of their experiences and expand the understanding of migration beyond economic, political, or social motives: "As agents of their own lives, men, women, and their children look not only for material security but for emotional-spiritual-intellectual security as well" (Hoerder, 2002, 15).

The first story challenges specific, firmly anchored images about Slovenian migration in the period from the mid-19th century until the end of World War II: that it was primarily men who were on the move, that the leading destination was the United States of America, that migrations were one-time moves to a different place, that women stayed at home and that migrations were linked to difficult circumstances, emotional distress, and feelings of loss. In 1896, a young woman, Helena Turk, first emigrated from Inner Carniola/Notranjska to Brazil, returned home and went to Egypt, returned again and emigrated to Germany, and from there moved to her final destination in the United States. She recounted her story of multiple moves and returns, rich experiences, curiosity, and exuberance in an autobiographical newspaper account.

The protagonist of the second story is another young woman who migrated on her own to the United States from her native Lower Styria/Spodnja Štajerska/Untersteiermark in 1906, planning to earn enough money to pay for her education at home. Marija Černe/Marie Prisland—like the vast majority of migrants in this migration period—intended to return home, but she never did. The story of the accomplishments of an orphan who, soon after arriving in the United States, fully committed herself to work in the Slovenian community, set up the first organization of Slovenian women immigrants, and left an indelible mark in both her homelands, has been collected from her extensive writing for magazines and her autobiography.

The greatest Slovenian American writer Lojze Adamič/Louis Adamic is another person who left a permanent mark in both homelands. He, too, set off for America as a young man, on his own, from his native Lower Carniola/Dolenjska in 1912, his motives being adventure and youthful exploration. His extensive body of work details the exceptional process of a creative, successful, and politically active migrant who simultaneously presents a construction of a hybrid identity on the intimate level and, on the public level, reveals the political division and ideological conflict of Slovenian migrant communities.

The political context is also the key to the story of Marija Ipavec from the Soča/Isonzo River Valley, who, as a young mother during World War I, experienced exile in Italy. Like many others caught up in the world's carnage, she most likely kept her intimate diary to make sense of the horrors of war events and share them with someone. The record that her descendants only recently donated to the Goriški muzej (Goriška Regional Museum) in Nova Gorica is scant and incomplete. Nevertheless, it provides invaluable insight into an individual's subjective experience of the mass suffering of refugees from the western part of the Slovenian ethnic territory and transcends the uniform understanding of the dramatic experience of being a wartime refugee.

The epistolary love story from the interwar period provides the most detailed insight into the emotional dimension of the migration process. The correspondence of a young couple reveals the complexity of keeping in touch during the years of separation, of strengthening emotional bonds, loyalty, belonging, and at the same time, the turmoil of doubts and worries defining the lives of those who leave and those who stay. In their words, Franc Peric and Felicita Koglot from the Goriška village of Bilje speak to us of love, duty, hope, and perseverance. In intimate letters traveling between Egypt and Slovenia in the 1920s, they also sketch the outlines of the changes that migration brings to family relationships, gender roles, and the traditional structures of the community of origin.

The final two stories portray two individuals' memories of the collective experience of mass forced migrations during World War II. They reveal the historical and political reasons for the ostensibly voluntary and forced migrations in circumstances where the racist war policies of the German Reich displaced millions of people and forced them into slave labor. The oral testimony by Terezija Černelič and the written autobiography by John Tschinkel help us to follow the different yet similar fates of two families, one that returns home and one that loses its home forever. Their experiences of exile in their youth more than 70 years ago connect past to present and warn about the common fate of humanity in which the routes of emigration and immigration—upon which, at certain moments, a community, a nation, an individual, and a family can find themselves—are forever interchangeable.

Multiple Mobility to Brazil, Egypt, Germany, and the United States of America: Helena Turk

As a young woman in 1896, Helena Turk traveled to Brazil—by herself. She, too, "caught the mania" of that time to go to the advertised promised land to earn money and be able to help her parents and herself. Her secret wish was to see the "beautiful world" and distant exotic lands. After staying in Brazil for five years and returning home, Helena Turk migrated again. She worked in Egypt for a few years, returned home, married, migrated to Germany, and finally went to the United States. She published the story of her multiple moves to different parts of the world in 10 short installments in the newspaper *Amerikanski Slovenec*. She started publishing her feuilleton "My Journey in the World" on September 23, 1925, beginning her story in 1896 when she first ventured into the unknown world. Her experience is important for discovering the forms, patterns, and modes of migration from the Slovenian ethnic territory, for seeing the role of gender in migration decisions, and for presenting a young female migrant's personal experience that, even a century later, sounds contemporary and modern.

THE OLDEST AND MOST POPULAR SLOVENIAN NEWSPAPER IN UNITED STATES OF AMERICA

AMERIKANSKI SLOVENEC

NAJSTAREJŠI IN NAJBOLJ PRILJUBLJENI SLOVENSKI LIST V ZDRUŽENIH DRŽAVAH AMERIŠKIH

PRVI SLOVENSKI LIST V AMERIKI.

GESLO: Za vero in narod — za pravico in resnico — od boja do zmage!

GLASILO SLOV. KATOL. DELAVSTVA V AMERIKI IN URADNO GLASILO DRUŽBE SV. DRUŽINE V JOLIETU, — S. P. DRUŽBE SV. MOHORJA V CHICAGI IN ZAPADNE SLOVANSKE ZVEZE V DENVER, COLORADO.

ŠTEV. (No.) 149. CHICAGO, ILL., SREDA, 23. SEPTEMBRA — WEDNESDAY, SEPTEMBER 23, 1925. LETNIK XXXIV.

Slovenski katoliški shod v So. Chicago

TAKO SIJAJNE SLOVENSKE MANIFESTACIJE SE NI VIDELA SO. CHICAGO. — TA VELIČASTNA KATOL. MANIFESTACIJA JE POKAZALA AMERIKANCEM, DA SMO SLOVENCI POŠTEN IN VEREN NAROD.

PETNAJST ITALIJANOV ARETIRANIH NA MEJI

KRIŽEM SVETA.

MEHIKANSKI BANDITI USMRTILI TRI OSEBE.

Iz Jugoslavije.

VELIČASTEN SHOD BIVŠIH BOJEVNIKOV NA BREZJAH. — VDELEŽBA OGROMNA, NAVZOČIH JE BILO 12,000. — DRUGE ZANIMIVE VESTI.

DRUSI PORAŽENI.

UMOR IN SAMOMOR.

TUDI NA ŠVEDSKEM NISO TLA ZA KOMUNISTE.

SIROVOST PIJANCA.

SMRTNA NESREČA; DVE ŽENSKI MRTVI.

COOLIDGE PROGLASIL TEDEN ZA PREPREČENJE POŽAROV.

SPLOVITEV PARNIKA, NOSILCA LETAL.

ZAŠITO SRCE.

DROBNE VESTI IZ STARE DOMOVINE.

IZ TRBOVELJ.

SIRITE A. SLOVENCA!

DENARNE POŠILJATVE

Picture 3.1: The front page of *Amerikanski Slovenec*, in which Helena Turk's feuilleton was published. Courtesy of: Digital Library of Slovenia – dLib.si.

Picture 3.2: Postcard of the *SS Rotterdam*. Courtesy of: Marjan Drnovšek.

For a long time, in understanding and studying the historical and contemporary aspects of migrations, women were considered merely passive companions of migrants or those who stayed at home. In the research on labor and the economy of migrations, they have, together with children, often been presented as dependants, dissociated from the sphere of paid work (Hoerder, 2015, 119–120). Only in the last three decades have studies begun to show that they were active economic migrants even in the past (Donato and Gabaccia, 2015; Hahn, 2015; Hoerder et al., 2015)—as they clearly are today when we talk about the feminization of migrations. Sheila Rowbotham explains why historical, statistical, and demographical overviews often skip the paid work that women, including migrants, have chiefly done:

> One reason for the lack of visibility has been the nature of female migrants' occupations. In many cases they went into domestic services or served as wet nurses—activities which have never been regarded within the prevailing definitions of "work" or the "economy" and have thus defied statistical reckoning. This is a gendered obscurity in a double sense. The women leave no traces because they are female and because the framework of who is to be seen has been biased towards the male. Another reason for the obscurity of the poor migrant woman, however, crosses gender. They belong to that great host of the ignoble who are not "known" to history (2001, xvi).

From the available data for the time before 1940, Marjan Drnovšek estimates that the share of women who emigrated from the Slovenian ethnic territory was about 35 percent of all migrants. This estimate changes drastically depending on the period and, particularly, on the region or administrative unit. In 1882, for example, the share of women among the emigrants from Carniola was 21.4 percent, but in the Kočevje region of Carniola, it was 31.3 percent. Slovenian women in Egypt in 1897 represented 96.3 percent of all Slovenian migrants, and these numbers remained stable until 1940 when the migration of women to Egypt stopped. Among the seasonal migrants from Prekmurje in the interwar period, there were 29 percent of women in 1929, 36 percent in 1930, and 40 percent in 1931. The German census of 1900 lists the share of women at almost 30 percent among those who selected Slovenian as their first language; in 1939, 45.7 percent of Slovenian migrants to Germany were women (Drnovšek, 2003a, 31–33). These estimates are related to different phenomena and phases of different migration patterns in different political and administrative frameworks.

Although women were less visible in Slovenian migrations, we know that in the Slovenian ethnic territory, there were markedly female migration routes along which women moved for employment and were often family breadwinners. Three distinctive types of work and their protagonists stand out before 1940: maids and care workers who left from the entire Slovenian ethnic territory to work in urban centers nearby or further away, such as Zagreb, Belgrade, Skopje, Rome, Udine, Milan, or to the faraway countries such as the United States, Belgium, Germany, France, and England (Kalc, 2004; Drnovšek, 2001a; Mlekuž, 2004); skilled artisans and traders, who are among the least researched and known of all women migrants—an example of such migrants were the straw-hat makers from Domžale, Mengeš, and the surrounding villages, who migrated seasonally to different parts of the Habsburg Monarchy and to cities in the United States (Roškar, 2014), and aleksandrinke, the most widely known and also represented in literature and film (see above; see also Makuc, 1994; Koprivec, 2013; Zorn, 2012; Milharčič Hladnik, 2015b).

The autobiographic description of Helena Turk's multiple mobilities confirms Annemarie Steidl's analysis of migration in the late Habsburg Empire when she claims that the majority of spatial movements were temporary rather than permanent. Even more, the story reveals diverse mutual interactions of internal, overseas, and continental migration patterns in which human mobility occurs within individual socioeconomic contexts. It is an example of Central Europeans who "participated in various spatial mobility patterns over diverse timescales, including medium-distance, long-distance, and overseas migration on a seasonal, temporary, or permanent basis. These patterns linked people from

Habsburg-ruled territories to different regions and provinces within the empire, to other European countries, and to destinations in the Americas as well as other continents" (Steidl, 2021, 170).

Brazil

In 1896, in a village in Inner Carniola, the young Helena Turk observed the neighbor's family who, along with many others, was preparing to migrate to Brazil. As she says herself at the beginning of her feuilleton, Brazil at that time was considered the new promised land:

> Their neighbors, too, had a diligent little daughter called Helenica. She, too, caught the mania to go to Brazil, earn money and help her parents and herself. She introduced this secret thought to her parents, who at first objected but later thought she'd go with familiar people from home, let her go, if the girl makes some decent money, she'd be able to help us there better than at home. [...] The ticket to Brazil wasn't expensive at the time. 15 guldens to Genoa and beyond paid for one person. Simon's daughter Helenica was so eager to go, to see this beautiful world and, of course, make a lot of money out there, and then return. That was what her heart decided (Turk, September 29, 1925, 2).

Migration to Brazil is a specific period in the mass migrations of the Slovenian population. At the end of the 19th century, in the form of a true "Brazilian fever," it mainly affected the countryside of Carniola, as well as Goriška and Istria. The impoverished rural population was the target of an intense propaganda campaign by the Brazilian state and from the Italian shipping companies: following the abolition of slavery, Brazil needed a workforce, and the migration mostly went through Genoa (Kalc, 2001, 57–58; see also the section Other Translatlantic Destinations). Aleksej Kalc describes the characteristics of the short but intense emigration to Brazil that lasted only a couple of years:

> Unlike the migration to North America, which was more about individuals, the South American migration was primarily family-style. It was part of colonization programs that, in the Brazilian case, often offered prepaid tickets for the transatlantic voyage and initial aid in setting up farms before the first crop. This support, as well as the possibility of gradually purchasing the land, was alluring bait for rural families with insufficient or no farmland who, due to financial distress, otherwise could not afford the decision to emigrate (2002, 19–20).

Kalc mentions all the elements that Helena Turk also describes in her feuilleton: many families were setting off on a journey, including the neighbors she wished to join. Her parents let her go because she was traveling with people she knew. The ticket was obviously subsidized because, in her own words, it did not cost much. Everybody counted on making good money and creating a better life.

Helena also wanted to see this "beautiful" world and help her parents with her earnings. In the continuation, Helena describes the voyage to Brazil and mentions other single women she socialized with on the journey and later in Brazil. An article in *Slovenski gospodar* from March 9, 1893, confirms that many single women migrated to Brazil. The unsigned editorial, which mentions migration from around Maribor, speaks harshly against the people who had decided to move to Brazil:

> Has this Brazilian fever turned your hearts into stone? Such contempt for oneself and one's homeland, such stubbornness, such horrific foolishness and defiance to oppose the most benevolent teachings, just how the Americans were, nobody believes that one could only stare and be silent. Who are these people? For the most part, tenants working in vineyards, *dvoraki*,[50] and unsatisfied workers, unlearnt, greedy, and without education along with their children; God knows why so many young girls joined them! (Dopisi, 1893, 76).

At the beginning of her feuilleton, Helena Turk answers this rhetorical question. In the time of mass migrations, the reasons for young women leaving, just like the reasons of other migrants, stemmed from the desire to improve their social and economic position and from the awareness—which spread among the people like a "mania"—that there are worlds where these desires could be realized. In the five years she spent in this "exotic country," she moved often and changed employers even more often. In the beginning, she worked on a coffee plantation only for a couple of days. She was then employed as a maid or a care worker in households of wealthy landowners and merchants and, for a while, in a restaurant. She had little to say about her work but often stressed her good relationships with her employers. She did not change jobs because she had issues with her employers but because, for example, she wished to live in less remote and isolated places. So, she left her job in a lonely place and moved to a big city, where she found other Slovenian migrants:

> Helenica almost squealed with joy when she found herself once more among her fellow Slovenians. She felt like she was home again. Here, she again heard the beautiful Slovenian language of her country people, she visited them often, and what is more pleasant abroad than a meeting with our compatriots. Soon, she became acquainted with two Slovenian girls in a restaurant where they worked as chambermaids. They were two sisters, Marija and Antonija Kastelic. They advised Helenica to stay with them. When they offered her a rather high salary, 40 milréis per month, she stayed. She liked the life in

50 *Dvoraki* is an old expression that probably describes the peasants working for aristocratic estates in semi-feudal conditions.

> the city, particularly as there were many Slovenians there, and she'd never before been to such a beautiful and big city (Turk, September 30, 1925, 2).

She also wrote that she only worked briefly in the restaurant because it was not a suitable work environment for a young girl. Unlike many bad experiences of the Slovenian migrants in Brazil, Helena Turk predominantly listed positive experiences.[51] Nevertheless, she did not forget to describe how she had doubts about the correctness of her decision in the first days after arrival. "I should have believed the wise men who told us that foreign lands are not quite as golden as the gilded tongues of cunning shipping agents would have us believe. But it is too late now" (Turk, September 24, 1925, 2). She described her acquaintances and friends, single young women like her, who traveled around Brazil "seeking their fortune," and the numerous Slovenian families and communities with whom she socialized. Her writing reveals a dense network of bonds and support that the young single women established in the vast country across enormous distances. She often mentioned socializing with other Slovenian women; she was surprised at how many there were. A few years later, she moved to an even bigger city, the capital:

> Once she arrived in São Paolo, she was pretty astonished at the big city, for it was the biggest city to which Helenica had ever been. Although she had the address for Marija Čelan, she did have difficulties finding her friend. After a long search, she found her. Luck did not abandon Helenica. She got a job in a big, rich house where Marija Čelan and two other girls worked. Now there were four girls together, all Slovenian, and that was something extraordinary in Brazil, far from the beautiful Slovenian homeland (Turk, October 1, 1925, 2).

She emphasized that she learned Portuguese because she wanted to communicate with the local population, even though her employers most often spoke other European languages, primarily Italian. From the following description, we can see, on one side, how self-confident she was because of her ethnic identity. On the other side, she had no second thoughts about integrating into the new environment. As she wrote, this integration also eased her homesickness and the lack of contact with other Slovenian emigrants: "Particularly, because she started practicing the Portuguese language. The better she could speak it, the

---

51 Andrej Kalc (1889) described an example of a bad experience that made a group of migrants decide to immediately return home after arriving in the depths of Brazil. The story was published in *Edinost*, the paper of the Slovenian political society in Trieste, under the headline "From Brazil—to Trieste" in five installments between January 16 and 30, 1889.

easier her communication with the people there became. She found it lovely, particularly when they called her *sinjorina* (miss), something she'd never been at home" (Turk, September 30, 1925, 2).

This brief description of the mainstream attitude toward a migrant helps us to discern what many autobiographical documents by Slovenian female migrants—in the later period of migration—also confirm. From Brazil and the United States to Egypt, female migrants report how the respectful and polite treatment of women differed significantly from the attitude toward them in their homeland. The Slovenian ethnic territory was ruled by rigid patriarchal relationships that made a returnee's reintegration into the home environment very difficult. The migrants returning to their hometowns came back with rich work experience and an experience of more equal relationships between genders (Drnovšek, 2003c; Škrlj, 2009; Koprivec, 2013). As Helena Turk writes several times, female migrants were also interested in learning about the "New World," which—to her—already when she decided to migrate was "this beautiful world." With her curiosity, learning new languages, self-confidence, and communicativeness, Helena is an example of a Slovenian woman migrant that we can also observe in some recollections by the aleksandrinke, straw-hat makers, and other Slovenian women migrants.[52]

Helena Turk stressed that her masters treated and paid her well. She never mentioned how much money she was sending to her parents, but it is clear from the narrative that she was very attached to them, particularly her mother. It was because of her mother and her advice that she returned home. When a young Slovenian man she was in love with proposed to her in Brazil, she asked her mother or parents for advice. Three months later, her mother's letter arrived in which she advised against the wedding and recommended she return home. Helena followed her advice and returned to her small village in Notranjska. But, as she describes in the continuation, not for long:

> She stayed home for two years, but a desire was born once more in her heart, just like the proverb says about travelers: When migratory birds bid farewell and leave for warmer climates, the traveler bids farewell and goes elsewhere. [...] Once more, she embarked on a journey, this time to Egypt. It was there that Slovenian girls went a lot at the time. She and her neighbor Marička decided to travel to the Holy Land. They read in the *Mohorjeva praktika* that Slovenian girls earn well in Egypt. And there were so many

---

52 Some testimonies and cases of similar experiences of Slovenian female migrants appear in the book *Going Places: Slovenian Women's Stories on Migration* (Milharčič Hladnik and Mlekuž, 2014).

there already, in Alexandria. Helena forgot about Brazil and decided to go to Egypt (Turk, October 2, 1925, 2).

Egypt

Seven years after her first departure from home, Helena Turk describes a new attractive destination for those young Slovenian women who wanted to work and use their earnings to help their primary families or wanted to start their own. In the second half of the 19th century, Egypt had become a promised land for women and, unlike Brazil, remained so until World War II. Because of the mass emigration of girls and women, predominantly from villages in Goriška and Primorska, different Church organizations were soon set up in Egyptian cities (Milharčič Hladnik, 2015b). In Cairo and Alexandria, the Slovenian Church opened institutions that received Slovenian migrants, helped them search for employment, provided for their religious needs, and exercised control over them. The Asylum of Franz Josef in Alexandria was among these institutions; it was where Helena and other women got their first lodgings and assistance when searching for work. After a couple of days, Helena got a job with a wealthy family, but because she could not cook, she had to wait for the next offer. It came from a wealthy Russian family, while her friend was placed with a French one. We learn little about Helena's job, but the following description hints that she had enough time to converse with the children's teacher:

> The teacher of the masters' only two daughters was a Frenchwoman who talked to Helena a lot. Still, because Helena didn't understand anything, they mostly communicated at the beginning using fingers and pointing. Helena would have loved to learn French because the family only spoke French, some Russian, and very little Italian (Turk, October 7, 1925, 2).

Helena Turk wrote less about her work and more about the environment, habits, social interactions, and her free time:

> In Alexandria, they also had a Slovenian reading room so that the then-Austrian migrants were relatively well provided for and didn't feel too abandoned. The reading room manager was a Slovenian priest who always made sure that Slovenians had enough good Slovenian reading material. Slovenian girls, particularly, were well organized (Turk, October 7, 1925, 2).

The feuilleton does not tell us how long Helena stayed in Alexandria. Helena returned, without hesitation, when her aunt wrote to her from the home village to come home because she was sick and wished to bequeath her property to her. Before returning, she visited the sights in Cairo, just as she had visited all the sights in Brazil and those on the way there, for example, in Genoa, Milan, and

Trieste, during her first migration journey. She describes them in detail in her feuilleton. Interestingly, Helena's other aunt (who, unfortunately, is only mentioned in passing) lived in Cairo. Since family or neighborhood networks are essential in migrations, we can deduce that her aunt was also an aleksandrinka (Milharčič Hladnik, 2015b).

Soon after her return, Helena married a miner and moved with him to Germany, where miners made good money at the time. As the historical part of this study mentions, migration to Germany was present in all Slovenian regions and increasing—on the one hand because of the general crisis within the agrarian economy and on the other because of the specific circumstances in mining, which was expanding at the time and needed a mass workforce. Despite that, Helena and her husband did not stay in Germany for long because, as she says, America beckoned as the real promised land:

> Helena and her husband stayed in Germany for a couple of years. Then they intended to migrate to America, which at the time was celebrated as the richest and truly real promised land. But it was always Helena's destiny to travel. Only here, in America, did she set up a permanent home. She often still remembers and, in her happy memories, flies back to those places that she traveled to and lived in her youth. She would love to see them again but probably won't because she now has a family. And thus ends the story of my travels around the world (Turk, October 8, 1925, 2).

Only here does Helena enter the traditional interpretation of migrations in which men are agents and women their passive companions. However, it is also possible that the young couple decided to migrate to the United States after joint deliberation. The life story of Helena Turk and her multiple migrations proves that, even in the past, women were the agents of migration, and many set off on a journey on their own and successfully found work abroad. It shows migrations as a system: multiple moves to different places in which the places of emigration and immigration are linked into a complex structure of migration experience and identities (Harzig, 2001, 16). It also shows how female migration encouraged the desire for emancipation, independence, freedom, and self-reliance. The very structure of her feuilleton proves the great importance she gave to her own migration experience as she allocated the largest part of her autobiographical writing to it and wanted to share it with other migrants. She describes the return from Egypt, marriage, the move to Germany, and then the move to the United States in the last brief installment, from which the following segment draws.

We do not learn from the feuilleton whether Helena was still employed in Germany and the United States. She claims that in Germany, miners earned well, yet we can still deduce that life was not quite comfortable, or they would not have decided to leave for the United States. A short description of their decision confirms what an important factor the circumstances in the labor market were for the move, along with the differences in pay for the same work and the images of the "true" promised lands. Perhaps she was not employed in the United States. Perhaps she only worked occasionally, but it is probable that she, like many wives of Slovenian migrants, took care of tenants—boarders who lived with the family and worked three shifts. That hard work lasted from 4 in the morning until midnight, consisting of care for full-board (including the food that workers packed for work) and laundry (clothes and linens) for the boarders and the family. Like many other forms of informal women's work, this work was paid and an important part of the family income (Planinsek Odorizzi, 1978, 31–33).

Helena Turk's story about migration routes and the work that she did in different parts of the world deconstructs the prevalent stereotype that women always stayed at home, did unpaid housework, and, in the case of a move, only ventured as far as the next big town, or joined their migrating fathers or husbands. Her subjective interpretation of the migration experience questions the prevailing image of the female migrant that was reinforced for centuries by the political, Church, ideological, literary, and media discourses about women who dared to migrate on their own and paid for their boldness with moral collapse (Hoerder, 2015, 120–121). Helena Turk's feuilleton describes a determined and independent young woman who searches and works in different unknown promised lands and moves within the migrant networks woven by communities, kin, and emotions. Her experience is a good illustration of Christiane Harzig's findings which she recorded after many years of historical research on global migrations:

> As migrants of past and present show, migration works within transworld migration systems which are formed not only by global capitalism but by well-informed global players—the migrants themselves— performing multiple gendered functions, deciding about their moves in transnational communities, considering their assets on the global labour market and relying on networks formed by the family, kin, friendship and neighbourhood/village. Within these transnational processes women are often decisive agents pursuing their own agenda at the local and global level, negotiating gendered strategies and options (2001, 25).

Helena Turk's autobiographical account is a small but significant contribution to understanding women's migrations and subjective perceptions of the migration processes. It illustrates the migration process as inherently rhizomatic, following Gilles Deleuze's description of rhizomes, which generate connections, heterogeneity, multiplicity, multiple entry points, and routes rather than roots (Deleuze, 1993). In the rhizomatic complexity of migrations, Helena Turk's story depicts a map of multiple journeys, mobilities, movements, connections, and destinations. The migration and mobility patterns of herself and her neighbors and friends are entangled, interrelated, and interwoven in similar processes of the same complex and dynamic spatial mobility. She describes multiple migrations to different continents and European countries as something ordinary and reminds us of the free movement of people over a century ago, which we have not experienced in a long time.

Organizing and Connecting Women Migrants in the Service of Two Homelands: Marie Prisland

Marie Prisland disembarked on Ellis Island in the United States in May 1906 as Marija Černe. The 15-year-old girl from the village of Rečica in the Savinja Valley had traveled on her own, with nothing but an address in Sheboygan, Wisconsin, sent to her by her neighbors who had emigrated years before. When she was three years old, her parents first went to Trieste to search for work, and two years later emigrated to Brazil and left her with her grandmother in Rečica. They were part of the migrant "Brazilian fever" that started in the 1880s and lasted until the end of the century. The migrants were primarily poor peasants from Carniola and the Austrian Littoral. Their migration routes took them "through a genuinely rough—compared to other migration destinations probably the roughest—collective life ordeal" (Kalc, 2001, 59). A year after arriving in Brazil, Prisland's mother died of yellow fever; her father remarried and never returned or took care of his daughter. Marija Černe, who wanted to become a teacher, went to the United States to earn money to pay for her tuition. Along with the address in Sheboygan, she took with her an independent mind, courage, determination, and a strong sense of national identity. Her life story can help us navigate the organizational structure of Slovenian communities in the United States and the richness of different traditions, particularly concerning interpersonal relationships, especially those between men and women. These encounters led her to radically rethink gender differences once she became one of the most important activists and organizers of the Slovenian communities in the United States (Milharčič Hladnik, 2007).

In Sheboygan, Wisconsin, Marija Černe found a lively Slovenian community. The community of several thousand Slovenian immigrants boasted active choral and theatre groups, brass bands, Slovenian shops, butcher shops, and taverns. In 1909, the community decided to build a church and establish a parish. The church was completed two years later, and in 1913, three new church bells, cast in Ljubljana, arrived from Slovenia. The church in tiny Sheboygan was the only one in all of the United States with bells from the old place. The community also raised money for the parish school, inaugurated in 1923, and a recreational center (Prisland, 1966, 182–190). In the first decades of the 20th century, they established the women's branches of the main benefit societies. The Slovenians in Sheboygan were a typical example of a highly organized community as benefit societies, a national home, an ethnic newspaper, a parish, Slovenian businesses, cultural institutions, publishers, and schools operated there.

The Slovenian Women's Union of America

Marie Prisland introduced a series of novelties into the operations of Slovenian communities in the United States. The most daring and far-reaching was establishing the first independent women's organization, the Slovenian Women's Union of America (SWUA), in 1926. How did the young Marie foster this idea for two decades after coming to an unknown country with an unknown language? She took up many activities in the bustling Slovenian community and thus came into contact with Slovenian organizations. The council of benefit societies first heard of Marie Prisland in the early 1920s when she was elected to the committee of the Kranjsko-slovenska katoliška jednota (Grand Carniolian Slovenian Catholic Union), or KSKJ, and became one of its vice presidents. In a large group of men, she was the only woman at this level. At that time, Catholic women throughout the United States were already actively organizing, and Prisland participated at one of their conventions as a delegate of her church. This event encouraged her to describe her experience in Slovenian papers. In one of the articles, published in *Amerikanski Slovenec* in October 1926, Prisland wrote:

> I thought how we could do this, how this would be possible among us, Slovenian women in the United States, to do the same and organize ourselves for the good of the many. With energy and talent as we have, we could have an organization that would enrich our cultural life, educate us to advance ourselves and our many young women, and make us proud to be a part of this new homeland we have selected for ourselves and our families (1926, 3).

She pointed out that the Slovenian women should learn from other immigrant women, for example, Czech, Polish, or Slovak women, who were well organized and had their own newspapers. In her opinion, Slovenian women had not achieved the same because of their lack of education and interest in organizing. Her firm intention was to establish an organization that would connect them and—because it would demand responsibility, education, and determination to work for the benefit of the community—make them visible and proud. The goal of a women's organization was to connect women from different scattered communities for the progress of their families and nation. Decades later, when remembering the events leading up to the establishment of the SWUA, she described the response to the article as favorable: women from all over expressed their support to "show their compatriots in the old homeland, that they are also aware of their Slovenian nationality, to get to know each other better and to be beneficial to themselves and their families" (Prisland, 1964, 210).

The Slovenian Women's Union of America was founded in December 1926 in Chicago, and Marie Prisland was its national president for 20 years. The main objectives of the organization were: "To unite the Slovenian women living in America, to assist in their social, moral and intellectual education, to foster American and Slovenian ideals, to encourage participation in American civic affairs, to help members to become American citizens, and to arrange an adequate interment for its deceased members" (Prisland, 1968, 78). In 1929, Prisland established the magazine *Zarja —The Dawn*, which became the official publication of the SWUA, a bond among members all over the United States. The organization is still active, and it still publishes the magazine. The magazine published educational articles, advice on housekeeping, and travel articles. It covered the activities of the many SWUA branches and published recipes. On the 30th anniversary of *Zarja*, Prisland wrote an article for the *Slovenski izseljenski koledar*, for which she wrote regularly. She mentioned that *Zarja* had always "advocated women's rights and encouraged members to fulfill their obligations as mothers, housewives, proud Slovenians, and loyal American citizens" (Prisland, 1960, 182). For Slovenian readers, she emphasized that the purpose of *Zarja* was to showcase everything good among the Slovenians in America and draw attention to the excellent work, commitments, and contributions of the SWUA members to the collective construction of the "American Slovenia." This self-perception is important for research on Slovenian migrant experiences and social constructions of migrant identities. It shows how Marie Prisland and other migrants saw themselves as a part of two nations, societies, and communities on both sides of the Atlantic.

Not all migrants shared these feelings or wanted to participate in ethnic activities. The communities of Slovenian migrants were heterogeneous and divided into at least two main ideological-political camps, and at the same time, depending on the local or regional provenance of the immigrants, also into smaller organizational units. The legacy carried over from their old homeland could be called "liberal and clerical," or "left and right," or "conservatives and progressives." All cultural, artistic, benefit, and social organizations were established and doubled around these divisions: choirs, dance and theatre groups, fraternities, newspapers and publishers, national homes, and recreation centers.

Picture 3.3: The portrait of Marie Prisland that hangs above her desk at the Slovenian Women's Union and Slovenian Heritage Museum in Joliet, Illinois. Author: Mirjam Milharčič Hladnik.

In this complex and polarized structure, we can place SWUA on the "Church, Catholic, conservative" side and the other women's organization, the Progressive Slovene Women of America (PSWA), established in 1934 in Cleveland, on the other. The PSWA was among the most active in collecting aid for Slovenia and Yugoslavia during and after World War II. For several decades they helped many health, cultural, research, and educational institutions in Slovenia and participated in American humanitarian actions.

In the introduction to her book *From Slovenia to America* (1968) about her migrant experience, Prisland wrote:

> "Recollections and collections" have been compiled with deepest respect for the early settlers who contributed their efforts and talents to the strength of America and its Slovenian communities. It is for their children and grandchildren that I have prepared this collection of Slovenian history so that they might have a small glimpse of the rough road traveled by their parents and grandparents to improve their own living standard and to give life to sons and daughters in a free and happy land (1968, 13).

Judging by these words, one would think that there were not many autobiographical elements in this book. It includes chapters like "Slovenian Womanhood," from which the above quote is taken, and "History of the Slavs," "Slovenia—The Beautiful," "American-Slovenians and Their Activities," "Slovenian Pioneers and other Notables," "Notable Newcomers Since World War II," and "Slovenian Communities." But the book also includes some invaluable autobiographical chapters, for example, "Impressions of America," "A Dream," "Memories of Our Old Stove," which paint vivid tableaus of impressions the young immigrant got upon her first contact with the American society, living conditions, and interpersonal relationships. In the chapter "Impressions of America," she presents a compelling interpretation of how the American views on gender and class equality, work ethics, and democracy impacted the European newcomers, including Slovenians and the author herself. Young Marija was the most surprised by the discovery of gender equality:

> The honor and the freedom which American women were enjoying was a marvel to me. This is not duplicated in any other country on the globe. A few married men, however, were of a different opinion. Used to European behavior, they thought that America was over-protecting the little woman. One complained: "In Europe, a man could mishandle his wife and nobody bothered him, but here, if a man beats his wife a little and the neighbors hear her cry, they quickly call the police! The man is taken to jail for something he believed it was his right to do. Isn't the wife his property? And is he not free to do with it what he thinks is right?" (Prisland, 1968, 31).

Prisland's descriptions of the impression that the substantially greater gender equality had on the newcomers from Europe reveal an important appeal that the American society had for women migrants. It is a partial explanation—apart from the significantly higher number of men of marriageable age than that of women—for the gender gap in the return migration as established in Annemarie Steidl's study:

> The journey from one continent to another covered a long distance, but during the early twentieth century, up to 40 percent of all migrants from both parts of the Habsburg Empire nonetheless returned to their country of origin within a few years of migrating to the United States. Men most often opted for only temporary stays across the Atlantic,

while for the majority of women who undertook journeys to the "New World," the move was more likely a permanent one (2021, 219).

Although the United States at the beginning of the 20th century was not an egalitarian society regarding the social status of women, the women immigrants were in a much better position than back in the country of origin. There were more economic and social opportunities to improve their status, and American society offered greater possibilities, even for the very traditional Italian and Slavic women. In a general study of ethnic history, *To Seek America* (1988), where women's roles were included for the first time, Maxine Schwartz Seller mentions how wife-beating declined when women discovered that abusive husbands could be reported to the police. In a unique conceptual insight into the connectedness of the private and public spheres, she adds that many Slavic and Italian women became active in their ethnic communities at this point. But she emphasizes that their activities were not "limited to their own communities. In Chicago, Slavic women helped organize a Woman's Civic League to register voters and campaign against a corrupt city administration. Women in other cities engaged in similar activities" (Schwartz Seller, 1988, 127).

The second important point is the gap between the private and public spheres, which is used as one of the most common methodological approaches in the studies of gender roles. However, if we try to study gender roles in society, this division is more of an obstacle than a valuable tool for understanding. The mentioned connection that Schwartz Seller makes between women's civic activities outside the home and their rebellion against abuses inside the home is a good example that shows how the private and public spheres are intertwined even when the private sphere is traditionally structured and protected from the influence of the dominant society by the closed ethnic immigrant community. It also shows that subjected women can use civic values and laws of the dominant society to empower themselves even if they simultaneously maintain their traditional role. The life and work of Marie Prisland is another good example of both points mentioned above. Her story shows the contradictions, ambiguities, complexities, and dynamics of a life lived in the Slovenian ethnic community in the United States.

The Position of Women in the History of Both Homelands

Many questions remain open regarding Marie Prisland's thinking and activities when all of the above is considered: how she was able to establish a women's organization that combined traditional Slovenian values and American civil liberties; how she combined religious and civic values; how she intertwined the goal of maintaining traditional women's roles with learning and exercising their political and human rights. Her rich body of work contains too few autobiographical texts to give us answers to these questions. From her many articles, lectures, and appearances, it is possible to understand the women's perception of their position in the world and their authority in their own society. Particularly meaningful is her strong determination to help Slovenian women in the United States achieve a position of authority and respect at home, in the community, in the "old country," and in the new homeland. She wanted precisely that: a position of authority and reputation for women gained through education, work, and activities at home and in the community. As much as she wanted to preserve the traditional gender roles, she also wanted to change them. She predicted that this single change alone would make respected Slovenian women with authority a part of the history of Slovenians in the United States and Slovenia (Milharčič Hladnik, 2007).

In the autobiographic parts of her book, in which she deals extensively with gender roles in the new country, she describes her experience of coming through Ellis Island:

> A group of Slovenian immigrants, of which this writer was one, arrived in New York from that part of Austria, which presently is the territory of Yugoslavia. It was a beautiful morning in May 1906. After leaving the French ship, *La Touraine*, we were transported to Ellis Island for landing and inspection. There we were "sorted out" as to the country we came from and placed in a "stall" with the letter "A" above us ("A" was for Austria).
>
> There were at least a hundred Slovenian immigrants. We separated ourselves, as was the custom at home—men on the right and women and children on the left. All of us were waiting to leave for all parts of the United States.
>
> The day was warm and we were very thirsty. An English-speaking immigrant asked the near-by guard where we could get a drink of water. The guard withdrew and returned shortly with a pail of water, which he set before the group of women. Some men stepped forward quickly to have a drink, but the guard pushed them back saying: "Ladies first!" When the women learned what the guard had said, they were dumbfounded, for in Slovenia, as in all Europe, women always were second to men. Someone dramatically explained it this way: "First comes man, then a long time nothing, then comes the woman."
>
> Happy at the sudden turn of events, one elderly lady stepped forward, holding a dipper of water, and proposed this toast:
>
> "Živjo Amerika, kjer so ženske prve!"
>
> (Long live America, where women are first!) (Prisland, 1968, 19).

It is easy to imagine that Slovenian immigrants are not often mentioned in the vast literature on US immigration history because they are a very small group within it. It is even easier to presume that Slovenian women migrants are paid even less attention. Yet one of the earliest researchers of women's experience of migration, Maxine Schwartz Seller, used Prisland's above description in the introduction to her influential book *Immigrant Women* (1994), adding:

> For Prisland, who later founded the *Slovenian Women's Union of America* and created a woman's magazine, *The Dawn*, the American dream became reality. Not all immigrant women were so fortunate. For many, life in the United States was bitter and the slogan, "ladies first," cruelly ironic. "Ladies" were first to be underpaid, unemployed, and abused (1–2).

Marie Prisland knew a lot about this cruel irony and included some stories about the hardships of Slovenian women in her book. She quoted Albina Novak saying: "Those who pine for the 'good old days' never knew what they were or never remembered what life had in store for the poor immigrants who came to America in the early years of this century" (Prisland, 1968, 75). The hard life of Slovenian immigrants at that time is also the subject of the books *Immigrant Woman* (1976) by Mary Jugg Molek and *Footsteps Through Time* (1978) by Irene Planinsek Odorizzi. In her book, Molek describes the life of her mother and other immigrants in the promised land, toward which she is critical and sharp:

> These young women were not "the tired, the poor" proverbially greeted with open arms by the Statue of Liberty. Neither were they descendants of uncultured, uncivilized lands. They were, instead, bearers of an already-established cultural heritage, centuries old. [...] Their composite efforts, dedication to work, indefatigable physical and mental labors, resourcefulness, self-pride, and aspirations toward nobler human potential and dignified activities helped to weave the fabric for the affluence that was to be (1976, 5).

Planinsek Odorizzi collected the life stories of 21 Slovenian immigrants. What they all had in common was a hard everyday life, which for them and their children only improved decades later. As one of the narrators, Mici from Jesenice, describes those hard years:

> I think that I remember those days the most because they were the busiest and the hardest ones. We put so much of our bodies and souls into living that the memory of these early days could never be forgotten. Every day it was work, work, and more work. Never a vacation and very little sleep. [...] When I look back now, I wonder how we immigrants ever managed in those early days. Then I realize that we were able to suffer those hardships because we knew we had to stay (Planinsek Odorizzi, 1978, 33–34).

The difficult life of Slovenian immigrant women was also why, in 1926, Marie Prisland founded the first Slovenian women's organization. Her determination to help Slovenian women gain authority and respect at home, in the community, and in society stemmed from two circumstances in her life. First, it arose from her youthful desire to become a teacher, which brought her to the United States and equipped her with the ability to do socially relevant things: establish a women's organization and magazine, write articles and publish them in Slovenian and American publications, write and edit books, take part in many organizations, and advocate for the importance of education and civic responsibility. The other circumstance was the society into which she migrated. American civic values regarding the social position of women were a great inspiration for the young Marija Černe. As a worker in shops, a chair factory, and a brewery in Sheboygan, she soon realized she would never earn enough to pay her teachers' college tuition. She never obtained her professional degree for the work she did all her life—teaching people through writing, organizing, and lecturing. However, American society gave her inspiration and the possibility to continue her educational path, although certain activities were not desirable in the Slovenian community.

In her book, Marie Prisland points out how difficult it was to establish such an organization because it was not welcomed in Slovenian communities, certainly not by men. For a person who valued ethnic community life as she did, it took extraordinary courage. About the beginnings of the SWUA, she wrote:

> This new organization was not very popular. Men eyed it as an intrusion into their domain and as something totally unnecessary. They maintained that a woman's place was in the home, taking care of the husband, the children, and the boarders which every home had to augment the family income. Even the women had little faith in the new society until it started to grow and then expanded beyond expectations (Prisland, 1968, 78–79).

Publishing *Zarja*, as well, was financially difficult and risky. Marie Prisland and her colleagues shared a vision of the importance of the publication, not only for the membership and the organization but also as an archive and a message for younger generations: "The entire course of history is recorded here. This paper will testify to late generations about the activities of Slovenian Americans particularly in the time when public life was limited due to the beliefs of the time that a woman belongs in the kitchen and nowhere else" (Prisland, 1960, 183–184).

Marija Černe left her tiny village in Carniola with ambitions for education and an awareness of the necessity of social engagement. She disembarked in a country where the women's movement for political rights was in its sixth decade of uninterrupted activity that proved successful in 1919. She settled in a multi-ethnic society where immigrants still valued their ethnic cultures and traditions. She was deeply involved in the Slovenian community and cared about Slovenian culture and tradition. Nevertheless, she recognized that the organizations of other Slavic women, their publications, and their activities were something Slovenian women should also have. She was inspired by the community in which she lived, other ethnic communities, the host society, and the homeland she had left forever.

Picture 3.4: The front page of *Zarja—The Dawn*. Courtesy of: Digital Library of Slovenia – dLib.si.

She succeeded because she linked the traditional ethnic culture of her country of origin with the modern citizen culture of the "New World." In her own way, she followed traditional ethnic habits and gender roles while also transforming them using citizenship values and political freedom. Her vision was for her work and the work of all the contributors to the organization and magazine she founded to become an important legacy of the country of origin and the new homeland.

A Multiculturalist Thinker and Transnational Fighter with a Pen: Louis Adamic

Louis Adamic, who moved from the Slovenian countryside to the United States in 1912 as a 14-year-old boy with no formal education, was an acclaimed and award-winning American writer by his early 30s.[53] The importance of his work made him a writer of national repute who was a guest of presidents Roosevelt and Truman, a winner of various awards, accolades, and writing grants (for example, the Guggenheim Fellowship in 1932, the Book-of-the-Month-Club selection in 1934, the Anisfield-Wolf Book Award for the best work on a racial question in the modern world in 1941). His works earned him an honorary doctorate from Temple University in Philadelphia (1941) and the Yugoslav Order of Brotherhood and Unity (1944). He enjoyed the private and public support of literary notables such as Upton Sinclair, Sinclair Lewis, Sherwood Anderson, and F. Scott Fitzgerald. Adamic's books were included in the White House Library and O'Brien's anthology of the best short stories (1931), listed on the reading syllabi at many American colleges, and among the literature distributed to American soldiers when the United States joined World War II (Žitnik Serafin, 2011).

Lojze Adamič, the son of a wealthy family from the village of Blato near Grosuplje and a student of a grammar school in Ljubljana, pondered his decision for some time while he listened to the exciting stories of the returnees from the United States. Young people who met returnees nurtured their plans for the future by observing their American clothes and their adopted habits and listening to their stories. Other factors significantly influenced his decision to emigrate. He was expelled from school because of his involvement in anti-monarchy demonstrations and his membership in a nationalistic student organization. He needed an escape to a place that offered different personal development possibilities and life choices. The United States of America, as a country of "permanent progress," attracted young people with its magnificent development, the images of freedom, and the position of individuals with

53 Practically all Slovenian, American, and other Adamic scholars list 1913 as the year of Adamic's emigration to the United States. However, the passenger list of the *Niagara* from the Le Havre Port in December 1912 and the information about his arrival to the port of New York that same month quoted in his Declaration of Intention to obtain American citizenship from December 1916, call these statements into question. I thank Janja Žitnik Serafin for this information.

different views on life and relaxed interpersonal relationships. Lojze Adamič thus disembarked on Ellis Island at the end of 1912 and, in the following decades, fulfilled his dreams: to learn about and experience this great land of opportunities. He became the most important Slovenian American writer and a political activist in both his homelands. Until his final breath, he fought for more equality and justice in society, freedom and coexistence between nations, and against fascism, colonialism, and imperialism (Enyeart, 2019).

When he disembarked in the United States as a young teen without any knowledge of the language, Adamic first followed the typical path of many Slovenian migrants. He took all sorts of menial jobs and moved from the east to the west coast of the vast county. During World War I, he joined the US Army and, in 1918, got American citizenship. After this time, he took a completely different route as Louis Adamic. Perhaps the key to his exceptional creative life was the determination and zeal with which he took to learning English and, once he mastered it, to describing the vast land of misery and riches he took as his new homeland. But it is more likely that his political and moral convictions decisively stirred his literary and social engagement as early as his secondary school years, convictions he never abandoned. He soon had such a good command of English that he translated a couple of Slovenian texts into English, in 1926, Ivan Cankar's *The Bailiff Yerney and His Rights*,[54] and started publishing in US magazines. In the "land of endless opportunity," he became a journalist, writer, translator, editor, historian, political activist, and a socially critical analyst of the complex, multicultural, and multiethnic fabric of American society, as well as an enthusiastic chronicler of historical and political events in his country of origin. Adamic's body of work contains 20 books, over 500 articles, translations, and lectures and fully conveys his political convictions and unyielding will as an active citizen of two homelands.

His first novel, *Dynamite: The History of Class Violence in America, 1830–1930*, was published in 1931 (Adamic, 1931). The following year he published the autobiographical *Laughing in the Jungle: The Autobiography of an Immigrant in America* (1932). The controversial—and banned in Yugoslavia—*The Native's Return: An American Immigrant Visits Yugoslavia and Discovers His Old Country* (1934) put him on the American literary map. It was followed by a book about migrants' experiences of division and uprootedness,

54 Ivan Cankar (1876–1918), a writer, playwright, essayist, poet, and political activist is regarded as the greatest Slovenian writer.

Grandsons: A Story of American Lives (1935), and then by *Cradle of Life: The Story of One Man's Beginnings* (1936), *The House in Antigua: A Restoration* (1937), and *My America 1928–1938* (1938). The decade of the 1940s began with a book about interethnic relationships, *From Many Lands* (1940), followed by *Two-Way Passage* (1941, *What's Your Name?* (1942), *My Native Land* (1943), *A Nation of Nations* (1945), and *Dinner at the White House* (1946). His last—also controversial—book, *The Eagle and the Roots* (1952), was published posthumously. The writer only published three books that could be strictly described as fiction: *Laughing in the Jungle*, *Grandsons*, and *Cradle of Life*. Narrowly categorizing Adamic's work as novels, travelogues, autobiographies, or historical reports is questionable. For the most part, they are a mix of different narrative and documentary elements, that is, literary journalism or documentary prose.

Picture 3.5: Louis Adamic. Courtesy of: Research Centre of the Slovenian Academy of Sciences and Arts (Slovenski biografski leksikon).

From the autobiographic elements in all the published works, we read how he experienced his two homelands and attempted to understand and influence their monumental political events. All his works show an obvious transformation of identity, which he often accurately described as compound. His answer to Oton Župančič's question whether he felt like a Slovenian or an American was quite explicit: "I feel one and the other" (Župančič, 1932, 518).

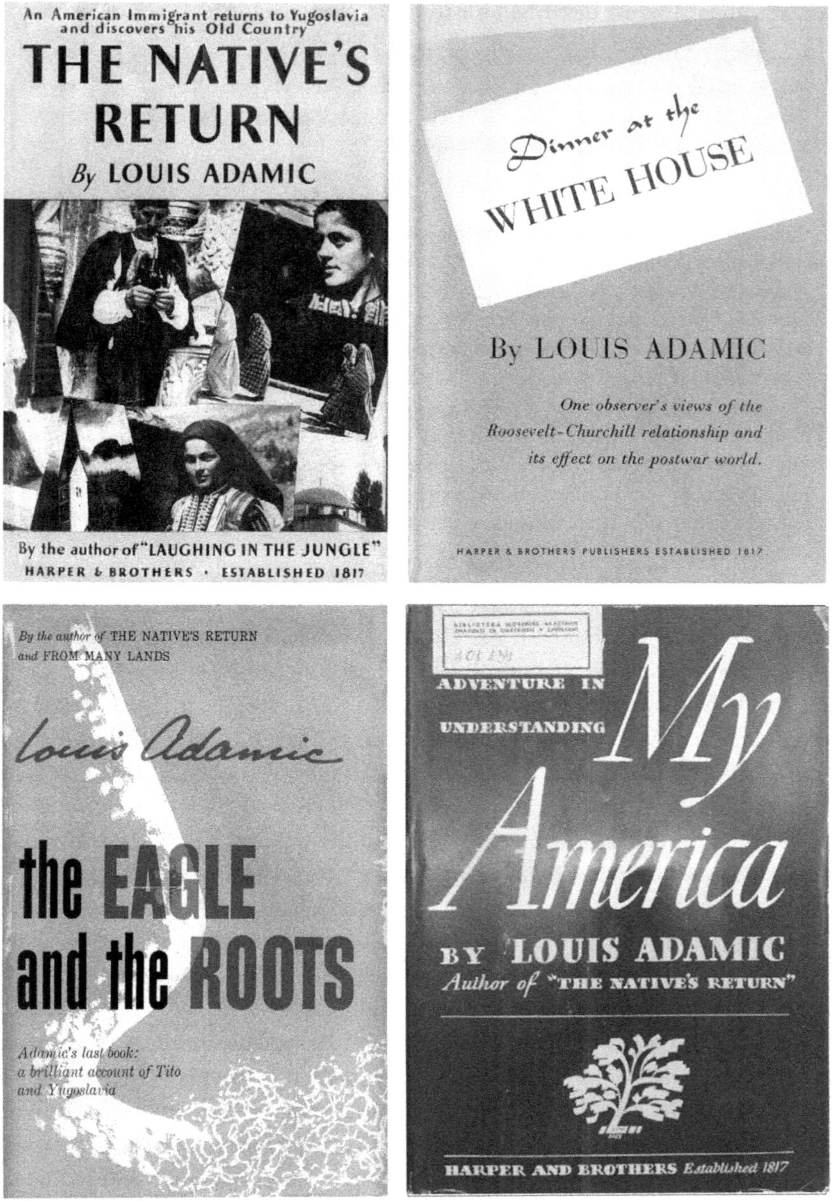

Picture 3.6: The covers of some of Louis Adamic's books. Courtesy of: Research Centre of the Slovenian Academy of Sciences and Arts, Slovenian Migration Institute.

In *Grandsons*, Adamic described his transnational identity:

> Now I was trying to get into America, become a part of her, a part of her essential life, realize myself. But I was careful. There were things in me (I was barely cognizant of them) which I did not want to see killed or harmed. I wanted to get into America and at her as a writer on my own terms. I didn't want the turbulent, superficial, temporary, non-essential America to get me altogether. She already had touched me too much. It was (as I see now) my Carniolan peasant egoism, of course, that stood in the way of my thorough Americanization in the current sense; but not that alone. There were things in contemporary America that just naturally went against my grain. Violently so. I wanted nothing to do with them except fight them if I ever got a chance; if I ever equipped myself to fight them. I was hysterical, but perhaps mainly because I was impatient with myself, bewitched and bewildered about America. I was twenty-eight, twenty-nine. Perhaps half of my life was gone, possibly more, and where was I? I was spending most of my energy fighting the forces of jitteriness in contemporary America that had put their hooks into me personally, for unwittingly, ignorantly, I had exposed myself to them. They held me, and I was uncomfortable; I felt like nothing at all, a shadow, no better than most people in America; but I would extricate myself from them. I was still partly an outsider in America. I had a chance to free myself (Adamic, 1935a, 80–81).

We can read about his positions, beliefs, views, and political actions in numerous books, essays, speeches, and letters. He kept his double position of the "insider" and the "outsider" throughout his life. It helped him understand the complex fabric of Slovenian and other ethnic immigrant communities, American society, and his birth country. From a privileged position of a free yet intimately engaged observer, he expressed clear political opinions and tried to "spite" the events that were, in his opinion, destructive for the individual and society. He was a ruthless critic of the American melting pot and the US postwar imperialist politics, as well as the prewar Yugoslav dictatorship and the postwar Tito personality cult. He directly and without any impediments described the closed-off and culturally self-conscious migrants and migrant communities. At the same time, he was a sharp critic of the majority society, which did not accept difference and diversity. He advocated that a society accept its population's ethnic and cultural diversity as a value and a fundamental principle of democracy.

In the United States, Adamic is still pertinent as one of the foremost harbingers of American ethnic studies and as one of the early proponents of the functional variant of cultural pluralism. He studied immigration questions and interethnic and intercultural questions in the United States during his many study trips, visits to archives and libraries, fieldwork, mass questionnaires, and in-depth interviews. He used the results of his research in articles, lecture tours, radio shows, and independent books about immigration. He was a founder and an editor of the magazine *Common Ground*, the official paper of the Common Council for

American Unity, which strove for multicultural education by including migrant content into public education and, in the last years of his life, he was the editor of nine books from The Peoples of America Series. His enduring pertinence in this field is confirmed by a surprisingly large number of American and other authors working on multiculturalism and the integration of migrants who still know and quote Adamic and comment on his thoughts (Shiffman, 2003).

The Extent of Creative and Political Activity

We can highlight three periods in Louis Adamic's versatile career when his work and political engagement heavily influenced the Slovenian immigrant community in the United States, his country of birth, and the broader American public. The political impact of his writing and work was the greatest during World War II and at the publication of his two books created after the two visits to his country of birth in 1932–1933 and 1949. Adamic undertook the first visit, which he extended by several months, with his wife Stella in 1932 when he received the Guggenheim Fellowship to travel to Europe for a year. In the description of his return, he disclosed what kind of person he had become since he left home:

> In those nineteen years I had become an American; indeed, I had often thought I was more American than were most of the native citizens of my acquaintance. I was ceaselessly, almost fanatically, interested in the American scene; in ideas and forces operating in America's national life, in movements, tendencies, and personalities, in technical advances, in social, economic, and political problems, and generally in the tremendous drama of the New World.
>
> Events and things outside of America interested me but incidentally: only in so far as they were related to, or as they affected, the United States. I spoke, wrote, and read only in English. For sixteen years I had had practically no close contact with immigrants of my native nationality. For three years I had been a soldier in the American army. [...] In the last few years I had become an American writer, writing on American subjects for American readers. And I had married an American girl.
>
> To Stella I had told but a few main facts about my childhood and early boyhood in the "old country"; and what little I had told her of my parents, and the village and the house in which I was born, had seemed to her "like a story." She scarcely believed me. To her I was an American from toes to scalp (Adamic, 1934, 3–4).

Just as he was "almost fanatically interested in the American scene" in the United States, the return to his place of origin reignited interest in life in Yugoslavia and Slovenia. He only explained the American society once he knew it well, "had roamed over half of the United States and had been to Hawaii, the Philippines, and Central and South America" (Adamic, 1934, 3). He did the same once the arrival to his old-new homeland stirred his curiosity. Instead of traveling through

Europe, he spent a lot of his 1932–1933 with relatives in his home village, criss-crossed the entire Slovenia and Yugoslavia, visited big towns and small villages, and spoke with many ordinary people, workers, and peasants, influential artists and writers, politicians in power and in opposition, and even the king. It was obvious that once he renewed contact with his homeland and people, he became "almost fanatically" interested in the situation and events in Yugoslavia. He wanted to see the people and their strivings and only then form opinions and write about them. After returning to the United States, he published *The Native's Return* (1934). In it, he described his native village, people, their habits, and the social and political situation in the Kingdom of Yugoslavia. He denounced the dictatorship, the persecution of political opponents, and the suppression of liberties. Because of Adamic's criticism of the dictatorship and King Alexander, for whom his book also foretold a violent death, *The Native's Return* was banned in Yugoslavia but saw several reprints in the United States and was selected as the book of the month in February 1934. Vatroslav Grill, the long-time editor of *Enakopravnost*, claims that by publishing it, Adamic became "some sort of unofficial, but morally de facto ambassador of the Yugoslav people in the United States and the entire informed American public hence saw him as such" (1979, 186).

As for the Slovenian diaspora, the book caused an irreconcilable rift between those who enthusiastically welcomed it and those who rejected it. It was received with opposite reviews that linked to the political and ideological positions of the two camps: the liberal and the conservative. The opposing and conflicting points of view of the two fundamental ideological and political orientations of Slovenian migrants in the United States had been increasingly strained since the establishment of the initial Slovenian communities. They had decisively influenced their organization and operation. Upon the publication of Adamic's book, the worldview divisions and ideological confrontations between the progressive/liberal and Catholic/clerical camps were obvious in the open support for the author and the mass purchase of the book on the one side and in the unrestrained shaming and mudslinging directed at the author on the other. Grill claims that the book was a literary sensation, which some papers of the Slovenian community nevertheless criticized, calling Adamic an "uprooted person," a "renegade," or a "janissary." *Ameriška domovina* even doubted that he wrote the book himself and suggested that it was written by his wife, Stella. The unexpected consequence of the vitriolic attacks on *The Native's Return* was the establishment of the women's organization Progressive Slovene Women of America, which defended Adamic and his work, and after his death even financially supported the buying of his literary estate and donated it to the library of Princeton University (Grill, 1979, 187–191).

Adamic's return to his birthplace and interaction with Slovenian writers also dramatically affected the Slovenian literary scene. After Adamic's lengthy visit to Slovenia, during which he had intense contact with Slovenian cultural workers and writers, Oton Župančič published the essay "Adamic and Slovenehood" in *Ljubljanski zvon*. In it, he compared Adamic's interesting, humorous, and attractive style of writing to the then-prevalent style among Slovenian writers and claimed:

> In our parts, bleakness is an aesthetic dogma, and dull writing is considered wisdom. Thus, so many of our writers twirl and twirl in some sort of mysterious fog and can't see a way out, while the poor reader can't see or find a way in—but this counts as deep insight around here. Into this eternal Slovenian sadness and misery of ours, laughter suddenly breaks! Slovenian laughter, from America, Adamič's laughter. Would he have retained it had he stayed home? I do not know. Hardly. To me, our entire literary output has sometimes seemed like the moaning of a feeble old man, tossing on a creaky bed, yet can't get comfortable in any position (Župančič, 1932, 516–517).

A large group of influential writers took offense at Župančič's writing and his evaluation regarding the "bleakness" of Slovenian literary production, which he compared to Adamic's attractive, interesting, and witty style. They left the central literary magazine *Ljubljanski zvon* in protest because the editor did not want to publish their reactions and established the new literary magazine *Sodobnost*. They disagreed with Župančič, who wrote not only an ode to Adamic in the contested text but also saw in his work the possibility for Slovenian cultural workers and writers to reconsider some of their positions on society and culture and change them:

> He reached above us all. He is the highest that our emigration gave us, this barely organized mass that searches for its survival in its new country: the chronicler and painter of our emigrants, theirs and our genius. A great writer on the world scale, the kind that makes a nation reevaluate its essential questions. I believe that after his book, which we are supposed to get in a lively translation soon, we will be forced to reorganize many beliefs about ourselves and shift many points of view (Župančič, 1932, 518).

After his return to the United States and the book's massive success, Adamic traveled and lectured a lot. Janja Žitnik Serafin describes how, in this time, the American public image of an author and activist who did not belong to any of the existing institutional frameworks was formed:

> It was Adamic's distrust of existing organizations and institutions and his passionate allegiance to the principles of individual human rights and liberties that brought him—despite his incessant emphasizing of the need to provide equal opportunities for development to all people—to significant disagreement with the ideas of American Marxists,

> while many extremists on the left and on the right saw in him primarily an anarchist (Žitnik Serafin, 1995, 14).

At this time, he intensively developed his idea to go beyond the prevalent American model of a "melting pot." He frequently wrote and lectured about the necessity of (self-)respect for the ethnic backgrounds of all people living in the United States. He considered it a foundation of democracy and a guarantee for the preservation of the American society that, in his opinion, was threatened by discrimination and violent Americanization.

Interwar and Postwar Activity and Writing

During World War II, Adamic was exceptionally active. Among other things, he was working for the US organization, the Common Council for American Unity, which advocated for the ethnic rights of immigrants. In his speech in Cleveland (1941), he explained the by-then well-developed idea of respect for the diversity of cultures and ethnic origins, which in the 1970s emerged as the policy of multiculturalism:

> We need to be trained, or train ourselves and help one another to be trained, in the direction of becoming creatively, positively interested in a man partly *because* he is different; because being different, he is apt to have something out-of-the-ordinary to offer to us personally and contribute to the evolving civilization and culture.
>
> Inviting diversity, being interested in it, will tend to produce unity in a democratic country; will tend to make it dynamic; will operate against the concentration-camp-like foreign sections and ghettos and restricted residential districts, and will encourage movement and dispersal, at the same time that it will work for harmony and fusion. [...] Inviting diversity brings out the basic sameness of people, just as the opposite results only in more and sharper differences. It breaks down both the superiorities and inferiorities of people, which are two ends of the same stick, equally productive of insecure, mediocre character and personality which cannot achieve any satisfactory function in a democracy (Adamic, 1941b, 14–15).

Because of his friendship with the US vice president (1941–1945), Henry Wallace, Adamic was involved in the work of different departments of the US administration. He was also active in the organizations of Slovenian migrant communities. In 1942, he became the honorary president of the Slovenian American National Council and, in 1943, joined the United Committee of South-Slavic Americans. Following his political convictions, which were based on the values of the national struggle for independence, the class and social struggle for greater equality in the society, and anti-fascism, he actively supported Tito and the National Liberation Army. He strove to explain the national liberation fight to the American public and convinced the US administration to support Tito's army and stop supporting Draža Mihajlović. After the war, he tried to convince the US administration to support the annexation of Trieste to Yugoslavia. Once he realized that his efforts were in vain because the US government decided otherwise, he dedicated himself to the fight against the US and global colonialism and imperialism.

Among the compatriots in Slovenian immigrant communities around the United States, Adamic's engaged activity was received with opposing opinions linked to the political and ideological stances of the two abovementioned camps. Despite the concern and aid collection for the homeland, ideological confrontations were also prevalent during World War II. Evaluating events in the homeland and responses to them coincided with the opposing attitudes toward Fascism, the prewar Yugoslavia dictatorship, and the national liberation fight and revolution so that the two camps found themselves in an irreconcilable conflict. The rejection of Adamic most definitely also stemmed from Adamic's writing and his descriptions of Slovenian emigrant communities. From his very first published works, he did not idealize them but described and criticized them without any embellishment. When he describes the community of Slovenian immigrants in the chapter "My Own People in America" in the book *Laughing in the Jungle*, we can imagine that such a realistic description did not earn him the mass sympathy of his compatriots. He calls the communities colonies and is particularly lucid in his description of their ideological and organizational splits:

> Most colonies were split into two antagonistic camps—the church members and the unbelievers, mainly Socialists. Each had its own lodges, clubs, singing societies, and social affairs. The unbelievers, of course, were in a minority and largely ineffective. On election day, if they happened to be citizens, they voted for the Socialist candidates, and on their club-room table lay such subversive publications as *Proletarec*, *Prosveta*, the *Appeal to Reason*, and *The Liberator*.
>
> In the other camp, life revolved entirely around the church. The alien Catholic parishes, with their parochial schools, were strongholds of mediaevalism. Unlike in the Old Country, where the priesthood included men of learning and superior character who commanded the respect of their antagonists, the Bohunk-American priests were usually of an inferior order (Adamic, 1985, 107–108).

Adamic's situation, both during the war and after it, was controversial. He supported Tito and the Partisans—the Yugoslav National Liberation Army as well as the Liberation Front of the Slovenian Nation and Slovenian Partisans—the anti-fascist resistance to the Axis powers. In this way, he "inevitably deepened the schism between the left and right wings of Slovenians in the United States, something for which many compatriots who thought differently than he did could never forgive him. Both camps of Slovenians were formed long before, but it was then that the right-wing camp started to gain reinforcement through the new political emigrants from Slovenia" (Žitnik Serafin, 1995, 18).

In 1948 and 1949, Adamic focused his writing, lectures, and media appearances on sharp criticism of the US internal and foreign policy. In the magazine he was editing, *T&T: Trends and Tides*, he spoke out about the undemocratic procedures in the United States, the McCarthyist witch hunt, and the US military interventions and imperialistic foreign policy. His other articles from that time also clearly showed his political engagement in support of the Yugoslav leadership during the Cominform, the struggle for independence, and the renovation and reconstruction of socialist Yugoslavia (Žitnik Serafin, 1993). Adamic's critical writing on US internal and foreign policy and his commending yet critical presentation of the postwar Yugoslav situation further complicated his controversial position in the Yugoslav diaspora. It turned out that in the acute ideological confrontation of the Cold War, the writer and active citizen lost both his homelands at once.

This loss became apparent when he was writing his second book on Yugoslavia, following an extended stay in the country in 1949. There, Adamic once again met with many people, the most influential politicians, including Tito, and saw "with his own eyes" the situation in postwar Yugoslavia, strongly influenced by the dispute with the Cominform in 1948. After his return, he worked on *The Eagle and the Roots* in the months of the worst escalation of the Cold War, encountering the double accusations of being "anti-American" in the United States and "anti-Yugoslav" in his country of birth. So a "strong, controversial work was created, which triggered a wave of reproach in the United States that it was pure propaganda because, in it, Adamic offers a black-and-white picture of the situation in Yugoslavia. Ironically, the book could not come out in Yugoslavia due to its criticism for another eighteen years" (Žitnik Serafin, 1993, 165). He never finished the book; after his death, his wife and the editor at Doubleday & Company in New York radically abridged and edited it.

The decision of the 14-year-old Lojze Adamič to emigrate by himself to the United States of America did not stem from poverty but rather from the youthful desire to get to know a new, different world and for self-realization. As he wrote in 1932 in his autobiographical work *Laughing in the Jungle*:

> I have come here for excitement and adventure. I have never been hungry for more than two days since I am here. The jungle has been vast and interesting. Too interesting. Sometimes it is overwhelming in its complexity and melodrama. And as I say, lately I find it difficult to laugh. But I stay and I intend to remain here (Adamic, 1985, 335).

And indeed, Louis Adamic succeeded in doing just that. He got to know a magnificent country, experienced excitement and adventure, was at the forefront of political and literary events, won awards, accolades, and fellowships, and was loved and hated, respected and despised—all that in a confusing and melodramatic jungle to which he once wanted to belong and which, in the end, in the whirlwind of heightened political tensions and persecutions, destroyed him. In 1951, he died in unexplained circumstances, aged 53. About this last phase of Adamic's life, Vatroslav Grill wrote:

> The final and most turbulent phase of Adamic's life was intimately linked to his country of birth. For a man of his spiritual disposition, this was just as natural as the love of a father and a mother for their children is natural. Rationally, he was deeply entrenched in the United States, but emotionally, he never stopped being a Slovenian, therefore a Yugoslav. It is also easy to imagine how he was attracted by the dynamics of the fight of his people under the occupation. This fight was a unique phenomenon, sparking the imagination of the entire freedom-loving world. He created and worked with the energy that went far beyond human endurance and got involved in political controversies for which he paid a high price—in the end, with his life, regardless of whose hand fired the gun that was found in the lap of his corpse (Grill, 1979, 215–216).

A Refugee Mother during World War I: Marija Ipavec

The mass exodus of the population from the Soča Valley began with the Italian declaration of war on Austria-Hungary on May 23, 1915. The 90-km-long front line established with lightning speed on the left and right banks of the Soča River cut into the Slovenian ethnic territory. It continued uninterrupted for two and a half years and "left indescribable devastation in its wake, a sheer ecological and economic disaster" (Svoljšak, 2009, 34), as well as around 500,000 soldiers dead and around 700,000 wounded. The civil population—around 100,000 people—became refugees. The front line split Kanal ob Soči/Canale d'Isonzo/Kanalburg, where Marija Ipavec and her family lived into the eastern and the western part. The Italian army immediately occupied the western part, which included Marija Ipavec's home village of Gorenja Vas. The Austro-Hungarian army destroyed the bridge in Kanal on May 25 and thus completely cut off the district's western part from the eastern part. The population was left to deal with the Italian army, daily bombing, the chaos of war, and the inevitable route into exile.

Born in 1892 in Podgora, Marija Bajt married Franc Ipavec in 1914. Franc Ipavec lost his first wife and mother of his three children—Franc Jr., Emilija, and Silvester—in 1913. Marija took care of them after she married Franc. Their first child, Rudolf, was born in the early spring of 1915. Soon after his son Rudolf's birth, Franc, like most men, was drafted, and Marija remained alone with her mother-in-law, three children, and a baby on the farm in Gorenja Vas, which found itself on the front line upon Italy's entry into the war. In a short diary, which survived over a hundred years of neglect and partial destruction, she wrote about her experience on the front line and as a refugee in Italy. It was given to the Goriška Regional Museum in 2015.[55] Journal entries, which reveal the experience of the civilian population and exile during the Great War, are an essential source for understanding the personal, intimate experience of historical events. World War I was "an event that changed the (Western) world, an event that brought about a complete transformation in remembrance, and, finally, the first 'global' event that left in its wake an abundance (if not overabundance) of material through which one can imagine certain situations almost to the last detail" (Luthar, 2015, 8). However, the most sporadic, and therefore so much more valuable, are such records and accounts by women who—in the otherwise extensive opus of personal testimonies from World War I—are the most rarely represented. In her account, Marija Ipavec first describes the period between May 24 and August 19, 1915, when some people persisted on their farms, then multiple stages of the refugee route, and the final settlement in the town of Vercelli near Novara in Italy.

When the front line opened along the Soča River in May 1915, the inhabitants gradually became refugees, moving in several stages and directions. The largest group of inhabitants from the eastern bank of the Soča River was deported to the inland of Austria-Hungary, to Carniola/Kranjska/Krain, Lower Styria, and Carinthia/Koroška/Kärnten, and the refugee camps in Lower Austria as soon as the Soška fronta/Isonzo Front started. There were 80,000 Slovenians among them. From the western bank of the Soča River, the population had to retreat to Italy. Among them were around 12,000 Slovenians who were dispersed throughout Italy, from Sicily to the French border. Later, when the front closed, and the war ended, migration went the other way; most refugees returned to their destroyed homes. Because of the annexation of Primorska to Italy in 1920, some did not

55 Marija Ipavec's son Jožef Ipavec gave the diary to Marjan Nanut to photocopy and give to the Goriška Regional Museum. The diary was transcribed by Dr. Marko Klavora; he was also the one who alerted me to them, and I would like to thank him for that. The 10 pages of the diary begin without an introduction and end without a conclusion, and there are some parts missing in between, which means that Marija Ipavec's diary is not preserved in its entirety. The original diary is kept by the author's grandson Miran Ipavec.

want to return home from the refugee camps and decided to stay in Slovenia in the Kingdom of Serbs, Croats and Slovenes (later the Kingdom of Yugoslavia). Their number is estimated to be around 15,000 (Svoljšak, 1991).

The experience of the civilian population from the Kanal district in World War I is "among the more tragic." Most adult males were mobilized into the army and the civilian population experienced direct attacks, so they had to leave their homes. Additionally, the shelling from both sides destroyed homes, infrastructure, and agricultural land. Marija Ipavec from Gorenja Vas, a stone's throw away from Kanal ob Soči, describes the tragic experience of war and exile of the population of her municipality in her diary. In those fateful spring and summer months of 1915, this "stone's throw away" simply meant that her refugee route led to Italy and not to inland Austria-Hungary where her compatriots from the east bank of the Soča had gone. Eventually, she went into exile together with the mother of her drafted husband, two children, France and Emilija (Milica), and baby Rudolf, who was a few months old when Italy declared war. In the part of the district that the Italian army had occupied, the "evacuation of the civilian population went faster and more thoroughly, but was carried out in haste and without prior notification of the civilian population." When the area was occupied, a no-go operative zone with a 500-meter radius was imposed on the inhabitants. The inhabitants of the area were moved to the hinterland by the Italian army for safety reasons. The Italian side was more efficient than the Austrian and "emptied most of the western part of the district by mid-June 1915" (Devetak, 2013–2014, 328–329).

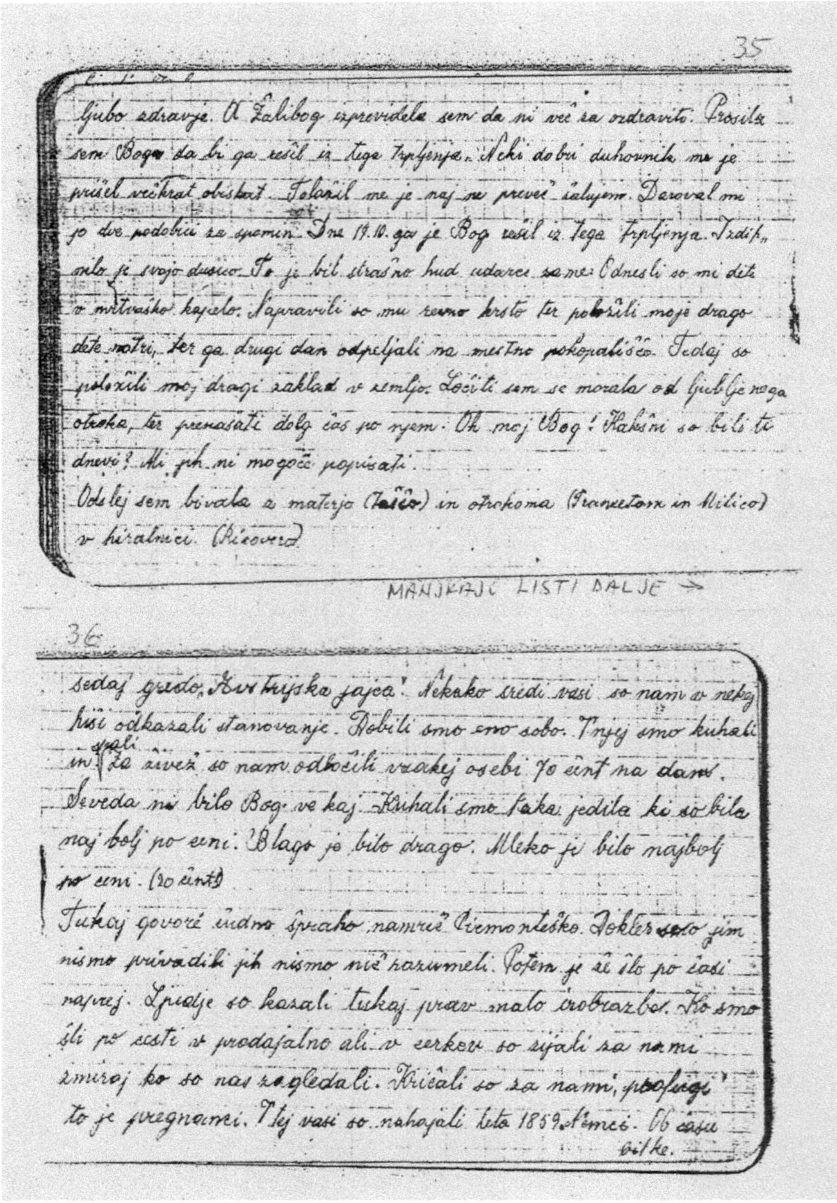

35

ljubo zdravje. A žalibog izprevidela sem da ni več za ozdraviti. Prosila
sem Boga da bi ga rešil iz tega trpljenja. Neki dobri duhovnik me je
prišel večkrat obiskat. Tolažil me je naj ne preveč žalujem. Daroval mi
je dve podobici za spomin. Dne 14.10. ga je Bog rešil iz tega trpljenja. Izdihnilo je svojo dušico. To je bil strašno hud udarec za me. Odnesli so mi dete
v mrtvaško kapelo. Napravili so mu krsto ter položili moje drago
dete notri, ter ga drugi dan odpeljali na mestno pokopališče. Tedaj so
položili moj dragi zaklad v zemljo. Ločiti sem se morala od ljubljenega
otroka, ter prenašati dolg čas po njem. Oh moj Bog! Kakšni so bili ti
dnevi? Ni jih ni mogoče popisati.
Odslej sem bivala z materjo (Tašča) in otrokoma (Francetom in Milico)
v hiralnici. (Ricovero)

MANJKAJO LISTI DALJE →

36

sedaj gredo. Avstrijska jajca. Nekako sredi vasi so nam v neki
hiši odkazali stanovanje. Dobili smo eno sobo. Tu smo kuhali
in spali. Za živež so nam določili vsakej osebi 70 cent na dan.
Seveda ni bilo Bog ve kaj. Kuhali smo taka jedila ki so bila
naj bolj po ceni. Blago je bilo drago. Mleko je bilo najbolj
po ceni. (20 cent)
Tukaj govoré čudno špraho, namreč Piemontesko. Dokler jim
nismo privadili jih nismo nič razumeli. Potem je že šlo po časi
naprej. Ljudje so kazali tukaj prav malo izobrazbe. Ko smo
šli po cesti v prodajalno ali v cerkev so sijali za nami
zmiraj ko so nas zagledali. Kričali so za nami, profughi
to je pregnanci. V tej vasi so nahajali leta 1859 Nemci. Ob času
bitke.

Picture 3.7: The diary of Marija Ipavec. Courtesy of: Goriški muzej (Goriška Regional Museum).

Picture 3.8: Marija Ipavec. Author: unknown. Courtesy of: Miran Ipavec.

Picture 3.9: Franc Ipavec. Author: unknown. Courtesy of: Miran Ipavec.

The Period of Persistence

But Marija Ipavec and some of her fellow villagers were not moved: despite the constant bombing and destruction of their houses and farm buildings, the wounded, and the dead, they could not decide to leave. Her diary draws a picture of the period of persistence, survival strategies, and the experience of a smaller part of the population, which was not exiled by mid-June 1915 and persisted on the front line until mid-August 1915. It tells us about the different survival strategies of the people who, in wartime circumstances, did not want to leave their homes and those with destroyed homes who tried to endure at least on their own land or in the local environment. Despite constant bombings, they worked their fields, harvested the crops, cared for the livestock, prepared wood for heating, and lay down to rest overnight, finding the occasional safe shelter in those buildings that had not yet been completely destroyed. Notwithstanding the hardships, they still endeavored to maintain their daily existence and, in preserving some continuity with the previous life through this everyday routine, they tried to defend themselves from the discontinuities of war—from its dangerous and cruel side. In her diary, Marija intertwines descriptions of the methods and directions of shelling and the deaths of fellow villagers with descriptions of looming starvation, lack of basic foodstuffs, and everyday life in the vise-like grip of relentless fear and confusion:

> Between May 24 and August 19, we slept in nine places. On May 25, they blew up the bridge. Now we are left like savages, without a priest, without a church. In a word, without

> anything. We have very little food. For a time, we made do while we still had homegrown crops. But once that was finished, we started feeling the lack, almost hunger. In the first days of June, two Italian soldiers came to check the trenches of our soldiers. They held their revolvers in their hands when they spoke to people because they feared somebody might attack them. They were sure that our soldiers would be there, hiding. People told them there were none. They could barely believe them. We were at first terrified of these "'Talians." But we had to get used to them. Very many followed these two. They took root in our hills. If one went into the forest to fetch wood or feed the livestock, one was in danger of being taken away to Italy. They took several people from our village. On the day of the Corpus Christi on June 2, the Italians started shooting from Ravnica Hill towards the trenches of our soldiers. The gunfire was constant, back and forth, from 9 in the morning to 3 in the afternoon. All of us were confused. We went into a shelter and stayed there for several hours. Once the shooting stopped, we went home. Those were difficult, sad days for us. In addition to the war, there was nothing to put on the stove. As soon as the potatoes stopped blooming, we were ready to dig them so that we could eat. I was so tired of this sad life, I had to carry my sick baby in my arms and its cradle on my back wherever I happened to go to sleep. And in the morning, return home.[56]

The motifs of hunger, shortages, and starvation are recurrent themes in the entire corpus of World War I testimonies. The civilian population—on the front line, in the hinterland, and in refugee camps—and the soldiers of the Austro-Hungarian Monarchy reported hunger, which affected all classes of the population and the soldiers. The touching descriptions from soldiers' journals about the starved and feeble soldiers who could "barely walk" (Jurkovič, 2018, 198) are complemented by the notes of mothers who, in the hinterland or in exile, watched their children die of hunger (Verginella, 2010, 82–84).

The following two motifs, the concern about the mobilized men and trying to find contact with them and searching for news about those who stayed at home or were in exile, are also a part of Marija Ipavec's diary. In its very first pages, we learn how the scarce news about her husband cheered her up:

> Sometime in mid-March, I got a letter from my uncle in Kanal, in which there were a few words of comfort written and a card included from my husband, which he wrote to my uncle in Loge. I was so happy to learn where he was. At the time, he was in Ljubljana. So I wrote back to my uncle that he should greet him if he'd write.

In June 1915, Marija and her fellow villagers persisted "in the battlefield"—as she described their situation. Persistence on the battlefield took its bloody toll. At the

56 Unless stated differently, the quotes in the text are from Marija Ipavec's diary: Marija Ipavec (1915), diary entries (photocopies). The collection of documents of the history department, Goriški muzej (Goriška Regional Museum), (ZG2/9/), Nova Gorica.

end of the month, precisely on July 31, a bullet fired by Austrian soldiers hit and killed Silvester, one of Franc's sons from his first marriage. In the chronicle of the municipal secretary of the Kanal ob Soči, Anton Bajt, the entry for July 30, 1915, states that "ours shelled the railway warehouse in Gorenja Vas. Rumor is that Silvester Ipavec from Gor. Vas 28 was wounded during the shelling. Another rumor is that he has died" (Sedmak and Bajt, 2006, 61).[57] From Marija's diary, we learn what death meant in such circumstances when the deceased could not be given a proper funeral. Her description of the endeavor to lay him to rest is shattering:

> Oh, what those days were like, I find it impossible to describe them. We needed to take care of his internment. We didn't dare to ask anybody as we knew full well that nobody likes to wade into danger. I went to Gorenje polje to order the gravedigger to dig a hole for him. People told me that the gravedigger doesn't like braving the danger because he could be seen from the trenches, and our soldiers shoot there. So I decided to take him to the graveyard myself with the help of the neighbor's girl (Betina). When I saw we couldn't dig a hole, I came back home with a sad heart. Oh, my God, what am I to do, I thought to myself. And then I remembered that a hole had been dug in Tišler's field. It was made for the late Štefan Sovič (Barba). But they didn't need it because they took him to the graveyard. So I asked the owner if they'd let me move the deceased there. They did. In the late afternoon, we laid him to eternal rest. What were those days like? It's not possible to describe them.

The common refrain among all the records, literary works, and other memoirs about World War I is that the horrors are indescribable. The situation of soldiers and civilians who, on the one hand, go mute, lose their voice and ability to describe the horrors and atrocities they witnessed, and on the other feel the need to do so, to convey them to others who will, although they are indescribable, read about them in the future, is a paradoxical one. Such strategies help them survive, not only in the physical sense but also, primarily, in the psychological sense. Soldiers on the front lines and women on the home front eased their suffering by writing diaries and recording everyday events and thoughts. In their diaries, they were able to describe their real lives and their experiences of the horrors of war much more honestly than in their letters, which were censored (Luthar, 2000, 19–23).

57 It is hard to imagine the division between the left and the right bank of the Soča River when the war divided both the common living area into ours and foreign and the fighting sides into ours and foreign. From the point of view of Anton Bajt in Kanal (Sedmak and Bajt, 2006), Gorenja Vas was shelled by "ours," which means that they killed "our" civilians.

Soon after burying Franc's son, Marija Ipavec received her uncle's second letter, explaining that he visited her husband in Ljubljana and relaying his greetings to her. After that, no more letters arrived. At the beginning of August, grenades destroyed more houses in the village and heavily damaged theirs. When describing the period between May and August 1915, Marija never mentions the children or mother-in-law but does mention the baby several times. For example, when she describes the destruction of the house: "The saddest thing I found was the fact that I had no place to put the baby to bed. A baby that was ill and not even six months old. He had to sleep in the stable."

Exile

Marija Ipavec decided to go into exile on August 19, 1915, along with two other families from the village. Although they were tired of the constant shelling, the fear and hunger, the dangerous life without a roof over their heads, leaving was a difficult decision. Italian soldiers, who were encouraging them to retreat into the inland of Italy, helped them. In the first days, they helped them carry their luggage, took care of their food and lodging, and gave them rides in their cars. Marija describes an unforgettable action of a soldier at one of their many stops: "He saw that our little one was drinking water (I couldn't get milk). It wasn't five minutes before he brought me a bottle of milk for the child. I was so grateful. Yes, this good man will stay in my memory forever."

Marija's first stop was the camp in Savogna/Sovodnje. This tent camp was just a "stop on the way for the refugees and not a pleasant one. They were placed into tents, equipped with some straw, just enough for them to not sleep on bare ground" (Komac, 2017, 220). Once they reached this camp where, as Marija mentions, around 1,000 refugees from the villages of Trnovo, Žaga, and Srpenica lived in military tents, those with small children were allocated lodgings in the attic of a house and given straw for their beds. The tents were near the village Azzida/Ažla next to San Pietro al Natisone/Špeter Slovenov. Marija Ipavec cites other good deeds of the civilian population and soldiers they met in those early days of their exile. When she went to the nearby village with her two neighbors to buy dishes and spoons because they had taken none with them, a soldier spoke to them, asking why they were crying:

> We told him of our grief each. He consoled us not to grieve too much, that God would help us. We arrived at the shop. We bought dishes, but they didn't sell spoons in that shop. When the soldier saw that we didn't have spoons, he went into some tavern and bought us a spoon each. He also wanted to buy us food, but we refused. He asked us if we had our loved ones in the army. We told him we did. He felt sorry for us. He told us

> he would be deployed to the battlefield the next day. He shook our hands goodbye and told us not to grieve so much. This good soldier won't leave my memory.

Marija's notes offer an interesting description of life in the refugee camp which was a part of a military complex and where, as she writes, they were "like soldiers" as well. We cannot know if this means that they were treated better or not, but after they complained about the food, they "lived like gentlefolk. We got black coffee and bread every morning, at noon, soup and meat. And a minestrone at night. We had a church near our flat. And water for laundry. We also had a doctor if anyone got sick. He gave little Rudolf some medicine, and he was better."

They stayed at the camp for a month. They were vaccinated and prepared for the further journey through Cividale del Friuli to Milan and onward to their final destination, Vercelli. In Cividale, where they waited for the train for six hours outside in the cold, Rudolf caught a cold and then got dysentery. The train journey took two days. When they stopped, they got food and milk for their children. Marija describes how "all sorts of curious people" waited for them at the Vercelli train station. Then, they were taken to the seminary, from where, despite their pleas not to be separated, they were displaced throughout Italy. Twelve families, including Marija's, were placed in the city's poorhouse. For Marija, the worst part of exile began. She was in the hospital with her gravely ill baby for a month, but they could not cure him:

> I stayed in the hospital with the baby for almost a month. He drank different medications, but nothing helped. He screamed more and more, crying and moaning day and night. He would have made a stone cry. He dried out so much that he was skin and bones. And so smart, he was. Even if he was only five months old, I felt so sorry for him, to the bottom of my heart, because I saw him suffering so much but couldn't help him. I asked dear God to restore him to good health. But alas, I realized that there was nothing to restore. I pleaded with God to deliver him from this suffering. A good priest visited me several times. He consoled me, told me not to grieve too much. He gifted me two icons as keepsakes. On October 14, God delivered him from suffering; the baby parted with his soul. This was a terrible blow for me. They took my baby to the funeral chapel. They made a meager coffin for him, put him inside, and put a lid on him, and the next day, they took him to the city graveyard. There they laid my treasure into the ground. I had to part from my beloved child and longed for him for so long after that. Oh, my God! What were these days like? Indescribable.

The foremost expert on World War I in the Slovenian ethnic territory, Petra Svoljšak, points out that diverse and complex experiences and events of the population from the Soča/Isonzo Front must be:

> [D]iscussed from different points of view. It is certainly possible to trace some common or general guidelines of their stories, such as the fears on the eve of the Great War, the chaotic escape, the arrival to unknown places, thousands of troubles, hope, and hopelessness, first returns to their destroyed homes. And yet, different elements and problems emerge in this unified picture. Exile is primarily an experience of different losses. The dispersed population was exposed to severe pressures and tensions conditioned with the rather disorganized departure without material and organizational tools. They had to adjust to previously unfamiliar living circumstances, from which economic and social dependence originated (2010, 91).

The common thread in Marija Ipavec's notes is exile as an experience of different losses. From the loss of home and extended family to certainly the worst loss, the death of a child. The description of the loss of her child is followed by the statement: "Since then, I lived with Mother [mother-in-law] and the children [France and Milica] in a poorhouse." A couple of paragraphs are missing. She then continues when they get a room in the center of a village (we do not know which village) where they cooked and slept.

In this part, Marija describes the places of her exile, the people and their habits, the farm work, the nature. She mentions that the people speak a strange language, "that is, Piedmontese," a land of strangely grown rice, a land with a lot of water and frogs, which they catch and eat almost daily. She vividly describes cutting grass, what tools they use, how and how long they must weed rice, and how they protect themselves from mosquitoes. She observes her environment meticulously and curiously. She finds that the churches are empty on Sundays because there is so much work and because the "people are rather godless, almost all socialists." These pages, written in tiny handwriting, not only reveal the reality of everyday refugee life but also shed light on the image of their writer, her personality, her curiosity, and her ability to observe, present, and interpret her environment.

She ends her notes with a sentence characteristic of diary entries, writing: "What I forgot to mention before, so I mention it now." Here, she addresses herself as a confidante and a friend with whom she "shares" her hard experience, which is, therefore, made a little bit easier for her. Thus, two complementary functions—the writing and the reading self—merge in one person:

> The diary is a written form of inner speech and conversation, of addressing the other, the hidden self. In the public space, it is a zone of privacy; in the situation of human reduction to function, it is a zone of individuality and emotionality. In a time of ongoing violence, it demonstrates an intact humanity; in a world of destruction, it demonstrates individual vitality; in a time of loneliness and abandonment, it expresses a longing for beauty, freedom, and love (Novak Popov, 2010, 116).

At the same time, because diaries are often addressed to someone, most often to family members who will read them in the future, she also addresses her potential readers. Marija "forgot to mention" that she was searching for her husband, as she and he found themselves in two different warring countries. She wrote to the International Committee of the Red Cross in Switzerland and asked them to "search for my husband." Petra Svoljšak asserts that in World War I, families had "a constant concern for those living in the trenches [...] It was this constant uncertainty that was regardless of the status common to millions" (2012, 11). Millions wrote daily to the International Prisoners-of-War Agency in Geneva, Switzerland, to get help finding any news about their relatives on the front line or in the hinterlands.

From the incomplete diary of Marija Ipavec, which ends here, we do not find out what happened to her husband. Her grandson, Miran Ipavec, who lived with his grandmother until she died in 1972, described the fates of the people mentioned in the diary. Franc Ipavec returned home, and he and Marija had five more children. Franc died in 1931, but Marija never remarried. She took care of the children by herself. Franc's children from the first marriage, who survived exile in Italy, emigrated in the late 1920s. Franc Jr. left to search for work in the Argentinean city of Córdoba; Emilija moved to Rome, where she worked as a maid.

Exile during World War I was often considered to have had a female character. Compared to the interwar period when, among civilian migrants, men were prevalent, there was a turn. Due to the mobilization of men, up to two-thirds of the refugees and evacuees were women. In the foreign environment, they became the primary carers and representatives of their children and often also of aged relatives since, in the rural Soča Valley, extended families were relatively common at that time (Strle, 2016, 47). Along with the horrors and devastation, World War I also brought "changed family relationships, changed roles of men and women" (Svoljšak and Antoličič, 2018, 295) when women took over the key roles in preserving lives, while men were forced to destroy them. Although women were prevalent among the refugees, they were long forgotten in historiography, invisible, unnoticeable, or merely blurred in the general descriptions of exile, which failed to consider the gender-specific structural positioning and experience of refugees (Wakounig, 2010).

For this reason, Marija Ipavec's diary is an even more valuable testimony of the trials and tribulations of women during World War I and their emotions and experiences, which will hopefully become a part of the collective memory. The Great War caused a hard and often tragic refugee experience for around 100,000 Slovenian exiles; most were women and children forced to leave their homes.

Many never returned. The Great War ended the epochal cycle of European and Slovenian mass migration. It forever changed migration patterns, mobility types, and the directions of intra-European and overseas migrations. Most European nation-states closed their borders and introduced restrictive migration regimes with strict identification requirements.

Negotiating Love, Gender Roles, and Family Care between Slovenia and Egypt: Felicita and Franc Peric

Felicita Koglot migrated to Alexandria, Egypt, in 1921 when she was twenty-two years old.[58] She was a Slovenian from Goriška, a contested region inhabited by peoples with complex identities and loyalties. She left from the nearby port of Trieste, which had, together with the westernmost part of the Slovenian ethnic territory that included her village, just been annexed to Italy after the end of the disastrous Great War.[59] Felicita's decision to emigrate was not motivated by the hardship of war or political instability but by love. She was engaged to Franc Peric, the oldest son of a big family from the same village, Bilje.[60] Like many other young women, Felicita wanted to marry with enough

58 This story has been previously published under the title "'People Cannot Live on Love Alone': Negotiating Love, Gender Roles, and Family Care between Slovenia and Egypt" (see Milharčič Hladnik, 2021). Slight changes have been made in the text to fit the style guide of this book, and some footnotes have been updated.

59 The Slovenian ethnic territory was a part of the Habsburg Empire for centuries. It was divided after World War I: the biggest part became a constitutional entity of the new Kingdom of Yugoslavia and the western part (including Goriška), became part of the Kingdom of Italy with the signing of the 1920 Treaty of Rapallo between the Kingdom of Italy and the Kingdom of Serbs, Croats and Slovenes. The population suffered greatly under the forced "Italianization" and repression by the Fascist regime from 1921 to 1943. This involved economic, social, and cultural measures: devastating tax policy, the prohibition of the use of the Slovenian language, name changes, the banning of all Slovenian organizations, prosecutions of intellectuals, forced migration, imprisonments, and killings. After World War II, the western territory (including Goriška but without the city of Gorizia/Gorica, which remained a part of Italy) became an integral part of Slovenia, the northernmost republic of the Socialist Federal Republic of Yugoslavia, and became an independent state in 1991.

60 Bilje is a relatively large village with orchards and vineyards and centuries-old brickyards. Many buildings were damaged or destroyed during the fights on the Isonzo Front. For political and economic reasons, migration was widespread. There were hardly any households that did not have one or more of their female members working in Egypt.

money saved to provide essentials for the couple's life together. Egypt offered lucrative opportunities, and so Felicita joined numerous women from the region and embarked for the port city of Alexandria. Felicita entered the tight network of translocal and transcultural life of the aleksandrinke, joining this region-specific women's migration circuit motivated by general political and economic reasons prevalent in the Goriška region in the interwar period, but above all, she aspired to meet family obligations and to follow intimate plans. Through an analysis of the correspondence between Felicita and Franc during their separation, this story highlights the importance of letters in maintaining intimate and collective ties in contexts of migration. In this correspondence, love emerges as a crucial feeling and a tool in preserving the relationship amid the challenges created by separation and changes in personal, familial, and gender roles. The analysis that follows explores how love defines subjective experiences of migration in relation to self, loved ones, family, and community. In addition to the relationship of the letter writers, this epistolary love story illuminates the broader family and community systems of migration at play in the sociopolitical contexts of two very different, yet closely connected worlds—the villages of Goriška, Slovenia, and the urban centers of Egypt in the 1920s.

Felicita and Franc: An Epistolary Love Story

Felicita migrated to Egypt twice. She first left in 1921, and through the migration network quickly found a job as a nanny for a family in Alexandria. She saved money, returned home after three years, married Franc, moved into his family's household—as was the custom for new brides—and gave birth to a girl named Danila.[61] In 1928, when Danila was three years old, Felicita and Franc decided she should migrate again. For the next three years, she worked as a nanny with a family in Cairo, occasionally visiting Bilje while traveling Europe with her employers, and returned home for good in 1931. This kind of migration cycle was nothing unusual in a region in which most families had someone abroad: be it in Egypt, Argentina, the United States, or Switzerland—the latter close enough for Bilje's construction workers, Franc among them, to go there for seasonal work.[62]

61 In the letters, her name appears in different loving variations: Neli, Dani, Danilka, Nila, Nini, etc.

62 Franc migrated for seasonal work in 1929, in the context of a worsening political situation and a generalized economic crisis.

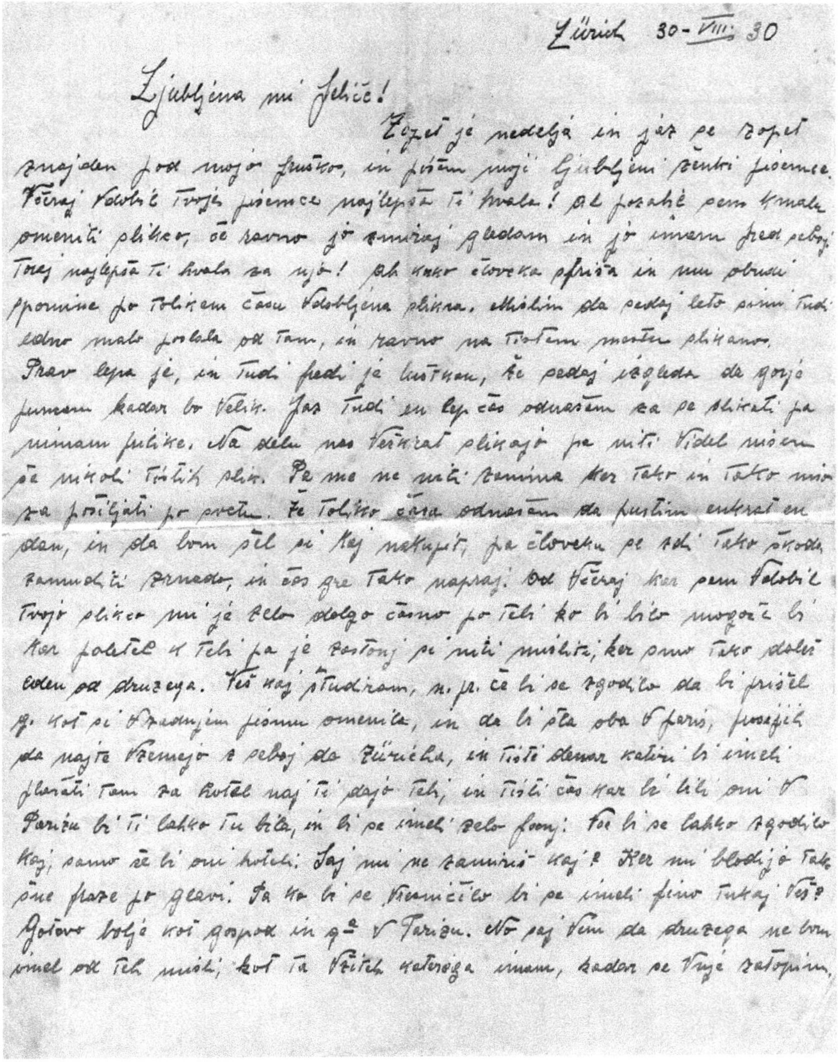

Zürich 30-VIII-30

Ljubljena mi Felice!

Picture 3.10: Letter from Franc to Felicita Peric. Courtesy of: Rusjan family archive.

Felicita and Franc discussed a wide range of topics in their correspondence—among them, the intensifying political repression at home and political tensions in Egypt, relationships with employers, her difficulties getting paid in a timely fashion, the care she received from her employers when she was ill, the exchanges of gifts and the dispatches of goods like grapes, wine, and ham for the Egyptian family, the activities of local ethnic organizations established by the Slovenian Catholic Church in Egypt, whose events Felicita attended regularly, and changes in everyday life in Bilje, Alexandria, and Cairo. However, I will focus on the intimate world that this correspondence illuminates. My selection of letters reveals the epistolary love story spanning six years in two periods of migration and highlights four topics that are relevant to understand the emotional dynamics experienced by the couple: love and longing, doubts and concerns, gossip and rumors, and family relationships. I have chosen these topics of epistolary conversation to present the main goal of the correspondence: to preserve and strengthen social and family solidarity, and above all, the young couple's intimate relationship.

Picture 3.11: Felicita Koglot. Author: unknown. Courtesy of: Rusjan family archive.

Picture 3.12: Franc Peric. Author: unknown. Courtesy of: Rusjan family archive.

I view the preservation and strengthening of this relationship through the perspective of the (temporary) rearrangements of the normative gender roles and attitudes created by migration. Apart from the challenges to emotional ties caused by the years of separation, Felicita and Franc's experience was characterized by two different familial dynamics and gender roles' constellations. In the first three years, Felicita was engaged, but single, with no children and also no work experience outside home. During her second stay in Egypt, she was a more confident and experienced worker, a wife, and the mother of a small child. She established an intense relationship with the family and with the children for whom she cared, Marika and especially Fredi. In the first period, Franc was young, single, and engaged to a migrant woman. In the second period, however, he was a married man and a father; yet having an absent wife and mother of your child was not a traditional and prescribed family ideal. Though it was common in these villages for wives and mothers to migrate and, in many cases, be the main breadwinners, rearranging gender roles and negotiating embedded sensibilities (Wickberg, 2007, 669) of femininity and masculinity was not a simple task. The correspondence exchanged by Felicita and Franc offers us a glimpse into how they negotiated their new roles and navigated the challenges created by the tension between socially prescribed gender expectations and their transformation in the migration process.

Correspondence in the Research of Emotions and Gendered Mobility

We must understand the preservation of the emotional ties and the negotiation of the normative gender roles visible in the epistolary dialog between Felicita and Franc in the context of the controversy that female labor migration caused at the individual, family, and community level (Gabaccia, 1994; Anthias and Lazaridis, 2000; Passerini et al., 2010).[63] From the second half of the 19th century until 1940, when it came to an end because of World War II, aleksandrinstvo, as the phenomenon of female migration to Egypt is called, represented an important financial resource for the Goriška region (Barbič and Miklavčič Brezigar, 1999). Migration to Egypt was spurred by the economic prosperity created by the opening of the Suez Canal in 1869 and by the profitable Egyptian cotton industry. Prosperous Egyptian cities witnessed an influx of cotton merchants and other wealthy families as well as many migrant workers from Europe, including aleksandrinke (Biancani, 2015).

A complex phenomenon, aleksandrinke encompassed diverse patterns of women's migration, re-emigration, circular and chain migration, and, above all, different personal experiences. The emotional underpinning of experiences varied considerably according to the type of work and the migrants' personal circumstances. There were clear differences between women who worked as nannies and wet nurses and women who did other types of work; between a few women who had their children with them and the majority who left them at home in Goriška; between single women and those who were wives and mothers. Regarding children, there are a few testimonies of truly abandoned children whom nobody cared for, but many more testimonies of less traumatic childhoods (Koprivec, 2013).[64] The different responses of men, spouses, and fathers to the mass migration of women varied from a caring

63 Suzanne M. Sinke conceptualized and defined the process of rearrangements of gender roles with the term "social reproduction." In her own words, she uses this term "[t]o describe how people put those roles back together, sometimes replicating, sometimes revising, sometimes totally reformulating their ideas of what women and men should do" (2002, 1).

64 We also need to consider the emotional experiences of the children the aleksandrinke cared for, and their testimonies of how the Slovenian nannies marked their lives and identities (Douek, 2004).

role they took without hesitation or performed with the help of their kin, to a retreat to the most reckless behavior of drinking and spending the money sent to them by their daughters, fiancées, or wives.

Even if the decisions to go to work in Egypt were always the result of family deliberations, the fact that women became breadwinners threatened prescribed gender roles and caused what Rhacel Salazar Parreñas calls "the gender paradox": the more changes of traditional family and gender roles are necessary because of the absent women, the more persistent everyone involved is to preserve them (Salazar Parreñas, 2001, 2005). The fact that women were far from the tight social control of family and village prompted public opinion and policy makers to attack them regularly. The Catholic Church and local priests condemned the practice as a threat to morality and used gossip about the supposedly immoral behavior of aleksandrinke to preach against the migration of women. They linked women's migration to the danger and fear of moral corruption, ignoring the economic and political causes of migration.[65] The institutions and discourses that supported the denigration of women migrants were far from limited to the Church. Dirk Hoerder argues that the discourse of disdain and moral condemnation of female work and migration was deeply rooted in 19th- and 20th-century nationalism and has not yet been overcome. The reduction of women to the role of "mothers of the nation" was intense in all political and ideological varieties and it was institutionalized on the legislative, cultural, and ideological levels. For this very reason, the concrete experiences, achievements, contributions, and importance of female migrants had to be silenced, overlooked, forgotten, or wrapped into traumatic intimate memories (Hoerder, 2015).

For women who migrated to Egypt, where they were mostly well respected and well paid, migration had a profound impact on their public image as unconventional female characters whose reputation fluctuated between "silent thankful adoration and loud moral condemnation" (Milharčič Hladnik, 2015b, 12). As such, it brought with it rearrangements of the traditional roles within families, of the value of women's work within the rural economy, and of the emotional system of the community. The concept of emotional community (Rosenwein, 2010) applies, in this case, to a social community of families from the cluster of

---

65 Moralistic accusations and condemnations of women migrants from Goriška to Egypt caused a collective trauma of families in the region that lasted until recently. For more on the recent attempts to reevaluate this specific women's migration by the local population of Goriška and the descendants of women migrants (see Milharčič Hladnik, 2015b, 11–38).

villages in the Goriška region where all were affected by this specific female migration. This concept allows a researcher "to uncover systems of feeling, to establish what these communities (and the individuals within them) define and assess as valuable or harmful to them (for it is about such things that people express emotions); the emotions that they value, devalue, or ignore; the nature of the affective bonds between people that they recognize; and the modes of emotional expression that they expect, encourage, tolerate, and deplore" (Rosenwein, 2010, 11). Autobiographical sources that convey experiences and emotions of individuals, of ordinary people and their everyday lives help us gain this perspective.

Since the 1980s, the importance of emotions has been recognized not only in histories of emotions, but also in migration studies (Redy, 2009). The use of diaries, auto/biographical texts, letters, photographs, and life narratives has been extensive in migration studies in the past few decades (Bailey and Ramella, 1988; Zempel, 1991; Cancian, 2010; Elliot et al., 2006; Thomson, 2011; Hladnik, 2014). Samuel Bailey and Franco Ramella assert the research importance of letters when they state: "We must use letters for what they can uniquely document—personal insights and feelings. Letters provide the subjective perspective on the immigrant experience" (1988). Personal correspondence reveals the complexity of individual and family decisions, everyday choices, cultural negotiations, and social experiences, and, above all, personal interpretations, intimate feelings, and the infinite series of emotions triggered by decisions about separation among those who left and those who stayed behind.

There are several methodological limitations regarding letters as a source. Susan J. Matt mentions the unclear relationship between words and feelings and the question of audience: "Writings, even private scribbles and diaries, are composed for an audience. That fact shapes what is included and what is left out, what is accented, and what is unstressed" (Matt, 2011). The chronology of the letters is also never an easy issue. There might be letters that should have arrived earlier but did not due to various circumstances; some are without dates, some with the date but not the year, and there might be missed letters or unanswered ones (Zempel, 1991, xiii–xiv). David Gerber calls our attention to another methodological challenge, which he terms the "epistolary masquerade," describing it as a problem of silence and untruths. He stresses that "immigrant letters are not principally about documenting the world, but instead about reconfiguring a personal relationship rendered vulnerable by long-distance, long-term separation. It is in the service of that goal that the letter develops its content, so that goal, profoundly illusive in its own way because it springs from the most profound recess of human needs and emotions, needs to be understood as the source of the aspirations of the parties involved" (Gerber, 2006, 143).

Nevertheless, correspondence reveals "the voice of the past" (Thomson, 1988) and the emotions of their migrating protagonists. As Sonia Cancian argues, "Letters take us inside the minds and hearts of ordinary people whose personal and family identities and circumstances were most affected by the realities of international migration" (2010, 5). This is exactly what we find in the correspondence of Felicita Koglot and Franc Peric: it is rich in content and extensive in the number of letters, postcards, photographs, and numerous references to packages that made up the exchanges of emotions, concerns, longings, doubts, and hopes in the decade between 1921 and 1931 from both sides.[66] I had the privilege to read and analyze a part of this vast correspondence: 110 letters written by Felicita to Franc and 122 letters sent by Franc to Felicita. During my research on the aleksandrinstvo phenomenon, I met many relatives and descendants of aleksandrinke, and one of them was Neda R. Bric, the granddaughter of Felicita and Franc Peric, who let me use her well-kept family correspondence for research purposes and provided additional background in several conversations.[67]

Uniting through Letters: Dreaming and Longing

The motif of an engaged couple's romantic love filled the correspondence immediately after Felicita left in February 1921. In the letter sent to Franc just "before Easter," Felicita wrote:

> Oh, how happy I was tonight, dreaming that Zorka [her roommate and friend] and I returned home and that you all came toward us on the road, but when we were only a few steps apart and I'd already thought how we would greet and kiss, Madam called me to go to the baby. Such a pity! Let's hope that this, like all in the world, passes; these three years pass and if God preserves our health we also live to see that happy reunion. How often do I picture that happy moment when the steamship approaches the dock. Ah, I mustn't ponder, because this goal is far, far away. You know, now I understand the saying that abroad more tears are shed than water is drunk. Just know that at times my heart wants to break with homesickness! (Felicita, before Easter 1921, s. d.).

Franc wrote his answer on Easter Sunday, pouring his sadness into words:

66 This is rarely the case as usually only one side of correspondence is preserved (Elliot et al., 2006, 3).

67 I would like to express my gratitude to Neda R. Bric for access to this correspondence, long conversations, her help in understanding family relations, and her trust. All presented letters are from her family archive. For my previous work about the experience of this migratory experience (see Milharčič Hladnik, 2015).

> To ease the bitterness of today I decided to write you this letter. First, please accept thousands of heartfelt greetings and kisses. It is a holiday, everything is beautiful, the procession at the resurrection mass was beautiful. Everyone is happy, everyone is brave, only my heart is terribly sad because it has lost its greatest treasure that it loves (Franc, Easter day, 1921).

These lamentations went on in the correspondence throughout the whole year. In August 1921, Felicita wrote: "Please know that when I was reading your letter I cried bitter tears. Why is God punishing us so much; I am truly afraid for you all, that you will suffer hunger. My God, every time I sit down to eat, I think of you, that here I have [food], and you over there, God knows" (Felicita, August 18, 1921). A month later her beloved replied even more inconsolably: "I don't know what would happen if I sent you any of those letters I write when I think of you too much, when I'm sad, when I can find no solace. If you felt like me in your heart you'd cry like I do, or you'd say, this man is stupid" (Franc, September 29, 1921). However, besides the hurting, longing, and sadness they also exchanged joy, expectations, and details about their dreams, especially Felicita. In the same letter from August 1921, alongside tears and fears, there were also encouraging words:

> How we used to watch the moon and the stars thinking of the days to come. Know that even now I look at the moon and the stars and I think that perhaps you are watching them at the same time and think we see them the same way. Unfortunately, we're far apart from each other, but know that in our thoughts we're always together, aren't we? I often see you in my dreams, and I find some consolation in this (Felicita, August 18, 1921).

Far from unique, the topic of looking to the same stars and the recounting of the dreams seem to be common motifs in the correspondences between lovers and couples separated by migration. Cancian has found references to the same topics in her study of Italian migration to Canada after 1945: "It seems that the act of dreaming functioned both to trigger the imagination of correspondents and to bridge the distance between lovers through a recounting of the experience" (Cancian, 2010, 136). It also seems that the recounting of dreams served the purpose of expressing erotic longing and fueled further desire for the absent lover.[68]

68 However, in his extensive research of epistolary practices of Irish immigrants in Australia and their loved ones in Ireland, David Fitzpatrick found the reporting of a dream also between friends and relatives. In the dreams, they were happily united again or "the dream signified not a happy conjunction but a supernatural warning of disaster or death, traveling faster than the promptest telegram" (Fitzpatrick, 1994, 494).

Separation also created uncertainty. Only nine months after the departure and a month after words of sad passion and longing were exchanged, a very different topic appeared in the couple's correspondence. A letter expressing doubts, suspicions, and jealousy arrived from Franc, and Felicita obviously felt compelled to answer it decisively:

> As for me, have no worries, because I have neither the freedom nor the parties, nor anything at all to get myself in any danger. And the thought of home and my beloved encourages me [...]and I cannot wait for that moment that we see each other again. What, have you already tried it, as you write that you feel when you hold a girl or even hug her, it feels like hugging a tree? I'm curious. I've not sullied myself not with the slightest thing, and God make it stay this way (Felicita, November 1, 1921).

The personal, intimate doubts Franc expressed in his letters were penned in the context of a general suspicion about aleksandrinke that resulted from a combination of intense moral condemnation voiced by priests and widespread gossip within the tight-knit communities of the villages. Franc wrote about gossip extensively as well as about other events and social changes at the level of local communities and individuals. He wanted to produce epistolary closeness and simultaneity with his beloved fiancée.[69] The letters from November 1921 provide good examples of this double objective: Franc reported to Felicita about nasty gossip but also sent endless confessions of love. He wrote that on a rainy Sunday, he had done nothing else but read her letters over and over: "My beloved Felička, one of my favorite letters of yours is the one you gave to me that last Sunday in our field when we went to Gorica, do you still remember? I will never forget you, nor the places where we were together" (Franc, 7 November 1921). Felicita also wrote back sweet words: "I was pleased to get your lovely letter this week again. Thank you from my heart. I'm glad that you respond to me regularly. I hope that now that long winter evenings have arrived you will have more time to unite with me through letters" (Felicita, 30 November 1921). Nevertheless, the romantic "uniting" through correspondence turned again to the bitterness of moral suspicions and gossip. Felicita wrote at the end of 1921:

> I was pleased to receive your letter, but as I was reading it such sadness came over me that I couldn't be happy at all. Because your writing reveals that you're very upset [with me] and are even doubting [me]!! I think that nobody can say a bad word about me for leaving. Do not curse Egypt, because people cannot live on love alone, and where there's

69 Many studies have helped us to understand that this "simultaneity" is an essential part of the subjective experience of the migration process—of those who leave and those who stay (see Levitt and Schiller, 2004, 1002–39).

> poverty it's often a cause for arguments in married life, and praise God that He gave me a chance to earn some money [...] As I understand from your writing, it might appear that because we're in a so-called bad city, we're all the same. You're wrong, when we're free, we go to the nuns, and there we play, sing, and that's all the joy we have.[70] My heart aches, because I think if we saw each other, you'd look at me with suspicion and not like before (Felicita, December 26, 1921).

In defending herself against suspicions and insinuations, Felicita writes in this letter about the tight connection between emotions and interest. She understands money as a prerequisite for marital happiness and well-being because love is not enough to guarantee happiness, and poverty can destroy it. Despite realistic and mature epistolary conversations about this problem, letters from 1922 brought more gossip. Another difficult topic appeared in these letters, namely the complicated relationship between the members of the two families, especially between Felicita's mother and Franc. Over the years, even when already married, the tension between Franc and his mother-in-law would grow as we will see in the letters from the second period of separation. In August 1922, Franc wrote a harsh letter regarding this matter:

> As for your picture, I swear I didn't get it. If you care to, resend it!? Further, you're insulting [his underline] me if you say things like that regarding your family, that I probably don't go there very often, or even not at all. Now that you haven't written to me in a while, I was there every week, I asked if they had anything from you. Finally, on Saturday they received your letter which your mother let me read. And when I reached the lines that were about me, I felt more than one knife in my heart. I'll leave these things be for today, perhaps more about them next week. I will just tell you this, you listen to people too much, and I think there is no uglier sin in the world than these old chin-waggers from Bilje (Franc, August 5, 1922).

Despite the tension created by distance, time of separation, gossip, the public accusations about aleksandrinke, and family difficulties, the following letters show that the love between Felicita and Franc prevailed. Deploying a familiar strategy, in September 1923, Felicita yet again expressed her love and doubts simultaneously: "Please know that although I love you with all the love that one can love with, if I knew that you'd be happier not marrying, of course, I'd be

70 On their free days, Slovenian women got together at the "nuns'" place in Cairo and Alexandria, where they participated in different cultural events and maintained a tight network of support. The Slovenian Women's Christian Union, the School Sisters of St Francis of Christ the King (known as the School Sisters, they took over managing the Asylum of Franz Josef, now St. Francis Asylum), and the Catholic Society of Saints Cyril and Methodius had been active in Egypt since 1908.

miserable, there are no words to describe it, but I'd please you" (Felicita, September 18, 1923). But shortly after, in a letter from October 3, she shared her happy dreams with him: "The night before last I dreamed that I came home and I saw you from afar plowing with the horses, I thought to myself 'He'll come,' and you did, but just to say hello [...] And then I told you, see, three years passed in a moment, and I'm here again. It was such a pity to wake up from dreams like that" (Felicita, October 3, 1923).

Optimism and happiness predominated in the correspondence as Felicita's departure from Egypt and return home came closer. Felicita summarized the three-year separation in positive, hopeful terms, in a letter from February 1924:

> So today as I write this, three years have passed since I stepped on Egyptian soil. Three years and yet they've gone by quickly, no? A lot of water under the bridge, a lot of tears, but our love is still firm, like a rock. Ours is that ditty: everything in this world passes, everything comes to an end, only love goes without changing, faithful love, forever there. So three years, now I'm calculating another three months, and if God grants us health, we'll pick the first cherries together (Felicita, February 1924, s. d.).

They did pick cherries together that spring, got married, and had a baby girl, Danila, the next year. As we can understand from the correspondence in the second migration period and from the oral testimonies of their granddaughter, they were happy. However, the Italian state's forced Italianization of the Slovenian population gradually worsened the political and economic situation in the Goriška region. This was undoubtedly one of the main reasons that Felicita and Franc decided she would leave again.[71]

Enduring through Letters: Weeping and Consoling

The correspondence started again on a sad day in February 1928, when Franc returned home after watching Felicita set sail from Trieste for the second time. He described his suffering vividly in his first letter, written at the moment that the second period of separation began:

> You have no idea how sad I came home. I couldn't even speak when I stepped into the house or when I saw Danilka. I thought I'd not make it until the hour when it's time for bed. But with a lot of attention and suffering I managed until 7 o'clock and me and

71 Her case was far from unique. Village and family life adapted to the migration of women. Family, relatives, friends, and neighbors took care of children when mothers (and fathers) left and, even in the case of those who went as wet nurses, the community had a system of support that worked well.

> Danilka went into our once so dear, but now lonely room. Oh, if only she could tell you, once you return, how many tears were shed that night, that we could both wash in them in the morning, Danila and I. The day was better, the second evening was still a little bad, but now I'm good. Danilka is fine, don't worry, you have no idea how many kisses she sends you every day (Franc, February 14, 1928).

At the intimate, emotional level of couples and children, the migration of women complicated life as much as it made it easier. Women's everyday physical and emotional care was absent but with the money they sent from Egypt there was enough food on the table, taxes and loans that put farms in danger were paid, houses were repaired or enlarged, and a new field or a vineyard purchased. Working abroad was a "blessing" but it all came at a high emotional price on the individual and family levels. Felicita described this complexity in her letter, addressing her daughter and husband, in December 1928.

> Honestly, I don't know if we'll ever be happy again. Like you write, perhaps if we cover our ears and fuse our eyes so we don't feel what's going on around us. In this way, we could still savor our previous happiness. Still, I wish indescribably to hold you both in my arms in the embrace of marital happiness. My heart hurts when you write about Nila that she asks for her mother when she comes home. Oh, child, if you only knew how your mama wants to see you and hug you and love you fiercely, but it's all in vain. God knows how much time will pass before this happens. Because the times are getting worse and blessed are the families who have members around the world (Felicita, December 5, 1928).

Felicita and Franc were loving parents who adapted to the challenges created by migration and family separation. In their correspondence, we see how Franc as a father and husband did not have a problem with the rearrangements of traditional and prescribed gender roles or with the expression of emotions.[72] He took care of their daughter lovingly, and understood Felicita's dire situation as a mother, being so far away from her child. In March 1929, he wrote to her the following compassionate lines:

> Truth be told, mommy, it hurts me that you cannot hear and see how our beloved daughter is growing. [...] it's impossible to tell you how I yearn for you, my beloved

72 The fact that women and men are equally articulate and passionate in conveying emotions on paper is also found in Cancian's analysis (2010, 142). I have argued the same in the analysis of the Udovič-Valenčič-Hrvatin family correspondence. In this correspondence, the brother and the sister who migrated to Cleveland in the United States expressed equally moving emotions in the letters to their mother and family in the Slovenian village of Jelšane (Milharčič Hladnik, 2014, 21–67).

> wife, and how Danila is also all pensive and upset when I tell her about you, her precious mommy, and how mommy loves her, and how she cares and sacrifices herself. [...] My love for you, I hope I don't have to convey it anymore, because surely you don't doubt that I love you with all my might and that I yearn for you as the thirsty yearns for water (Franc, March 18, 1929).

It is obvious that being parents did not mean that they stopped expressing their love and passion as a couple. Yet, it all became more difficult as the gossip continued and family problems intensified. In 1929, Franc joined the seasonal migration to Switzerland where construction workers from Bilje and neighboring villages worked every year. Thus, Danila stayed with his family and relatives. His 1929 work season ended in December, and he reported to Felicita very bad news from home:

> Please note that they all talk about your mother, how she is the way she is, and everybody's surprised how she can be like that. I've never spoken ill to her, I've put up with it patiently, I've cried, you and I, we've both cried to console ourselves, but now I'm so fed up with everything that I don't know if I'll be able to keep quiet about everything when I go over there. Come what may! She's gone too far, I don't know what would happen if someone else was in my place. And I'm advising you, too, not to write to her, not to send her anything, because, mother or not, she's not worth it.
>
> I know that this insults you, Felička, but I've heard from reliable people that she wants to alienate you from me, and that she wants to disgust my entire family toward you without any of us doing anything bad to her. These days, I've not had a happy hour. All days and nights I am wrought with worries. And instead of looking forward to your return, it scares me, our much-awaited reunion! And this is the truth, mommy, I know very well, that there will be no peace (Franc, December 4, 1929).

Apart from the growing problems with his wife's mother, Franc had difficulties being back in the village and having to readapt to its traditional ways. Perhaps he felt what so many women felt when they returned home from Egypt—unexpected alienation and isolation. Instead of happiness, aleksandrinke often felt that they no longer belonged there. The difficulties upon returning home were caused by changed identities brought about by the new experience of independence and dignity, by the acquired knowledge of foreign languages and an urban, higher-class lifestyle, and by traveling the world and Egypt. Upon arriving home, women had to again accept the prescribed gender norms, the required social control, and the rural, farming way of life. For aleksandrinke there was another problem: the adjustment to the interpretation of their experience as morally questionable, which remained as a collective trauma deeply embedded in the community until recently.[73] Feelings of inadequacy were also common

73 This trauma has been collectively dealt with through the Society for the Preservation of

among returned seasonal workers like Franc. These feelings were compounded by the absence of wives, in the case of married men whose wives were working abroad. Franc confided to Felicita his feelings of isolation: "I feel very bored, no company, I like it best to keep at home. My old mates, they don't fit right." And he mentioned the sorry fact that hard-earned money from Egypt was trickling down the throats of ungrateful, indifferent, or even frustrated and suspicious husbands in the local bars. He indignantly described his peers as "only drunkards and scroungers" (Franc, 4 December 1929).

Franc's detailed descriptions of Zurich in the next year's correspondence, when he was again a seasonal builder, are particularly noteworthy. In the midst of describing to her the beauty of Zurich's surroundings, he dreams about seeing her soon:

> Since I got your picture yesterday I've missed you so much; if it were possible I'd fly to you, but it's in vain to even think about it when we're so far apart. You know what I've been thinking, for example, if it happened like you mentioned in your last letter that they'd both go to Paris [the couple she worked for], ask them to take you with them as far as Zurich, and give you the money they'd spend for the hotel over there, and for the entire time they're in Paris you could be here and we'd have a good time (Franc, August 30, 1930).

They did not meet and stay together in Zurich. Their reunion had to wait until another cherry-picking season in Bilje in 1931, when Felicita finally came home and never left again.[74]

---

the Cultural Heritage of the Alexandrian Women. In 2005, the descendants of aleksandrinke in Slovenia organized themselves into this society, started to collect testimonies and material heritage and, a year later, opened a participatory museum in the village of Prvačina, in the Goriška region, in order to establish the perception of aleksandrinke as courageous, successful, and determined saviors of their families. See the website of the museum and the society: http://www.aleksandrinke.si/eng/ (See also, Milharčič Hladnik, 2015b, 3).

74 In many cases, the love for the children in the care of an aleksandrinka and the love of those children for her did not cease with her departure. The existing love and affection created translocal emotional ties that, in many cases, lasted until the death of the aleksandrinka and even through generations, with her descendants. Upon her arrival home, Felicita put Fredi's photograph on a shelf in the dining room and it stayed there. Bric remembers that for them, Felicita's grandchildren, Fredi was their "cousin" (Neda R. Bric, e-mail message to author, October 15, 2016).

Conclusion: "Our Love Is Still Firm, Like a Rock"

Correspondence is considered one of the central documents to study "history from the inside out" (Matt, 2011) and to understand the solidarity, bonds, experiences, and emotions of individuals involved, whether they stay or migrate. The analysis of correspondence offered in this text provides an insight into the complexity of feelings of love as driving forces behind the decisions and everyday negotiations in translocal and transcultural settings. The correspondence Felicita and Franc kept during the years of separation helped them maintain and strengthen their relationship. The selected epistolary topics convey expressions of love, loyalty, and devotion that were exchanged to preserve the relationship even in times of disagreements, resentment, and misunderstandings. These hard times were further complicated by the couple's family problems as well as by external issues such as public gossip and attitudes instilled by the Church and community, and by the social, economic, and political realities in Europe and Egypt at this turbulent time. Testimonies of their "epistolary emotions" (Seymour, 2010, 148) reveal how gender and migration were interconnected in the correspondence of a young couple involved in the phenomenon of aleksandrinstvo. The selected letters offer a glimpse of the subtle and complex emotional landscape resulting from the rearrangement of gender roles and of the embedded sensibilities of femininity and masculinity caused by migration.

Apart from the relationship of the writers of the letters, my main goal was to show how their love story was part of the larger family and community system of migration in the social and political context of two different yet connected worlds—Goriška villages and Egyptian cities. They were connected by women and men who loved, cared, hoped, and feared on both sides of the Mediterranean. They were aware that no matter how strong their love was, migration and separation were a necessary challenge of their pursuit of happiness. As Felicita put it in her letter to her beloved Franc, "people cannot live on love alone."

Forced Migrations and Childhood Experiences of Exile during World War II: Terezija Černelič and John Tschinkel

Following the German invasion of Yugoslavia on April 6, 1941, the territory of Yugoslav Slovenia was divided among Italy, Hungary, and Germany. The regions of Lower Carniola/Dolenjska and Inner Carniola/Notranjska, as well as the capital, Ljubljana, and its surrounding area, were occupied by Italy and named the Province of Ljubljana. The greater part of the Prekmurje region on the eastern side of Slovenia was occupied by Hungary. A few villages on the Slovenian–Croatian border were occupied by the Nezavisna Država Hrvatska (Independent State of Croatia, or NDH, the Ustashe-led Nazi puppet state). The largest part—Upper Carniola/Gorenjska, Lower Styria/Spodnja Štajerska/Untersteiermark with a part of Lower Carniola, the Meža Valley, and the remaining northwestern part of the Prekmurje—was occupied by Germany. In all the occupied territories, the occupying forces were to carry out ethnic cleansing.

The program of ethnic cleansing consisted of ethnic and racial classification of the population, internment, and destruction of national institutions, the expulsion of the majority of the population, and the colonization of the territory by settling the Italian, German, and Hungarian populations (Gombač and Gombač, 2013, 18). In the Province of Ljubljana, around 30,000 people were either exiled or jailed. In Prekmurje, the first to be deported were nationally conscious Slovenians who moved there in the interwar period after the annexation of Primorska to Italy. Later, they deported other nationally active Slovenians and, in 1944, the entire Jewish community. The deportation of the Jews started in April 1944, and most of them were murdered in Auschwitz in that same month. With that, the largest and most flourishing Jewish community in Slovenia was destroyed (Toš, 2010, 645–646). Germany had the most ambitious ethnic cleansing plan for the largest segment of the occupied Slovenian territory: expel the majority of the population, that is, between 220,000 and 260,000 people, and "Germanize" the territory (Luthar et al., 2013, 416–421).

Map 3.1: The occupation of Slovenia 1941.

The Collective Experience

The Nazi Germany planned to make this territory "German" within three or four years. "No other occupied territory experienced a deadline for total Germanization quite as short as the Nazis set for the occupied Slovenian regions, particularly Lower Styria" (Ferenc, 1971b, 6). They removed all the Slovenian signs and replaced them with German ones. They abolished and expropriated all Slovenian organizations and institutions, Germanized personal and geographical names, confiscated and destroyed all books, closed Slovenian schools, and, as early as August 1941, brought in German teachers who started to teach in German (Kokalj Hočevar, 2010, 139). Immediately, they started an intensive expulsion, first of educated Slovenians, then the rest of the population. Before the expulsion, they expropriated the people and racially

> categorized them, divided them into categories suitable for Germanization. During World War II, the Germans deported about 63,000 inhabitants from these regions: 45,000 to Germany, 7,000 to Serbia, and 10,000 to Bosnia and Herzegovina and Croatia. They also expelled about 1,000 Slovenians from Carinthia and 8,000 Slovenians from Lower Styria and Upper Carniola, the relatives of partisan fighters and assassinated hostages (Ferenc, 1993, 76).

The expulsion was the most meticulous in the regions of Posavje (Lower Sava Valley) and Obsotelje (Sotla Valley): between October 23 and December 9, 1941, 32,367 people were deported, the rest by the end of July 1942 (Ferenc, 1971a, 48–49). Within a year, the Nazis expelled about 43,000 people from Posavje and Obsotelje, around 83 percent of all the people that lived there at the 1931 census (March 31, 1931), the last before World War II (Ferenc, 1971b, 16).

The exiles had the status of Absiedler, that is, emigrants, and constituted a segment of the plan for the ethnic cleansing of the territories to be colonized by ethnic Germans and, later, annexed to the German Reich. The population of these regions was transported to German labor camps or farms as slave labor (Ferenc, 1971a). Slovenian exiles were placed into more than 300 camps and moved frequently. Their distribution to the camps or workplaces depended on whether they were—before they were exiled—racially classified as capable or incapable of being "Germanized." A special program was in place for young girls regarded as particularly suitable for Germanization, so they were taken from their families and placed in large German families. A similar fate befell some children who were separated from their parents and handed over to German families or sent to special re-education camps. They have become known as "stolen children." There were allegedly more than 600, but their exact number is impossible to establish (Repe, 2010, 254). Children and children's experiences in migrations in general, and particularly in forced migrations, are less well studied (Milharčič Hladnik, 2015a), so adults' testimonies in which they focus on their childhood spent as exiles or deportees are particularly important (Gombač and Gombač, 2013; Strle, 2015b).

Slovenian exiles were part of an immense mass of people who were forcibly displaced from their homes throughout Europe and brought to the territories occupied by the German Reich as forced laborers. The editors of the book *Hitler's Slaves: Life Stories of Forced Labourers in Nazi-Occupied Europe* estimate that there were around 13.7 million of them; of these, 8.4 million were forced laborers-civilians, while the remaining were prisoners of war and concentration camp prisoners who did slave labor. Without their work in agriculture, mines, quarries, the chemical and metal industries, and the war industry, the economy of the German Reich would have collapsed long before the military defeat (Plato et al., 2010, 3–6). These millions of long-term forced and slave laborers of the German Reich form the least researched and least "heard" part of the mass experience of World War II victims (Ogris, 2011). Therefore, the abovementioned book is all the more relevant, as it was written based on a large international research project that ran in 2005 and 2006 in 27 European countries, including Slovenia. The project's purpose was to record the testimonies or autobiographical stories and memories of forced and slave laborers who were almost silent, their suffering unknown and forgotten until then. The Slovenian contributor was Monika Kokalj Kočevar (2010), who collected 20 stories of Slovenian exiles. The testimonies and other documentary materials are available on the project portal "Forced labor 1939–1945."[75]

The largest collection of personal testimonies by slave and forced laborers in Slovenia has been curated by the Slovenian Exiles' Association 1941–1945 (Društvo izgnancev Slovenije 1941–1945), established in 1991. The association prepared an extensive documentary collection which included lists of forced and slave laborers and more than 6,000 German factories where they worked, the names of the camps and records of war damages, and minutes of the sessions and procedures for establishing the status and rights of Slovenian deportees (Žnidaršič and Kaplan, 2012, 7). The association published three comprehensive books which present 488 autobiographical stories of women, men, and children with rich photographic material and poems written in exile (Žnidaršič and Kaplan, 2003, 2006, 2012). In addition to historical, archival, and documentary materials, today we finally have insight also into the personal stories and memories of people who, in Slovenia as well as in other parts of Europe, were only allowed to describe their experience and co-create the collective memory of forced migrations 70 years after the end of the war (Novak, 2016).

Among these people was also Terezija Černelič (née Cizel), born in Globoko near Brežice in 1928. She answered the question of how she remembers the day of the expulsion with: "I will never forget this."[76] The Cizel family was expulsed in the first days of the swift and efficient deportation of the population in the Lower Sava and Sotla Valleys at the end of October 1941. In her narrative, Terezija lists all the aspects

75 The international project was financed by the German Foundation Remembrance, Responsibility and Future (EVZ Foundation), which was established in 2000 in order to pay the reparations to the victims of forced and slave labor in the German Reich and made sure that the memory of their suffering is preserved. The National Museum of Contemporary History was the Slovenian participant in the project.

76 I talked to Terezija Černelič (née Cizel) on May 7, 2019, in Ljubljana. The recording is with the author.

that are mentioned in the collected and published testimonies of Slovenian deportees during World War II: they deported entire villages, only a few that Germans needed were allowed to stay; the expulsion was swift, they transported them to the German Reich in a few days. Lodging and food were insufficient. They suffered from hunger, cold, and deprivation; frequent moves from camp to camp and the type of forced labor depended on racial checks and evaluations; and they worked for free or for minimum pay, and every transgression was brutally punished.

The Personal Experience

> I will never forget this. We knew there would be deportations; people knew ever since the Germans came in April 1941. I was 13 years old, my brother was two years older, and my little sister, six years younger. We were deported on October 28, first to Rajhenburg and then straight on to Germany. That day I went on a cart to the mill with my godmother and other children, and on the way back, by the time we reached school, our parents were already there, having been forced from our home. They intercepted us there. I still feel like crying when I think of it. God knows how much my parents must have cried when they were taken. It's a good thing we weren't there to see it. That same evening, night, we went on. People were saying they'd give us new houses up there, but they packed us, just like the migrants today, wherever there was room: castles, monasteries, massive buildings. We were in four camps. I remember their names to this day: Schlauphof, Mittelsteine, Templin, and Sternberg. In the first one, we were all packed into a single large room, all the families, the air was such that you couldn't breathe, and the winter was harsh that year, so very harsh. We were very hungry there. There was hardly any food, a rutabaga here and there, and two slices of bread per day. We were there for four months. Our papa was clever and left the camp to beg the baker for some bread; he hid it and brought it to us. At that time, the surveillance was not that strict. Then they moved us to a different camp. It was even worse there, starvation. We were so terribly hungry. But they already started to send us out to work, to farms and factories. First, a farmer took my brother, some 15 km away, and then they gave me to the next farm. Mother stayed in the camp, she had to look after her mother, who was also there with us, and my little sister and papa worked in the camp. My brother and I were alright. I helped on a farm, in the stable, whatever was needed. I wasn't hungry, and they didn't force me to work. My parents were happy that my brother and I were living on a farm because they knew we weren't hungry. We worked for them for free, but they were nice. I got used to it, and I was fine. When my mother came up to visit me, there was a train that took her up there. The little woman—she was older, kind—gave her a big loaf of bread and a stick of butter.

Picture 3.13: The Cizel family in the Templin camp, 1943 (Terezija second from the right). Author: unknown. Courtesy of: Terezija Černelič.

After four months they moved us again, I don't know why they moved us so much. When we were still at home, they racially labeled us. We were not *heimatbund*, we were not suitable for Germanization. We were the second class; we'd be sent to Siberia. So we got to Templin, in the north, above Berlin, and we were there for a year. It was a resort for postal workers, little huts in a pinewood by the lake. We got one hut, the kitchen was downstairs and the bedroom upstairs, my parents were very pleased. I got very lucky in Germany. First, I went to work for a rich family, but I only stayed for six weeks. I lived in a nice room for a maid, but the lady was stingy; she only ever gave me a small bread roll [a day], only one. From there, I went to another family; they were very good, I got a lot to eat, I only worked until three, then I went home. Not all Germans were for Hitler, some were good. Papa and brother had to go work in Berlin. They were in grave danger, all the time under bombardment, all the time afraid. My brother got into an argument once because they wanted to send him to the Russian front, and they put him into a *Straflager*. He stayed there for 56 days. When he came home, he was just skin and bones, he had a terrible time. My parents didn't want to tell me what they did to him, they beat him, poured water over him, kicked the dish in which he got food, but my father somehow got him out.

Picture 3.14: Sisters Jožefa and Terezija (right) in the Templin camp, 1943. Author: unknown. Courtesy of: Terezija Černelič.

Then we went to the fourth camp, 30 km east of Frankfurt an der Oder, today this is Poland. We stayed there the longest, one year and a half. We had a small room, and I worked in the barracks in the canteen. It wasn't bad, and the lady was nice, she loved me, we wrote to each other for a long time [afterward]. There was enough to eat, and she always piled my plate. She paid me and gave me food every night to take to my mother

> in the camp. Papa and my brother were still working in Berlin. My brother knew the war was over. He escaped from Berlin, but they sent him back. But papa was with us when the Russians came. We were so afraid of this, the Russians and the front. We were scared.
>
> The Russians came in February 1945 for Candlemas, but the front line passed us. When the Russians came, it was bad because they chased women. These are ugly stories, my, I cannot speak. Luckily, nothing happened to me, I was terribly lucky. We had a very tiny little room. At one time, six officers came and started fighting and arguing, pushing each other around. I was hiding on the bunk bed, and one of them stepped on a chair and started talking to me. He was young and sober, and he was showing me an icon of Mary and hoping that she would help him through the war unharmed. And then they left. That was lucky.
>
> The way home was hard. First, we went on foot, for a very long time, to the station. I still remember it was Wolsztyn/Wollstein. That was so bitter! When we walked, there were dead soldiers, one next to the other, and dead cattle. We crossed the battlefields on foot. There, they put us on wagons, there were 120 wagons, and then they took us down, and on other wagons, two stations, and down again. It was a harsh winter, and we were in open carts, the four remaining families. Oh, it was hard. There was no food. The Russians gave us *komis* [Komissbrot], it was hard as bricks, and dry vegetables and tea. And then they attached us to those 120 wagons, and we went across the Carpathians and arrived in Hungary. In Szeged, my grandmother died. We arrived in Pančevo for Easter and buried her there. In Pančevo, there were many Slovenians, and they brought everything to the train, ham, potica, meat, sausages. They didn't know where to put us, but that was spring already. They put us up in the empty houses of Banat Germans, whom they'd expelled. Entire villages were empty. It was just like at home when they expelled the Gottscheers. They moved into our houses, and when we returned, they expelled them.[77]

The Gottscheers that Terezija mentions at this point in her narrative are known in the Slovenian language as Kočevarji (or kočevski Nemci, the Gottschee Germans) and have a special place in this drama of forced migration and expulsions. German-speaking settlers had moved to the Kočevje/Gottschee region in the southeastern part of Slovenia around 1330 from Tyrol and Carinthia. They established and maintained their specific Gottschee identity, customs, and language and coexisted with the Slovene-speaking residents of the Kočevje region for centuries. "During the rise of nationalistic fervor in the second half of the 19th century, the local identification 'Gottscheer' (i.e., inhabitants of the Kočevje region) became tinged by imposed nationalism, and gradually the division into Gottschee Germans and Gottschee Slovenians began to take hold" (Moric, 2020,

77 It is unclear whether Terezija meant Mittelsteine, a subcamp of Mittelbau-Dora, located in Woffleben, Germany, or Mittelsteine, a subcamp of Gross-Rosen/Sachsenhausen, located near today's Ścinawka Średnia, Poland.

141). After the collapse of the Austro-Hungarian Monarchy, Gottschee Germans became an ethnic minority, and their status and situation worsened. Some of their community leaders embraced Nazism already in the 1930s.

As the Kočevje region became part of Italy under the 1941 partition plan, the Gottschee Germans were "offered" to leave their homes and move to their German "homeland." In fact, the designated resettlement area for them was not a place in the German Reich itself but the region along the Sava and Sotla rivers in Lower Styria. The ethnic cleansing plan was immediately put in motion with the expulsion of the Slovenian population from the designated region, swiftly followed by the resettlement of the Gottscheers in the emptied villages:

> Following the intense propaganda campaign of their leadership, one hundred and thirty-five trains transported the Gottschee Germans to their new homeland in mere nine weeks between November 1941 and January 1942. According to their census, almost 95 percent of the around 12,500 people, inhabitants of around 170 villages, were moved. The majority—11,509 persons were resettled in Štajerska (Lower Styria), a small part—571—in Germany, while a tiny number of people—66 persons—were not allowed to migrate. They had to leave all their immovable property at home. Their property was taken over by the German moving company *Deutsche Umsiedlungstreuhangesellschaft* which sold it on behalf of the displaced persons to the Italian real estate institute *Emona* (Ferenc, 2010, 529).

The Gottschee Germans were resettled in the Sava and Sotla river valleys in the houses from which Slovenian families had previously been expelled. Trains had taken the racially inappropriate Slovenian population to be enslaved in the German Reich and its occupied territories. Their homes were colonized by the "German" defenders of the Reich's southern border, who, in turn, were resettled from the region in which they had lived for centuries. The Gottschee Germans cohabitated in this linguistically and culturally mixed region with others until the nationalist ideology of exclusion and, consequently, the war aiming to realize nationalistic—and Nazi—ideas destroyed this symbiosis (Ferenc, 2005). As Terezija Cizel mentions in her narrative, their relocation had disastrous consequences because they were expelled to Austria at the end of World War II. From there, they migrated all over the world. Among them was Marija's peer Johann/John, whose testimony presents the subjective experience of yet another war-torn childhood.

Map 3.2: Kočevje/Gottschee and the resettlement region 1941–1945.

John Tschinkel, born in Grčarice/Masern in the Kočevje region in 1931, described the drama of the ethnic community in great detail in the autobiographical work *The Bells Ring No More* (2012). The program of "return" to the Reich was created after the partition of Slovenia, which had allocated the Kočevje region to Italy. The German authorities then promised the Gottschee Germans collective resettlement to the territory of annexed Poland: the largest territory of the new German *Lebensraum*. The promise was fake. The actual destination for resettlement was the region along the Sava and Sotla rivers. As Tschinkel wrote, the young, zealous community leaders knew that they could only convince the people to resettle in the homes of expulsed Slovenians by fraud:

> To succeed in their mission, the young Gottscheer fanatics, now fully committed to the objectives of the Reich, decided to keep the resettlement destination a top secret from the population. At least until the deed was done. With this decision, the betrayed became the betrayers. They managed to keep the secret and with their effective tactics of persuasion during the remainder of 1941 the mission was successfully completed in February 1942 (Tschinkel, 2012, xix).

The people from Grčarice/Masern left in early December 1941. They left their homes, full stables, granaries, and cellars unlocked, as instructed. And when young Johann's—John Tschinkel's—family arrived at their new "home" in the village of Veliko Mraševo, they found the house unlocked. They were totally shocked that they were not taken to the Reich but had been resettled in the homes of Slovenian families who had clearly been expelled:

> Signs of this were everywhere, in the unmade beds, the open doors of the half-empty wardrobe, the pulled-out drawers of the chest from which most contents were removed. After a quick look about, we settled in the kitchen where the unwashed dishes of that day of hasty departure were still on the table and the water in one of the pots on the stove frozen to solid ice (Tschinkel, 2012, 298).

Immediately after they arrived, the mother sent the father to check the stable and the barn. "But she did this mainly to spare him his crying in front of the children, something he could do freely outside but tried to hide when I joined him in a little while" (Tschinkel, 2012, 298). Those Gottscheer who had learned from the ones already resettled in the emptied Slovenian houses what their real destination was tried to stay in their homes, but it was too late. Their property had already been seized in full.

At the end of World War II, the six-century-long existence of the Gottschee German community in Slovenia ended. On May 8, 1945, the Tschinkel family, along with their fellow villagers, joined the escape of the endless mass of soldiers and civilians who tried to flee from the victorious National Liberation Army:

> And when permission to leave finally arrived that morning, it took only a little time to get going.
>
> Few, if any, in the wagons now clogging the road had an inkling of what destiny had in mind for us. No one yet knew that the war had come to an end and with that our promised future in the Reich. The journey that began in 1941 was now on an unpredictable final stretch toward an unimaginable future. The folly of having pulled up, only three and a half years ago, the anchor to a place established centuries before, was still to become fully apparent in the tumult of the terrifying days yet to come (Tschinkel, 2012, 418).

Tschinkel describes their desperate attempts to flee to the north in the narrow roads jammed with hundreds of thousands of German soldiers and other military formations, military vehicles, carts, and carriages, along with thousands of civilians fleeing from the National Liberation Army. It was total chaos. After four days of agony, the partisans stopped them near Radeče and made them set up camp. They then moved them to a camp near Brežice. On May 24, they took them by train to the Austrian border near Maribor, where they ordered them off the train:

> The bolt was lifted, the door rolled open and a Partisan, standing off to the side yelled: "Raus!"
>
> The train had stopped at the edge of a broad downward sloping field and then after a flat part, upward toward the road a few hundred feet from the tracks. On the other side of the road was a long, 20 foot high stone embankment, keeping in place the hill into which the road had been constructed.
>
> On the stones of the embankment was written in huge white capital letters, plain to all spilling from the boxcar to see:
>
> "WIR DANKEN UNSEREM FÜHRER"
>
> Plain for all to contemplate, in silence, the four words of now-bitter irony, words we had been made to utter so often in mass rallies during the summer of 1941 and the years after that without really knowing why (Tschinkel, 2012, 455).

In spring, on May 25, 1945, John and his family crossed the border into Austria. In front of them lay a path into the unknown and the search for a new home, so different from the route of the Cizel family and Terezija, who knew where they were returning. Terezija described the homecoming:

> In Banat, they put us up in a tiny little house, without any doors or windows, but outside it was so beautiful. Papa borrowed tools and cut grass and immediately planted salad. Mama baked bread in a brick oven. We were there for four months because we couldn't go home, there was no bridge yet. My little sister was taken in by nuns so that she could learn something, and Papa and I went to work for peasants, and they gave us food. We bought everything before we went home: flour, corn, twenty chickens, even a piglet. We put everything on the train because we knew there was nothing at home.
>
> Once we got home, we were fine. My brother was already home. He came via another route. Papa found his way around, and we got well soon. The house was new before we went, for those times it was a good house. One Gottscheer lived in it, and he kept it nice. He didn't destroy anything. But there was nothing inside, it was empty. The Gottscheer was a tailor, a good man, and he stayed in the village for a long time afterward. He lived above the grocery store, and they said he'd supported the partisans. They took everything from our house the minute we were expelled—our home was cleared out, but nobody wanted to admit who took what: the cows and the calves, the hens, everything. This was before the Gottscheer arrived, it was the locals that took it.

> After the war, I went to a hospitality school. Then I worked for the local council. Everybody could get a job. My sister was a trained shop assistant, and her husband was an upholsterer. They lived in Celje and had good jobs, she, in a nice shop and he, in upholstery. Our mother told her that a colleague of hers had escaped and was already in Canada. Then her daughter—my sister, that is—asked for the address, and then they escaped as well. I guess that colleague told her money was better in Canada. They found something in Austria where he had family, and then they escaped to Germany, where they stayed for three years. They could have had a house here because they were skilled, they had jobs, but they escaped. That was in 1957.[78] In 1961 they then went to Canada. At that time, quite a few people defected. My brother, as well, to Germany. He went to visit my sister and her husband and stayed. He's also buried in Germany.

Terezija and John were peers, born three years and 70 kilometers apart, in the same country and similar rural environment. They belonged to two different ethnic groups, which had lived together for centuries and whose members had mixed and mingled for centuries. In the German Reich, John belonged to the superior, "pure Aryan race" (if only conditionally, as his mother was Slovenian). Terezija was of a "Slavic race" designated for slave labor and starvation to death. Both suffered the same fate: deportation, exile, and return home, but in a different way. Terezija had to leave her home in Globoko near Brežice because of the same National Socialist politics that made John leave his home in Grčarice; the Cizel family were displaced by force, the Tschinkel family by deceit. They both spent three years in the German Reich that, for John, was supposed to mean "home" and, for Terezija, inevitable extermination. In 1945, Terezija and her family embarked on a long and dangerous journey home from exile. At the same time, John and his family embarked on a long and dangerous journey from home—which never felt theirs because they lost theirs at the beginning of the war—to exile. A large majority of the 83 percent of the displaced population from the Sotla Valley returned home after the war. The large majority of Gottschee Germans lost their homes and homeland forever.

78 After 1945, the Yugoslav borders were heavily controlled and many young people who could not get passports crossed them illegally. A more liberal migration and mobility regime of "open borders" started in 1962.

Concluding Remarks

This book offers an insight into migrations within and from the Slovenian ethnic territory between the mid-19th century and World War II and places them onto the larger map of European migrations in that same period. The small Slovenian space at the junction of Central Europe, the Northern Adriatic, and the Balkans has been a place of intensive and diverse migration processes. Such intensity and diversity reflect the density and complexity of the factors and circumstances linked to the geographical traits and administrative structures, socioeconomic circumstances and political events, and historical geopolitical shifts from which these migrations were born.

One prominent trait marks the main character of the Slovenian story: emigration. Within the countries to which the Slovenian ethnic territory belonged and, at times, even on the European scale, it was one of the regions with the highest levels of out-migration. Such a massive population transfer significantly reduced the natural demographic potential and systemically conditioned the population's growth dynamic and structural characteristics. Adhering to the classic interpretative categories, we could define the Slovenian territory, structurally, as an emigration region, which, before World War I and during the interwar period, recorded a continuing loss of its population due to out-migration. The reasons lie in the socioeconomic structures of the (predominantly agrarian) Slovenian Lands, which were a part of the rural Slavic periphery of the Austro-Hungarian Empire. The rationalization of agriculture, the advancement of non-agrarian economic sectors, and other modernization processes occurred very slowly in the Slovenian territory. These processes were thus unsuccessful in preventing unemployment and social marginalization and fulfilling the population's needs and social aspirations. Since the mid-19th century, the population thus looked outwards en masse and entered labor markets in faster-growing economic areas within and beyond the borders of the Habsburg Monarchy. A similar trend continued in the interwar period, even though Slovenia at that time was the most developed region of the new Yugoslav state and experienced accelerated industrialization and substantial growth of the labor market. However, this progress was not enough to eliminate the pre-modern structural elements that hindered the economic and social modernization of the Slovenian countryside.

From the mid-19th century onward, Slovenia, like the rest of Central Europe, experienced a permanent increase in spatial mobility. This mobility first appeared as moves within and between the lands and provinces of the Austro-Hungarian

Monarchy; such moves remained the most numerous throughout the entire period. From the 1880s onward, transatlantic migration brought on the expansion of long-distance moves, which peaked in the decade before World War I. The scope, orientation, and patterns of migration processes differed from region to region. The configuration of migration geography reflected the gravitational links with the traditional and new labor markets. It was influenced by the historic administrative and political division of the Slovenian Lands. The migration routes from Lower Styria thus led primarily to the mining and industrial centers of Upper Styria and Lower Austria. Trieste attracted a large share of migrants from Goriška and other areas in the western Slovenian ethnic territory. The inhabitants of Prekmurje in the northeast left to work deep in the agrarian regions of the Kingdom of Hungary. Transatlantic migration, directed almost exclusively toward the United States of America, was more or less present throughout the Slovenian territory. Still, the epicenter was in the southern regions of Carniola. Migrants to the United States and Germany found a way to join the established transnational corridors and, through them, a path to the more favorable US and German labor markets, primarily in mining and heavy industry and, with time, other industrial sectors.

To a lesser extent, Slovenian migrations also turned toward several other foreign destinations, from European countries to the South American continent and Egypt. With this, Slovenian migrations joined international migration systems and, consequently, globalization processes. In the interwar period, migration forms and directions adjusted to the political reorganization of the new Europe and the shift in the international migration regime from liberal to selective and controlled migrations. After North America ceased its open-door policy and established national barriers to transnational movement, migrants turned partly to South America. However, they mostly remained within Europe and other parts of the Yugoslav state, of which Slovenia was now a part.

In the Slovenian ethnic territory, like elsewhere in Central Europe, the spatial mobility and migration in the studied period became an increasingly more pronounced element of the economic and social strategies of individuals, families, and communities faced not only with material needs and pressures but also with planning lives and pursuing aspirations. The book highlights the repertoire of the migration routes and forms we encounter in the small Slovenian territory. People moved in diverse ways: short and long-distances, within the Slovenian territory and abroad, temporarily or permanently, back and forth. Particular migration patterns were more characteristic in, and sometimes specific to, particular areas. Often, they were a continuation of the already existing practices and routes. However, many were established anew following the new possibilities and opportunities in the internal and international labor markets. One noticeable aspect everywhere was the coexistence of different migration forms and systems and people's transitions between them. For example, the peddlers from Kočevje, the *suha roba* (dry ware) sellers from Ribnica, and the Resian itinerant knife grinders expanded their circular routes. At the same time, people from Resia and Benechia started leaving in large numbers for seasonal work in construction. The seasonal migration of agrarian laborers from Prekmurje expanded and structurally changed when the shift from growing wheat to cultivating commercial crops affected the traditional employment destinations; at the beginning of the 20th century, the transatlantic labor market was included in the mix as a destination. Young women from rural areas, who traditionally found employment as domestic servants in regional towns and larger cities, ventured farther, sometimes as far as Vienna, with those from Goriška often going to the Egyptian metropolises. The Jesenice ironworks, the tobacco factory in Ljubljana, and the straw-hat-making workshops in Domžale attracted a migrant workforce as well as foreign specialists from beyond the immediate surroundings. At the same time, female straw-hat makers from Domžale left to work as skilled workers and craft tutors in large European and US cities.

A characteristic trait of the great migrations in the period covered was the transition from rural to urban environments. For Slovenians, this transition from rural to non-rural activities and social classes typically occurred through migrations beyond the Slovenian ethnic territory. However, a considerable share of the movements was not of this type and consisted of rural-to-rural moves and relocations; this type of migration involved not only seasonal agrarian workers and farmhands but also redistributed the rural population in general. In Lower Styria, for example, farmers from Prekmurje and elsewhere, who either did not have land or had too little land, took over the agricultural lands of those who had migrated to the industrial centers in Upper Styria. Moves from farmland to farmland also occurred when longer distances were involved, for example, to the United States and Brazil. In the United States, many immigrants from the Slovenian countryside ended up taking over farms after a period of work in industrial sectors and proletarianization and thus returned to farming.

A constitutive part of the migration activity throughout the entire period was, of course, also immigration. Immigration processes were partly linked to the deficits in technical and administrative staff and skilled workforce, which the Slovenian mining industry attracted from elsewhere. Many immigrants were entrepreneurs. For example, the straw-hat-making craft in Domžale was developed into an industry by Tyrolean entrepreneurs who attracted both a local workforce and Tyrolean laborers into the town. In many places, immigration resulted from emigration and the need to replace the workforce. The districts in Lower Carniola, where the number of inhabitants dropped because of the emigration to the United States, faced an agricultural labor shortage; new laborers had to be attracted from neighboring Croatia and paid significantly higher wages. Likewise, the Slovenian mining regions attracted workers from other parts of the Austro-Hungarian Monarchy and abroad to replace the departed Slovenian workers.

The Americas radically altered the geography and physiognomy of the migration processes in the Slovenian ethnic territory. With its pre-World War I programs of agrarian colonization, South America only attracted small groups of farming families who yearned for larger estates and more fertile soil. The United States, however, with its expanding labor market, at least from the 1890s onward offered significantly greater employment possibilities to the unskilled population from the Slovenian countryside than they could have expected within Austria-Hungary. Differentials in wages were also key in directing the movements to the German Ruhr Valley and Westphalia, where the workers were paid better than they were in Austrian mining centers. Another characteristic is that migration to Germany went predominantly from the Slovenian mining areas and consisted of skilled and semiskilled miners and other mine workers. The US labor market mainly attracted members of the peasantry and agrarian proletariat from all the Slovenian regions, who, for the most part, had no experience in the industrial sectors that employed them. The "American myth," which, from the 1890s onward, permeated the society in the Slovenian territory and incited an image of a "better life" also propelled them across the ocean.

The book underlines that migration in the Slovenian ethnic territory must not be studied only in terms of the transfer of the people into other social milieus (as mainstream historiography often does). In addition, our research reveals that, primarily, these moves did not happen in "one step" but through a long and multi-phased mobility process. We wished to show the multiplicity and complexity of migration patterns and their persistence or changeability over time and space on the community, family, and individual levels. Migrations—short- or long-term, continental or transcontinental—occurred, with some differences in logistics, timing, and practices and in the service of specific needs, plans, and socioeconomic strategies. Agrarian laborers from Prekmurje, masons from Goriška, and foresters from Inner Carniola moved within precisely defined temporal and geographical frames to combine income from the seasonal migration with agricultural work at home. The integration into the mining regions and industrial centers in the Austrian Crown Lands or Germany did not adhere to such strict timelines. Migrations could be several years long and iterative. Migrants could also remain in the place of their employment for extended periods or settled there permanently. Transatlantic migration had a similar character, as 40 percent of migrants to the United States returned after a short or long employment and often made the journey several times. The departures and returns depended not only on individual plans and choices but also on the labor markets' changeable conditions and conjunctures. These circumstances influenced the macrodynamic of departures and returns and the multi-phased migrations and mobility within the destination countries. In the mining towns in the Mesabi Iron Range in Minnesota, Slovenian communities were among the largest. However, their numbers oscillated considerably because their members moved to other mining centers in times of crisis. Similarly, in Europe, miners moved between Westphalia, Alsace, Belgium, and the Netherlands. Even towns in Slovenia experienced the dynamic and multi-phase nature of these processes. Their demographic growth in the decades before World War I was supported predominantly by immigration. Celje, Ptuj, and Maribor in Lower Styria were at the forefront among the Austrian cities in population turnover.

It is also worth highlighting the structural factors of diversity in the directions and patterns of migrations. The book shows male and female migrations as complementary practices, equally embedded in social and economic systems and processes. We wanted to emphasize the position and trajectories of male and female migrants, which were, due to different social roles and specifics of the labor markets, often divergent and asymmetrical. The migrations to mining and industrial centers were predominantly male, with women following men. However, some routes saw more gender balance, while others still were predominantly, or even exclusively, female-dominated. Moves to North America were about 80 percent male until the ratio shifted because of family reunifications and the expansion of employment possibilities in immigrant settlements. Women, particularly those with special professional profiles, for example, straw-hat makers, found their individual paths to the American labor market. Circular migrations of the peddlers from Resia and the Kočevje region were almost exclusively male, while the seasonal agrarian work in Prekmurje included many women. In Trieste, the statistics consistently show more Slovenian women immigrants than men, a fact connected to the labor market for domestic workers, which young women temporarily entered before they married, with many later settling down in the city. Migration to Egypt was a true female specialization, in which, opposite to the case of the United States, men sometimes functioned as "migration consorts." *Aleksandrinstvo* (the name given to this female migration at the time, retained in historical memory) was also about the economic role of mothers who, because of the more affluent Egyptian labor market, took on the role of the breadwinner and provided the majority of the income to support and develop their families back home.

Neither can we ignore the less obvious but no less lively career mobility, for example, the moves and itineraries of public servants, technical personnel, and intelligentsia. The immigrant communities in the United States, for example, attracted people from the "old place" who had a certain level of education and then took leading positions in the migrant social, economic, and cultural organizations. Priests, as well as seminarians who still needed to complete their studies, crossed the continent and the seas to serve immigrant communities and their parishes. Experts in particular fields went all the way to South America. At the same time, after World War I, the newly established Yugoslav space became a frequent destination not only for laborers of both genders but also for middle-class people active in trade and commerce, state and private white-collar jobs, and highly educated people from engineers to teachers, and others practicing skilled professions.

Because the sources are scarce, it is not easy to measure and discuss the economic effects of migration in the areas of origin. As a consequence, this field has been largely neglected both in economic history and migration studies. Generally speaking, migration, in one form or another, allowed the satisfaction of material needs through access to opportunities that were not available or were lacking in the original environments. On the macroeconomic level, labor migration systemically eased the burden of the weak socioeconomic structure and created a significant source of income, which, on the one hand, had positively affected the population's living conditions. On the other hand, it exacerbated those imbalances which triggered the need to migrate in the first place. In Prekmurje, seasonal work in intensive agriculture on estates in combination with subsistence farming represented one of the pillars of the economy. By the end of the 19th century, leaning onto this external source of income increased the population. It also accelerated fragmentation and decreased income from smallholdings. Therefore, at the beginning of the 20th century, an increasing number of Prekmurje inhabitants turned toward the more robust US labor market. In some districts of Lower Carniola, the epicenter of the transatlantic migration, the remittances represented a structural source of income for most families. Local authorities looked favorably on their residents' employment in the US industry because that took the edge off the local unemployment, eased the burden on social services, and because the influx of US money had a positive effect on monetary circulation and commercial exchange and the municipal budgets themselves. Among the indirect effects of the "emigration economy" was the increase in savings and local savings institutions in the most emigration-affected areas. Money from abroad enabled a slightly higher standard of living: houses were built or renovated, farming estates were enlarged or revitalized, sometimes new economic activities were started, and consumption generally increased. Often, however, remittances were simply used to pay off debts and bridge difficult economic situations. The community benefited as well, as the emigrants, through organizing collections, generously supported the initiatives for the public good in their places of origin. This dynamic mostly held true for the transatlantic, "rich" migration, which provided a significant influx of means despite the end of mass migrations and the economic crisis between the two world wars. From the macroeconomic development point of view, however, work migration in all this time had no visible impact because the absence of an adequate financial system, an economic policy, and entrepreneurial initiatives meant that the remittances were not transformed into capital investments that would transform socioeconomic structures and speed modernization processes.

The economy is, of course, but one explanatory framework of the processes under scrutiny. Through the micro perspective of life stories and individual experiences presented in the book, we show the complex reasons and motivations that led people to migrate. The decision to migrate may have been based on material—or at times political—causes, but mobility was also spurred by personal ambitions and aspirations, a thirst for knowledge, the desire for freedom from the traditional shackles of family and community, and a yearning to participate in processes of social and cultural modernization. Imitation also played a significant role. Young women from Goriška waited impatiently for when they, too, could leave for the cosmopolitan Alexandria. The schoolchildren in Bela Krajina were keen to learn about the geography of the United States and dreamed of their own transatlantic routes while listening to the returnees' stories. Young teachers hoped to get jobs in larger centers and cities, where they could be a part of the rich cultural and social life.

A particular aspect of the great migrations in the Slovenian ethnic territory is their correlation with the "national question." Since the early stages of its nation-building process, the mass departures of the young generations had evoked concern for the future of the small Slovenian nation, surrounded by large and historically dominant ones. The Slovenian political and national cultural rhetoric compared emigration to "a cancerous wound on the Slovenian national body." Together with assimilation, it depleted the nation's vital forces and then abandoned them in service of foreign locales. The doyens of the Slovenian national idea even had reservations about people moving into cities within the Slovenian territory because they promoted the idea of Slovenians as an inherently rural nation, and cities were the place of assimilation into the German or Italian culture. In ethnically mixed regions of the Slovenian territory, both on its edges and in the center, migrations were a source of national conflict. This conflict shifted to another historic phase after World War I, when the Slovenian ethnic territory was divided between the countries of the newly drawn map of Europe, and the question of ethnic minorities grew in importance. The forced displacements from the territories along the Soča/Isonzo Front during World War I were now followed by political migrations, a consequence of the nationalization policies of the countries striving for the ethnic homogenization of their territories. Denationalization policies affected the Slovenian minority in the Republic of Austria and particularly in Fascist Italy, in which one-quarter of all Slovenians lived. However, German and Hungarian communities in Yugoslav Slovenia experienced it as well.

The book also addresses migration regimes and the state's attitude toward migration and its diaspora. We have highlighted the continuous stance of the Slovenian politicians who, in principle, opposed the laissez-faire approach to migration processes, which left them in the hands of the market logic of liberal capitalism. However, they did not outright reject the freedom of migration and emigration as fundamental rights. They were proponents of regulated processes congruent with the "actual needs" of the population and the socioeconomic and national-political interests. Until World War I, these principles existed only on paper because, in its constitutional law, Imperial Austria enshrined free migration within and outside its territories, and political endeavors to limit it in any way failed. Restricting emigration without offering concrete alternatives was unrealistic, considering the socioeconomic circumstances. Political and administrative authorities in the Slovenian ethnic territory tended to be lenient even when enforcing the existing restrictive regulations (particularly regarding draftees), as strict enforcement would increase illegal departures from the country. The Church strove to stand by the migrants' side, en route and in their new communities, in the religious, moral, social, and national senses. After World War I, the newly created Yugoslav state, in the spirit of state interventionism, finally established direct control over migrations, intending to regulate them according to its national vision and interests. However, the new state had little leeway to do so because migration policies had to be adjusted to the changed international migration regimes. The changes no longer accommodated free transnational access to labor markets which the states protected with restrictive and selective immigration policies. Yugoslavia concluded bilateral agreements on posted workers and secured their social rights with those European countries that were workforce importers. Nevertheless, the position of Slovenian and other Yugoslav workers in France, Belgium, the Netherlands, and Germany reflected the negotiating weakness of Yugoslavia in international economic relations.

The political agenda of the Kingdom of Serbs, Croats and Slovenes particularly emphasized the national-political aspects of migrations. This agenda was reflected in the differentiating migration policy, which was restrictive towards the three constitutive nations (Serbians, Croatians, and Slovenians) but lenient, or even encouraging, towards the members of non-Slavic and non-Christian nations and minorities. The relationships with the emigrant communities proved a parallel issue. During the war, this diaspora—inherited by the new country from its predecessors, particularly Austria-Hungary—provided substantial material and political support for establishing the Yugoslav state. The richer diaspora remained the source of precious foreign currency. At the same time, during the peak of nationalisms and nation-states, Yugoslav politics recognized its role as an element and factor of the nation- and state-building process. It strove to integrate it—culturally and nationally—into the Yugoslav nation as an extraterritorial component. Slovenian and Croatian emigrants, who represented the central part of the Yugoslav diaspora, thus became the primary addressees of this politics. Nevertheless, they were active creators of the relationships with the homeland and often shaped them according to their independent views, bypassing the wishes of Yugoslav politics.

We have highlighted these and other traits of the Slovenian diaspora by presenting the activities of Slovenian migrants in organized community life, cultural and political activity, and creativity. Many Slovenian migrants possessed a strong sense of regional identity and retained the traits of their region. However, through migration, they underwent a process of association, recognition, and integration through which they came to embrace the national language, culture, and traditions. The ways Slovenian migrants organized reveal their ethnic separation from other migrants, their attitudes toward integrating into the dominant societal, linguistic, and cultural environment, and the shaping of specific group identity. Although most protagonists of these processes were less educated and poorly paid, they nevertheless succeeded in creating, in their new environments, social structures and complex institutional systems for mutual aid, the support of cultural and religious life, and the social and economic progress of the community. They built churches, national homes, schools, and recreational centers, published newspapers and books, and established educational, religious, cultural, political, economic, humanitarian, and sports organizations, some of which continue to operate today. All this testifies to the Slovenian diaspora's worldviews, political and ideological heterogeneity, identity transformations, and cultural hybridizations in the processes of retaining its ethnic and linguistic identity. The story of the organization of Slovenian migrants and migrations is a story of how the small Slovenian world geographically expanded and how its people adjusted to the new environments they encountered. It is also a story of integrating into host environments, retaining a close connectedness with the old place, and transferring the social values and cultural patterns that helped shape migrants and their communities, their places of origin, and the Slovenian ethnic territory.

The intertwining of the aspects and complexity of migrations with the general historical events over a long period is also shown through the concrete experiences of individuals. Life stories are not merely illustrations but also instruments of analysis that offer an in-depth understanding of the more generally described processes through a "human-centered approach." They introduce those who leave—always simultaneously emigrants and immigrants—and those who stay as agents and interpreters of their experiences and expand the understanding of migration beyond economic, political, or social motives. Through their subjective perceptions and individual interpretations, the participants in the migration processes present us with a kaleidoscope of individual narratives about complex decisions and everyday negotiations in transnational and translocal worlds. The subjective experiences of those who left and those who stayed indicate the establishment of translocal bonds and transnational families. Likewise, they show the dynamics of preserving national identities, traditions, and migrant communities. Through stories, it is easier to recognize the multi-dimensional circumstances of moves and experiences of individuals.

The book's timeline ends with World War II, but only certain aspects of the war are covered in its second and third parts. The war and its consequences meant new historic breaks that started new phases in Slovenia's migration history. War and occupation divided and separated the Slovenian ethnic territory between the German Reich, Italy, and Hungary. This period saw expulsion, deportation, forced acculturation, and the destruction of the Slovenian population all in line with the plan to violently change the ethnic landscape of the Slovenian territory. The victory of the liberation movement, the new post-war division between Italy and Yugoslavia, with which Slovenia gained most of the western part of the Slovenian ethnic territory, and the implementation of the socialist socioeconomic order were the sources of new, nationally and ideologically conditioned moves and transfers of the population. At the war's end in 1945, members of the groups and formations that collaborated with the occupiers fled from the Partisan Army to Italy and Austria, as did political and ideological opponents of the new regime, as well as those who did not see their future in the socialist regime and those who got caught up in the mass relocation of peoples after World

War II. The Gottschee Germans and members of other German communities were expulsed. The territories in northern Istria, integrated into Yugoslavia at the abolishment of the Free Territory of Trieste in 1954, were deserted by almost the entire Italian-speaking population and, along with it, a part of the Slovenian. Because of the prohibition of free emigration, tens of thousands of people fled to the West illegally in search of an environment with more freedoms and better socioeconomic opportunities. Illegal departures lasted until 1962, when Yugoslavia was the only socialist country to open its borders and reinstate prewar labor migrations based on new bilateral agreements with economically developed countries. Migrants turned predominantly toward Germany, Austria, France, and the Benelux, as well as Sweden and Australia. Migrations to other parts of Yugoslavia and those within Slovenia also continued.

An essential shift in the postwar migrations in Slovenia is that the Slovenian ethnic territory changed from a predominantly "emigration territory" to an "immigration territory" as well. Immigration, as the book also points out, has always been present. However, in the 1970s, the massive labor migration from other parts of Yugoslavia became a structural element of the demographic and socioeconomic development of the Socialist Republic of Slovenia. This trend has been retained in the independent Republic of Slovenia, which, in the last two decades, has consistently recorded a positive migration balance. In the 1990s, forced migrations once again accompanied the disintegration of Yugoslavia and the wars connected to it. Forty thousand refugees from the war zones in Yugoslavia found refuge in Slovenia. In 2007, Slovenia entered the Schengen Area of free mobility; it became a closely guarded "border mark" and the entry point of the Balkan refugee route to the fortress of Europe until Croatia replaced Slovenia in this role in 2023.

The picture of migrations in Slovenia from the end of World War II reflects the numerous similarities and continuities compared with the period that the book covers. Migrations with different patterns and structural characteristics still link Slovenia with places that have long been traditional destinations for the Slovenian workforce and, from the 1980s onward, also career mobility. In the first place, we must mention Austria and Germany, and, from the perspective of two-way migration connections and the transition of migration processes, the other Yugoslav republics and post-Yugoslav states, primarily Bosnia-Herzegovina. The case of Bosnia-Herzegovina can be used as a symbol of continuities and breaks in the long migration history. After many Slovenian foresters and, later, miners, railway workers, administrators, teachers, and tradespeople started going there in the mid-19th century, Bosnian workers have been arriving in Slovenia in large numbers from the 1970s onward; they and their families represent the largest immigrant group in Slovenia. Another discernible element of continuity in the history of migration in Slovenia is the changeability of the political space. Using the famous statement by Klaus Bade, we can say that migrations in the Slovenian ethnic territory are an emblematic case of a double process: one of people moving across state and administrative borders and another of borders moving over people.

List of Tables

List of Graphs

List of Pictures

List of Maps

Cartography by Mateja Rihtaršič.

Key Historical Dates and Events

14th century–1918: The Slovenian ethnic territory is part of the Habsburg Monarchy. In the period covered in the book, this territory comprises Carniola, parts of the Austrian Littoral (further divided into Trieste, Istria, and Goriška), Carinthia, and Styria. Prekmurje, the smaller, easternmost part across the Mura River, belongs to the Kingdom of Hungary, and in the west, Benechia and Resia are part of the Kingdom of Italy.

1848: The Springtime of Nations and the formulation of the political program for the unification of Slovenians into a United Slovenia.

1857: The Austrian Southern Railway opens between Vienna and Trieste.

1866: Benechia and Resia pledge loyalty to the unified Kingdom of Italy at a referendum.

1867: The Dual Monarchy of Austria-Hungary is created following the Austro-Hungarian Compromise; Prekmurje now belongs to the Hungarian half of the Monarchy.

1878: The Congress of Berlin and the Austro-Hungarian occupation of Bosnia-Herzegovina.

1908: Austria-Hungary annexes Bosnia-Herzegovina.

1912–1913: The First and the Second Balkan Wars take place.

1914–1918: World War I, general mobilization of the male population.

1915–1917: The Twelve Battles of the Soča/Isonzo Front lead to a mass exodus of the population from this overwhelmingly Slovenian region, the number of Slovenian refugees reaches 100,000.

1918: The State of Slovenes, Croats and Serbs is established on October 28, 1918, and merges with the Kingdom of Serbia and the Kingdom of Montenegro to form the Kingdom of Serbs, Croats and Slovenes on December 1 of the same year. In 1929, the country is renamed the Kingdom of Yugoslavia. Its Slovenian part is now the Drava Banovina, with Ljubljana as its administrative capital.

1920: With the Treaty of Trianon, Prekmurje, the eastern part of the Slovenian territory, becomes part of Slovenia/Yugoslavia.

1920: With the Treaty of Rapallo, Primorska (Istria, the Trieste hinterland, and the entire Goriška region), as well as parts of Upper and Inner Carniola, are annexed to Italy as its new eastern province, the Julian March.

1920: When the border between the Kingdom of Serbs, Croats and Slovenes and the Republic of Austria divides the former Austrian Crown Land of Carinthia, a referendum is held, and a large of the contested territory becomes part of Austria.

1921–1922: The Emigration and Immigration Act is adopted, and the Yugoslav General Emigration Commissariat in Zagreb is established as the national body to manage migrations and ties with the diaspora.

1928/29: The Emigration Office of the Drava Banovina is established in Ljubljana.

1941: Following the German attack on Yugoslavia, the occupying forces of Italy, Germany, Hungary, and the Independent State of Croatia divide the Slovenian territory.

1945: The Federative People's Republic of Yugoslavia is established, the precursor to the Socialist Federal Republic of Yugoslavia, established in 1963.

1947: With the drawing of a temporary border with Italy, Slovenia/Yugoslavia acquires most of Primorska, the Slovenian-populated former part of the Julian March.

Bibliography

10. Jahresbericht der Oesterreichischen Gesellschaft von New York. (1908). New York: Austrian Society of New York.

"A Report on the Migration of Persons Living in Knežak to the United States between January 1 and December 31, 1900." Ahiv Republike Slovenije, 136 Okrajno glavarstvo Postojna, Občina Knežak, box 34.

Adamič, France (1983). *Spomini in pričevanja o življenju in delu Louisa Adamiča.* Ljubljana: Prešernova družba.

Adamic, Louis (1931). *Dynamite: A History of Class Violence in America, 1830–1930.* New York: Viking.

Adamič, Louis (1932). Američan v svoji stari domovini. *Ljubljanski zvon* 52/7, 444–466. https://www.dlib.si/stream/URN:NBN:SI:DOC-JTEGBRS8/e257c5af-6312-4207-9b1d-4dd43fac9992/PDF, November 11, 2022.

Adamic, Louis (1932). *Laughing in the Jungle: The Autobiography of an Immigrant in America.* New York: Harper & Brothers.

Adamic, Louis (1934). *The Native's Return: An American Immigrant Visits Yugoslavia and Discovers His Old Country.* New York: Harper & Brothers.

Adamic, Louis (1935a). *Grandsons: A Story of American Lives.* New York, London: Harper & Brothers.

Adamic, Louis (1935b). *Struggle.* New York: Tomorrow Publishers.

Adamic, Louis (1936). *Cradle of Life: The Story of One Man's Beginnings.* New York, London: Harper & Brothers.

Adamic, Louis (1937). *The House in Antigua: A Restoration.* New York: Harper & Brothers.

Adamic, Louis (1938). *My America 1928–1938.* New York: Harper & Brothers.

Adamic, Louis (1940). *From Many Lands.* New York: Harper & Brothers.

Adamic, Louis (1941a). *Two-Way Passage.* New York: Harper & Brothers.

Adamic, Louis (1941b). *On Unity and Uniformity.* Cleveland: The Cleveland Press Building.

Adamic, Louis (1942). *What's Your Name?* New York: Harper & Brothers.

Adamic, Louis (1943). *My Native Land.* New York: Harper & Brothers.

Adamic, Louis (1945). *A Nation of Nations.* New York: Harper & Brothers.

Adamic, Louis (1946). *Dinner at the White House.* New York: Harper & Brothers.

Adamič, Louis (1951). *Vnuki: Zgodba iz ameriških usod*. Ljubljana: Cankarjeva založba.

Adamic, Louis (1952). *The Eagle and the Roots*. Garden City, New York: Doubleday and Company.

Adamič, Louis (1962). *Vrnitev v rodni kraj*. Ljubljana: Cankarjeva založba.

Adamič, Louis (1983). *Smeh v džungli: Avtobiografija ameriškega priseljenca*. Ljubljana: Borec.

Adamic, Louis (1985). *Laughing in the Jungle*. Salem, NH: Ayer Company. (Originally published by Harper & Brothers, New York, 1932).

Adamič, Louis (2018). *Iz mnogih dežel*. Ljubljana: Sophia.

Adlešič, Juro (1909). Organizacija slovenskega izseljeništva. *Čas* 3/4–5, 170–189. https://www.dlib.si/stream/URN:NBN:SI:DOC-PLI8WD8C/20f66a21-1359-4a24-aa48-f15f19c701f2/PDF, November 11, 2022.

Agstner, Rudolf (2012). *Austria (-Hungary) and Its Consulates in the United States of America since 1820*. Vienna: LIT Verlag.

Alberti, Mario (1911). *Il costo della vita, i salari e le paghe a Trieste nell'ultimo quarto di secolo*. Trieste: ettore Vram Editore.

Alešovec, Jakob (1883). *Ne v Ameriko! Slovencem v poduk*. Ljubljana: self-published.

Amar, Marianne, Milza, Pierre (1990). *L'immigration en France au XXe siècle*. Paris: Armand Colin.

Amerikanci (1909). *Narodni List*, February 25, 1909, 1. https://www.dlib.si/stream/URN:NBN:SI:DOC-SQ3MHYAM/8745c729-3bdf-4757-9122-b852d244f363/PDF, November 11, 2022.

Anić, Josip (1988). Jugoslavenski iseljenici u Brazilu. *Migracijske teme* 4/4, 395–411.

Anthias, Floya; Lazaridis, Gabriella (eds.) (2000). *Gender and Migration in Southern Europe: Women on the Move*. Oxford: Berg. https://doi.org/10.5040/9781474214810.ch-001.

Antrag des Abgeordneten Dr. Krek und Genossen betreffend das Auswanderungsgesetz (Eingebracht in der Sitzung des Abgeordnetenhauses von 23. April 1912). (1912). *Der Auswanderer* III/6–7, 99–100.

Arhiv Republike Slovenije, 136 Okrajno glavarstvo Postojna, box 27, 7013/1908.

Arhiv Republike Slovenije, 136 Okrajno glavarstvo Postojna, box 34, January 1, 1901.

Arko, Mihael (1931). *Zgodovina Idrije*. Gorica: Katoliška knjigarna.

Arnež, John (1966). *Slovenci v New Yorku*. New York: Studia Slovenica.

Avsenak, Ivan (1939). *Spominski zbornik Slovenije: Ob dvajsetletnici Kraljevine Jugoslavije*. Ljubljana: Založba "Jubilej".

Bade, Klaus (1992). "Billig und willig" —die "ausländischen Wanderarbeiter" im kaiserlichen Deutschland. *Deutsche im Ausland, Fremde in Deutschland: Migration in Geschichte und Gegenwart* (ed. Klaus Bade). München: Verlag C. H. Beck, 311–324.

Bade, Klaus (2001). *L'Europa in movimento: Le migrazioni dal Settecento a oggi*. Bari: Laterza.

Bader-Zaar, Brigitta (2003). Foreigners and the Law in Nineteenth-Century Austria: Juridical Concepts and Legal Rights in the Light of the Development of Citizenship. *Migration Control in the North Atlantic World: The Evolution of State Practices in Europe and the United States from the French Revolution to the Inter-War Period* (eds. Andreas Fahrmeir, Olivier Faron, Patrick Weil). New York: Berghahn Books, 138–152.

Bailey, Samuel L.; Ramella, Franco (eds.) (1988). *One Family, Two Worlds: An Italian Family's Correspondence across the Atlantic, 1901–1922*. New Brunswick: Rutgers University Press.

Bajec, Jože (1980). *Slovensko izseljensko časopisje 1891–1945*. Ljubljana: Slovenska izseljenska matica.

Balch, Emily G. (1910). *Our Slavic Fellow Citizens*. New York: Charities Publication Committee.

Balkovec, Marjetka (1994). *Občina Metlika: Etnološka topografija slovenskega etničnega ozemlja —20. stoletje*. Ljubljana: Znanstveni inštitut Filozofske fakultete, 45–47.

Banko, Catalina; Mouzakis, Pablo (2003). Slovenska emigracija v Argentini med obema svetovnima vojnama. *Dve domovini / Two Homelands* 18, 145–158.

Baraga, Frederic (1850). *A Theoretical and Practical Grammar of the Otchipwe Language: For the Use of Missionaries and Other Persons Living among the Indians*. Detroit: Jabez Fox Printer.

Barbič, Ana; Miklavčič Brezigar, Inga (1999). Domestic Work Abroad: A Necessity and an Opportunity for Rural Women from the Goriska Borderland Region of Slovenia. *Gender, Migration, and Domestic Service* (ed. Janet H. Momsen). London: Routledge, 164–177. https://doi.org/10.4324/9780203452509.

Bartsch, Franz. (1911). Einfluß der Wanderbewegung und des Fremdenverkehrs auf die Zahlenbilanz Österreich Ungarns. Mitteilungen des K.K. Finanzministeriums, 125–183.

Bartulica, Milostislav (1929). *O novom iseljeničkom zakonu*. Zagreb: Savez Organizacija Iseljenika u Zagrebu.

Bartulović, Alenka (2001). Slovenci v Srbiji. Seminar paper (unpublished). Ljubljana: Univerza v Ljubljani.

Bearzatto, Francesca (2012). *Sopravvissuto a Buchenwald e al Vajont*. Portogruaro: Nuova dimensione.

Becker, Peter (2010). Governance of Migrations in the Habsburg Monarchy and the Republic of Austria. *National Approaches to the Administration of International Migration* (ed. Arnold E. Peri). Amsterdam: IOS Press, 32–52. https://doi.org/10.3233/978-1-60750-599-0-32.

Becker, Peter (2019). The Practice of Control and the Illusion of Evidence: Passports and Personal Identification in Cities of Habsburg Austria. *Migration Policies and Materialities of Identification in European Cities Papers and Gates 1500–1930s* (eds. Hilde Greefs, Anne Winter). New York: Routledge, 217–242 https://doi.org/10.4324/9780429435065.

Beguš, Ines (2018). Integriranje dohodkov v kmečki ekonomiji v zgodnjem novem veku: Historiografski pregled po slovenskih pokrajinah. *Preživetje in podjetnost: Integrirana kmečka ekonomija na Slovenskem od srednjega veka do danes* (eds. Aleksander Panjek, Žarko Lazarević). Koper: Založba Univerze na Primorskem, 51–74. https://doi.org/10.26493/978-961-7023-81-7.51-74.

Benvenuti, Silvano (1982). Da "peccatori" a "depravati": Note sull'emigrazione dalla provincia di Gorizia (1878–1891). *Qualestoria* 3, 57–107.

Bernik, Franc (1923). *Zgodovina fare Domžale*. Domžale: self-published.

Bernik, Franc (1939). Zgodovina fare Domžale. Domžale: self-published.

Bertuzzi, Carlo; Fait, Francesco (2010). *Un secolo di partenze e di ritorni. L'emigrazione dal Friuli Venezia Giulia verso l'estero (1866–1968)*. Udine: Forum Edizioni.

Biancani, Francesca (2015). Globalization, Migration, and Female Labour in Cosmopolitan Egypt. *From Slovenia to Egypt: Aleksandrinke's Trans-Mediterranean Domestic Workers' Migration and National Imagination* (ed. Mirjam Milharčič Hladnik). Göttingen: V&R unipress, 207–228. https://doi.org/10.14220/9783737004039.207.

Bicha, Karel D. (1882). Hunkies: Stereotyping the Slavic Immigrants, 1890–1920. *Journal of American Ethnic History* 2/1, 16–38.

Bilec, Janez (1891). Slovenski drvarji na tujem. *Koledar družbe sv. Mohorja 1891*, 10–105.

Biondi, Neva (2001). Regnicoli: Storie di sudditi italiani nel Litorale austriaco durante la prima guerra mondiale. *Profughi, internati ed emigrati di Trieste, dell'Isontino e dell'Istria* (ed. Franco Cecotti). Gorizia: LEG, 49–69.

Blagotinšek Turk, Metoda (2001). Prvih deset let delovanja SNPJ. *Dve domovini / Two Homelands* 13, 101–110.

Blaznik, Pavle et al. (eds.) (1970). *Gospodarska in družbena zgodovina Slovencev: Zgodovina agrarnih panog 1.* Ljubljana: Državna založba Slovenije.

Bock, Gisela (1987). *"Drugo" delavsko gibanje v ZDA od 1905 do 1922: The Industrial Workers of the World.* Ljubljana: ŠKUC.

Bock, Gisela; Carpignano, Paolo; Ramirez, Bruno (1976). *La formazione dell'opraio massa negli USA 1898/1922.* Milano: Feltrinelli.

Bodnar, John (1977). *Immigration and Industrialization. Ethnicity in an American Mill Town, 1870–1940.* Pittsburgh: University of Pittsburgh Press.

Bogovič, Alenka; Canjko, Borut (1983). *Slovenci v Franciji.* Ljubljana: Znanstveni inštitut Filozofske fakultete v Ljubljani.

Borges, Marcelo J.; Cancian, Sonia (2021). *Migrant Letters: Emotional Language, Mobile Identities, and Writing Practices in Historical Perspective.* London: Routledge.

Brecelj, Aleš (1992). Slovenski etnični tisk v Argentini do druge svetovne vojne. *Dve domovini / Two Homelands* 2–3, 167–183.

Brecelj, Aleš (1997). Nekaj podatkov o izseljevanju iz Slovenskega primorja v Južno Ameriko, predvsem v Argentino. *Annales: Historia et Sociologia* 10, 215–236.

Brunnbauer, Ulf (2012). Emigration Policies and Nation-building in Interwar Yugoslavia. *European History Quarterly* 42/4, 602–627. https://doi.org/10.1177/0265691412458399.

Brunnbauer, Ulf (2016). *Globalizing Southeastern Europe: Emigrants, America, and the State since the Late Nineteenth Century.* Lanham, Md.: Lexington Books.

Bućin, Rajka (2017). Državna iseljenička služba od 1918. do 1941. godine: Ustroj i djelatnost tijela sa sjedištem u Zagrebu. *Arhivski vjesnik* 60/1, 37–60. https://hrcak.srce.hr/file/287102, November 11, 2022.

Bufon, Milan (2021). Slovenci v Trstu nekdaj in danes: težavno ohranjanje multikulturnosti na robu slovenskegaetničnega ozemlja. *Kronika* 70/1, 177–198.

Burjak, Matjaž (1996). Jugoslovanska katoliška jednota (JSKJ) oziroma Ameriška bratska zveza (ABZ) od ustanovitve do leta 1948. *Etnični fraternalizem v priseljenskih državah / Ethnic Fraternalism in Immigrant Countries* (ed. Matjaž Klemenčič). Maribor: Pedagoška fakulteta, 371–378.

Cafnik, Nataša; Cingerli, Bojana (1996). Progresivne Slovenke Amerike (PSWA): 1934–1956. *Etnični fraternalizem v priseljenskih državah / Ethnic Fraternalism in Immigrant Countries* (ed. Matjaž Klemenčič). Maribor: Pedagoška fakulteta, 379–394.

Cancian, Sonia (2010). *Families, Lovers, and their Letters: Italian Postwar Migration to Canada.* Manitoba: Manitoba University Press.

Cankar, Ivan (1976). *Očiščenje in pomlajenje.* Ljubljana: Državna založba Slovenije.

Cannistraro, Philip V.; Rosoli, Gianfaiusto (1979). Fascist Emigration Policy in the 1920s: An Interpretive Framework. *International Migration Review* 13/4, 673–692.

Caro, Leopold (1907). Unsere überseischen Auswanderer ind die Enquete vom Jahre 1905. *Zeitrschrift für Volkswirtschaft, Sozialpolitik und Verwaltung* 14, 529–560.

Caro, Leopold (1909). *Auswanderung und Auswanderungpolitik in Österreich.* Leipzig: Verlag von Dunker & Humbolt.

Cattaruzza, Marina (1979). *La formazione del proletariato urbano: Immigrati, operai di mestiere, donne a Trieste dalla meta del secolo XIX alla prima guerra mondiale.* Torino: Musolini editore.

Cecotti, Franco (2001a). Emigranti e marinai: I cittadini del Litorale trattenuti all'estero 1914–1919. "*Un esodo che non ha pari". 1914–1919: Profughi, internati ed emigrati di Trieste, dell'Isontino e dell'Istria* (ed. Franco Cecotti). Gorizia: LEG, 27–47.

Cecotti, Franco (2001b). Internamenti di civili durante la prima guerra mondiale: Friuli austriaco, Istria e Trieste. "*Un esodo che non ha pari". 1914–1919: Profughi, internati ed emigrati di Trieste, dell'Isontino e dell'Istria* (ed. Franco Cecotti). Gorizia: LEG, 71–98.

Cecotti, Franco (2001c). Trieste 1914–1919: La città spopolata, la città rifugio. "*Un esodo che non ha pari". 1914–1919: Profughi, internati ed emigrati di Trieste, dell'Isontino e dell'Istria* (ed. Franco Cecotti). Gorizia: LEG, 155–182.

Cecotti, Franco (2003). L'emigrazione dal Litorale austriaco verso Argentina e Brasile 1878–1903. *Un'altra terra un'altra vita: L'emigrazione isontina in Sud America tra storia e memoria 1878–1970* (eds. Franco Cecotti, Dario Mattiussi). Gorizia: Centro isontino di rocerca edocumentazione storica e sociale Leopoldo Gasparini, 15–58.

Cecotti, Franco (2010). Mobilità dei confini e modelli migratori: il caso della Venezia Giulia. Archivio Storico dell'emigrazione Italiana. https://www.asei.eu/it/2010/04/mobilita-dei-confini-e-modelli-migratori-il-caso-della-venezia-giulia/, November 11, 2022.

Cesar Nezbala, Polonca (1990). Rajska dolina. *Dve domovini / Two Homelands* 1, 83–106.

Chicaški Slovenci na noge. (1919). *Glasilo K. S. K. jednote,* June 18, 1919, 5, 7. https://www.dlib.si/stream/URN:NBN:SI:DOC-2JRWHSTB/0bc3c869-a009-435I-a156-474d7a89bf7b/PDF, November 11, 2022.

Chmelar, Hans (1974). *Höhepunkte der österreichischen Auswanderung: Die Auswanderung aus den im Reichsrat vertretenen Königreichen und Ländern in den Jahren 1905–1914*. Wien: Österreichische Akademie der Wissenschaften.

Colakovic, Branko Mita (1973). *Yugoslav Migrations to America*. San Francisco: R and E Research Associates.

Comstock, Lyndon (2013). *Annie Clemenc and the Great Keweenaw Copper Strike*. North Charleston: CreateSpace Independent Publishing Platform.

Corti, Paola (2003). L'emigrazione italiana in Francia: Un fenomeno di lunga durata. *Altreitalie* 26, 4–26.

Corti, Paola (2015). Migration as a Historical Issue. *The History of Migration in Europe. Perspectives from Economics, Politics and Sociology* (ed. Francesca Fauri). London, New York: Routledge, 39–50.

Cosattini, Giovanni (1903). *L'emigrazione temporanea del Friuli*. Roma: Bertero e C.

Cross, Gary S. (1983). *Immigrant Workers in Industrial France: The Making of a New Laboring Class*. Philadelphia: Temple University Press. https://doi.org/10.2307/j.ctv6mtdp3.

Cvirn, Janez (1998). Nemška manjšina v Kraljevini Jugoslaviji (1918–1941). *Die "Deutschen" in Slowenien (1918–1955): kurzer Abriß / "Nemci" na Slovenskem (1918–1955): kratek oris* (ed. Dušan Nećak). Ljubljana: Znanstveni inštitut Filozofske fakultete, 61–67.

Čačić, Jadranka; Kumpers, Josip (1989). Između Jugoslavije i Francuske (pogled na jugoslovanske migracije u Francuskoj). *Migracijske teme* 5, 287–306.

Čermelj, Lavo (1958). O ljudskem štetju v Trstu leta 1910. *Anali Jadranskog instituta* 2, 7–51.

Čermelj, Lavo (1972). *Med prvim in drugim tržaškim procesom*. Ljubljana: Slovenska matica.

Čermelj, Lavo (1974). *Sloveni e Croati in Italia tra le due guerre Trieste*. Trst: Založništvo tržaškega tiska.

Černy, Dragica (1953). Rast prebivalstva v Prekmurju: Pregled gibanja prebivalstva od leta 1869 do 1948. *Kronika* 1/2, 131–137. https://www.dlib.si/stream/URN:NBN:SI:DOC-H70YWNAT/e83cd01f-76c8-4f5e-a0d8-27f520598c7e/PDF, November 11, 2022.

Česen, Frank (1970). Spomini ob moji osemdesetletnici. *Slovenski izseljenski koledar 1971*, 196.

Čizmić, Ivan (1974). *Jugoslavenski iseljenički pokret u SAD i stvaranje jugoslavenske države 1918*. Zagreb: Sveučilište u Zagrebu, Institut za hrvatsku povijest.

Čizmić, Ivan (1990). Emigration from Croatia, 1880–1914. *Overseas Migration from East-Central and Southeastern Europe 1880–1940* (ed. Juliana Puskás). Budapest: Akadémiai Kiadó, 143–167.

Čizmić, Ivan (1992). Review of American-Yugoslav Relations in World War II. *FDR and His Contemporaries* (eds. Cornelis A. van Minnen, John F. Sears). Houndmills: Macmillan, 75–84, 224–226. https://doi.org/10.1007/978-1-349-21901-8_6.

Čizmić, Ivan (1998). Political Activities of Croatian Immigrants in the USA and the Creation of an Independent Croatia. *Društvena istraživanja* 7/1–2 (33–34), 5–25.

Čizmić, Ivan (2002). Conflict and Co-Operation: Clevelands Croatians, 1880s–1930s. *Identity, Conflict, and Cooperation. Central Europeans in Cleveland, 1850–1930* (eds. David C. Hammack, Diane L. Grabowski, John J. Grabowski). Cleveland: The Western Reserve Historical Society, 44–76.

D. (1927). O naši izseljeniški službi. *Slovenec*, March 10, 1927, 6. https://www.dlib.si/stream/URN:NBN:SI:DOC-2EKR9UCJ/4acc0096-2983-4d7a-afa9-dfe79e64e5ba/PDF, November 11, 2022.

Deleuze, Gilles (1993). Rhizome Versus Trees. *The Deleuze Reader* (ed. Constantin V. Boundas). New York: Columbia University Press, 27–38. https://doi.org/10.7312/boun93296-003.

Devetak, Robert (2013–2014). "Razkropljeni smo v širni svet, naš sled je vroča solza in bolest ...": Izkušnja prve svetovne vojne civilnega prebivalstva iz sodnega okraja Kanal. *Goriški letnik* 37/38, 323–338.

Devoto, Fernando (1989). Argentine Migration Policy and Movements of the European Population (1876–1925). *Estudios Migratorios Latinoamericanos* 4, 135–158.

Di Caporiacco, Gino (1969). *Storia e statistica dell'emigrazione dal Friuli e dalla Carnia, I.: Dall'età veneta al 1915*. Udine: Edizioni del Friuli Nuovo.

Die Auswanderung aus Gottschee nach Amerika (1913). *Der Auswanderer* IV/7–8, 74–75.

Die Auswanderungsfrage im österreichischen Abgeordnetenhause. (1912). *Der Auswanderer* III/4–5, 74–75.

Dolinšek Divčić, Marija (2001). Slovenci v Bosni in Hercegovini. *Slovensko izseljenstvo: Zbornik ob 50-letnici Slovenske izseljenske matice* (ed. Milica Trebše Štolfa). Ljubljana: Združenje Slovenska izseljenska matica, 301–306.

Donato, Katherine M.; Gabaccia, Donna (2015). *Gender and International Migration: From the Slavery Era to the Global Age*. New York: Russell Sage Foundation.

Dopisi (1893). *Slovenski Gospodar*, March 9, 1893, 76. https://www.dlib.si/stream/URN:NBN:SI:DOC-O9VW1JPX/820b2a4d-6c83-440f-a9a6-86dcf118bf8b/PDF, November 11, 2022.

Douek, Ellis (2004). *A Middle Eastern Affair*. London: Peter Halban.

Dražumerič, Marinka (1987). Izseljevanje iz Bele krajine od začetkov do druge svetovne vojne. *Traditiones* 16, 169–204.

Dražumerič, Marinka (1988). Gospodarsko življenje v Beli krajini med 1. svetovno vojno. *Zgodovinski časopis* 42/1, 85–102.

Drnovšek, Marjan (1978). O stališčih slovenskih socialistov v ZDA do vojne in jugoslovanskega vprašanja med prvo svetovno vojno. *Prispevki za zgodovino delavskega gibanja* 1–2, 75–96.

Drnovšek, Marjan (1989). Prispevek k zgodovini izseljevanja iz Gorenjske pred prvo svetovno vojno —s posebnim poudarkom na Škofji Loki. *Kronika* 37/3, 203–218. https://www.dlib.si/stream/URN:NBN:SI:DOC-9725QEE6/6f49bc15-c06d-4240-9273-8b2099e60ead/PDF, November 11, 2022.

Drnovšek, Marjan (1990). Delovanje Toma Brejca med slovenskimi izseljenci v Franciji v letih 1936–39. *Dve domovini / Two Homelands* 1, 179–204.

Drnovšek, Marjan (1991). *Pot slovenskih izseljencev na tuje*. Ljubljana: Založba Mladika.

Drnovšek, Marjan (1992). Izseljensko in drugo časopisje med Slovenci v zahodni Evropi do leta 1940. *Dve domovini / Two Homelands* 2–3, 265–316.

Drnovšek, Marjan (1993). Načrti in rezultati raziskovanja slovenskega izseljenstva do leta 1941 —s posebnim poudarkom na izseljevanju v Ameriko do prve svetovne vojne. *Dve domovini / Two Homelands* 4, 11–40.

Drnovšek, Marjan (1996). Slovenska izseljenska društva v zahodnoevropskih državah med obema svetovnima vojnama. *Dve domovini / Two Homelands* 7, 33–49.

Drnovšek, Marjan (1997). Izseljenci in Kolodvorska ulica. *Homo sum: Ivan Hribar in njegova Ljubljana. Zbornik Mestnega muzeja Ljubljana* (eds. Taja Čepič, Janja Rebolj). Ljubljana: Mestni muzej, 193–203.

Drnovšek, Marjan (1998a). Spomenica jugoslovanske izseljeniške katoliške akcije iz Aumetza v Franciji (1931). *Dve domovini / Two Homelands* 9, 113–135.

Drnovšek, Marjan (1998b). *Usodna privlačnost Amerike: Pričevanja izseljencev o prvih stikih z novim svetom*. Ljubljana: Nova revija.

Drnovšek, Marjan (1999a). Izseljevanje Slovencev v razvite evropske države do leta 1940. *Slovenska izseljenska književnost 1: Evropa, Avstralija, Azija* (eds. Janja Žitnik, Helga Glušič). Ljubljana: Založba ZRC, 29–60.

Drnovšek, Marjan (1999b). Okrajna poročila o izseljevanju iz Kranjske v letih 1892–1913. *Gestrinov zbornik* (ed. Darja Mihelič). Ljubljana: Založba ZRC, 413–432.

Drnovšek, Marjan (2000). Slovenian Emigrants and Europe: Their Dispersal in the European Area in the Past Two Centuries. *Ethnic and National Issues in Russian and East European History* (ed. John Morison). London: Macmillan Press, 323–330.

Drnovšek, Marjan (2001a). Kuharica Liza v New Yorku. *Izseljenec: Življenjske zgodbe Slovencev po svetu*. Ljubljana: Muzej novejše zgodovine, 43–46.

Drnovšek, Marjan (2001b). Slovenski izseljenci in Zahodna Evropa do druge svetovne vojne. *Slovensko izseljenstvo: Zbornik ob 50-letnici Slovenske izseljenske matice* (ed. Milica Trebše Štolfa). Ljubljana: Združenje Slovenska izseljenska matica, 281–290.

Drnovšek, Marjan (2002a). Kulturna in šolska dejavnost med slovenskimi izseljenci v evropi v 20. stoletju. *Ustvarjalnost Slovencev po svetu: 38. Seminar slovenskega jezika, literature in kulture* (ed. Boža Krakar Vogel). Ljubljana: Filozofska fakulteta, 149–162.

Drnovšek, Marjan (2002b). Ljubljana: Križišče izseljencev na poti v svet. *Kronika* 50/3, 363–374. https://www.dlib.si/stream/URN:NBN:SI:DOC-ZXQR214W/1d1e1049-465b-43d3-b209-70a93c7a7b9e/PDF, November 11, 2022.

Drnovšek, Marjan (2003a). Emigration of Slovene Women: A Short Historical View. *Dve domovini / Two Homelands* 17, 29–46.

Drnovšek, Marjan (2003b). *Franc Pirc (1785–1880), sadjar na Kranjskem in misijonar v Ameriki*. Naklo: Občina Naklo.

Drnovšek, Marjan (2003c). Pismo Apolonije Noč od Sv. Jožefa/St. Joseph, Minnesota (1855). *Dve domovini / Two Homelands* 18, 63–81.

Drnovšek, Marjan (2006). Jugoslovansko-nemška sporazuma o sezonskem delu med svetovnima vojnama. *Prispevki za novejšo zgodovino* 46/1, 219–234.

Drnovšek, Marjan (2007a). Krekova vestfalska pisma: Socialno-ekonomski pogledi in izseljenstvo. *Dve domovini / Two Homelands* 25, 161–185.

Drnovšek, Marjan (2007b). The Attitudes of the State and the Catholic Church Towards Slovenian Emigration: A Historian's View. *Historical and Cultural Perspectives on Slovenian Migration* (ed. Marjan Drnovšek). Ljubljana: Založba ZRC, 13–43.

Drnovšek, Marjan (2007c). Kočevarski krošnjarji in nacistična propaganda. *Prispevki za novejšo zgodovino* 47/1, 85–101.

Drnovšek, Marjan (2008). Društveni arhivi v Nemčiji: Med preteklostjo in sedanjostjo. *Arhivi* 31/2, 251–261.

Drnovšek, Marjan (2009). Zakaj je bila za slovenske izseljence Amerika bolj privlačna kot Srbija in Rusija? *Kronika* 57, 467–482. https://www.dlib.si/stream/URN:NBN:SI:DOC-XP7TBH4T/ae904adf-68cc-4f95-89b1-3c2bbd6c8662/PDF, November 11, 2022.

Drnovšek, Marjan (2010a). *Izseljevanje, "rakrana" slovenskega naroda: Od misijonarja Friderika Barage do migracijske politike države Slovenije*. Ljubljana: Nova revija.

Drnovšek, Marjan (2010b). Nadzor nad komunisti-emigranti in cenzura. *Cenzurirano: zgodovina cenzure na Slovenskem od 19. stoletja do danes* (ed. Mateja Režek). Ljubljana: Nova revija, 67–88.

Drnovšek, Marjan (2012). *Slovenski izseljenci in zahodna Evropa v obdobju prve Jugoslavije*. Ljubljana: Založba ZRC. https://doi.org/10.3986/9789612543884.

Drnovšek, Marjan; Kalc, Aleksej (2014). Poklicne migracije Slovencev v jugoslovanskem prostoru med svetovnima vojnama. *Priseljevanje in družtveno delovanje Slovencev v drugih delih jugoslovanskega prostora: Zgodovinski oris in sedanjost* (ed. Janja Žitnik Serafin). Ljubljana: Založba ZRC, 91–118. https://doi.org/10.3986/9789612546687.

Dubrović, Ervin (2012). *Veliki val: Iseljavanje iz srednje Europe u Ameriku 1880–1914*. Rijeka: Muzej grada Rijeke.

Dular, Andrej (1996). Obrti in trgovina na Bovškem. *Traditiones* 25, 151–171.

Đikanović, Vesna (2012). *Iseljavanje u Sedinjene američke države: Jugoslavensko iskustvo 1918–1941*. Beograd: Institut za noviju istoriju Srbije.

Đikanović, Vesna (2014). More than just a Piece of Paper: Yugoslav Citizenship Policies and Immigration in the USA 1918–1941. *Staatsbürgerschaft und teilhabe: Bürgerliche, politische und soziale Rechte im östlichen Europa* (eds. Katrin Boeckh et al.). Oldenburg: De Gruyter, 83–93.

Đikanović, Vesna (2016). *Iseljenici u Severnoj Americi i jugoslovenska država 1918-1945*. Doktorska disertacija. Beograd: Univerzitet u Beogradu, Filozofski fakultet.

Elliot, Bruce S.; Gerber, David A.; Sinke, Suzanne M. (eds.) (2006). *Letters across Borders: The Epistolary Practices of International Migrants*. New York: Palgrave Macmillan.

Englisch, Karl (1911). Zu unserer Auswanderungsfrage. *Statistische Monatsschrift* 16, 89–120.

Englisch, Karl (1913). Die österreichische Auswanderungsstatistik. *Statistische Monatsschrift*, N. F. XVIII, 65–167.

Enyeart, John P. (2019). *Death to Fascism: Louis Adamic's Fight for Democracy*. Baltimore: University of Illinois Press.

Esteves, Rui; Khoudour-Castéras, David (2009). A Fantastic Rain of Gold: European Migrants' Remittances and Balance of Payments Adjustment During the Gold Standard Period. *The Journal of Economic History* 69/4, 951–985.

Evanish, Joe (1996). SNPJ History: A General Overview. *Etnični fraternalizem v priseljenskih državah / Ethnic Fraternalism in Immigrant Countries* (ed. Matjaž Klemenčič). Maribor: Pedagoška fakulteta, 59–64.

F. K. (1907). "O potnih listih," *Občinska uprava. Glasilo Kmetske županske zveze*, July 25, 1907. https://www.dlib.si/stream/URN:NBN:SI:DOC-9GAAHIJT/9520e9ed-1a2b-4117-974e-720c50d03f18/PDF, November 11, 2022.

Fassmann, Heinz (1985). A Survey of Patterns and Structures of Migration in Austria, 1850–1900. *Labor Migration in the Atlantic Economies* (ed. Dirk Hoerder). Newport: Greenwood Press, 49–93.

Fassmann, Heinz (1994). Emigration, Immigration and Internal Migration in the Austro-Hungarian Monarchy, 1910. *Roots of the Transplanted: Late Nineteenth Century East Central and Southeastern Europe* (eds. Dirk Hoerder, Ingle Blank). New York: Columbia University Press, 253–308.

Ferenc, Mitja (2005). *Kočevska pusta in prazna: Nemško jezikovno območje na Kočevskem po odselitvi Nemcev*. Ljubljana: Modrijan.

Ferenc, Mitja (2010). Posledice izselitve kočevskih Nemcev in neizvedene kolonizacije Kočevske med vojno in po njej. *Migracije in slovenski prostor od antike do danes* (eds. Peter Štih, Bojan Balkovec). Ljubljana: Zveza zgodovinskih društev Slovenije, 528–550.

Ferenc, Mitja; Repe, Božo (2004). Nemška manjšina v Sloveniji med obema vojnama. *Slovensko-avstrijski odnosi v 20. stoletju* (ed. Dušan Nećak). Ljubljana: Oddelek za zgodovino Filozofske fakultete, 147–160.

Ferenc, Tone (1968). *Nacistična raznarodovalna politika v Sloveniji v letih 1941–1945*. Maribor: Obzorja.

Ferenc, Tone (1971a). Izgnanci kot cenena delovna sila nemškega rajha. *Zgodovinski časopis* 25/1–2, 47–59. https://www.dlib.si/stream/URN:NBN:SI:DOC-Z5Q0PITD/4b65102f-96c7-4e4a-82f3-22d77b9951be/PDF, November 11, 2022.

Ferenc, Tone (1971b). *Muzej slovenskih izgnancev v Brestanici*. Ljubljana: Zavod za spomeniško varstvo SR Slovenije.

Ferenc, Tone (1993). Množično izganjanje Slovencev med drugo svetovno vojno. *Izgnanci: Zbornik slovenskih izgnancev 1941–1945*. Ljubljana: Društvo izgnancev Slovenije, 19–107.

Feys, Torsten (2013). *The Battle for the Migrants: The Introduction of Steamshipping on the North Atlantic and Its Impact on the European Exodus*. St. John's

Newfoundland: International Maritime Economic History Association – IMEHA. https://doi.org/10.2307/j.ctt21pxhnf.

Finzi, Roberto; Panjek, Giovanni (Eds.) (2001). *Storia economica e sociale di Trieste. La città dei gruppi.* Trieste: LINT.

Fischer, Jasna (1984). Delavke Tobačne tovarne v Ljubljani v letih 1871–1914. *Prispevki za zgodovino delavskega gibanja* 1–2, 11–62.

Fischer, Jasna (1988). Prebivalstvo v obrti in industriji na Slovenskem od sredine 19. stoletja do prve svetovne vojne. *Prispevki za novejšo zgodovino* 1–2, 23–40.

Fischer, Jasna (2005a). Slovensko narodno ozemlje in razvoj prebivalstva. *Slovenska novejša zgodovina 1848–1992* (eds. Jasna Fischer et al.). Ljubljana: Mladinska knjiga, 17–21.

Fischer, Jasna (2005b). Zagate v kmetijstvu. *Slovenska novejša zgodovina 1848–1992* (eds. Jasna Fischer et al.). Ljubljana: Mladinska knjiga, 72–74.

Fischer, Jasna (2005c). Razvoj obrti, industrije in trgovine. *Slovenska novejša zgodovina 1848–1992* (eds. Jasna Fischer et al.). Ljubljana: Mladinska knjiga, 77–86.

Fitzpatrick, David (1994). *Oceans of Consolation: Personal Accounts of Irish Migration to Australia.* Ithaca, NY: Cornell University Press.

Friš, Darko (1991). Zgodovina časnika Amerikanski Slovenec in "verske vsebine", ki jih je obravnaval med leti 1881–1931. *Znanstvena revija: Humanistika* 2/3, 405–414.

Friš, Darko (1992). Delovanje šolskih sester kongregacije Sv. Frančiška Asiškega – Kristusa Kralja v ZDA med letoma 1909 in 1915. *Dve domovini / Two Homelands* 2/3, 371–389.

Friš, Darko (1994). Jugoslovanska katoliška jednota (1898–1920). *Dve domovini / Two Homelands* 5, 37–62.

Friš, Darko (1995). *Ameriški Slovenci in katoliška cerkev 1871-1924.* Celovec: Mohorjeva založba.

Friš, Darko (1996). Naši izseljenci na obisku v "stari domovini" (1836–1941). *Razvoj turizma v Sloveniji: zbornik referatov* (eds. Franc Rozman, Žarko Lazarević). Ljubljana: Zveza zgodovinskih društev Slovenije, 165–176.

Friš, Darko (1999). *Korespondenca slovenskih katoličanov v ZDA med leti 1882–1924.* Ljubljana: Arhivsko društvo Slovenije.

Friš, Darko; Kolar, Bogdan; Vovko, Andrej (1997). *Prvih sto let Kranjsko slovenske katoliške Jednote: Pregled zgodovine KSKJ 1894–1994.* Ljubljana: Založba Ilex-Impex.

Gabaccia, Donna R. (1994). *From the Other Side: Women, Gender, and Immigrant Life in the U.S., 1820–1990.* Bloomington: Indiana University Press.

Gabršček, Andrej (1934). *Goriški Slovenci: Narodne, kulturne, politične in gospodarske črtice. 1901–1924.* Ljubljana: Self-published.

Garroni, Maria Susanna; Vezzosi, Elisabetta. (2009). Italiane migranti. *Storia d'Italia, Annali, 24, Migrazioni* (eds. Paola Corti, Matteo Sanfilippo). Torino: Einaudi, 449–465.

Genorio, Rado (1988). Slovenci v delavskem gibanju Argentine v obdobju med obema vojnama. *Migracijske i etničke teme* 1/2, 187–193.

Genorio, Rado (1989). *Slovenci v Kanadi.* Ljubljana: Inštitut za geografijo Univerze Edvarda Kardelja.

Genorio, Rado (1992). Slovenski tisk v Kanadi. *Dve domovini / Two Homelands* 2/3, 343–351.

Gerber, David A. (2006). Epistolary Masquerades: Acts of Deceiving and Withholding in Immigrant Letters. *Letters across Borders, The Epistolary Practices of International Migrants* (eds. Bruce S. Elliott, David A. Gerber, Suzanne M. Sinke). New York: Palgrave Macmillan, 141–157. https://doi.org/10.1057/9780230601079_8.

Gestrin, Ferdo (1991). *Slovenske dežele in zgodnji kapitalizem.* Ljubljana: Slovenska matica.

Gestrin, Ferdo (1998). *Slovanske migracije v Italijo.* Ljubljana: Slovenska matica.

Gestrin, Ferdo; Vasilij Melik (1966). *Slovenska zgodovina od konca 18. stoletja do 1918.* Ljubljana: DZS.

Gilman, Rhoda R. (2000). The History and Peopling of Minnesota: Its Culture. *Daedalus* 129/3, 1–29. https://www.jstor.org/stable/20027645, November 11, 2022.

Gobbi, Olimpia (2011). Emigrazione femminile: Balie e domestiche marchigiane in Egitto fra Otto e Novecento. *Proposte e ricerche* 66, 7–24.

Gobetz, G. Edward (1979). Slovenian American Literature. *Ethnic Literatures Since 1776: The Many Voices of America*, Part 2 (eds. W. T. Zyla, Wendell Aycock). Lubbock, Texas: Texas Tech University Press, 547–568.

Gobetz, G. Edward; Donchenko, Adele (eds.) (1977). *Anthology of Slovenian American Literature.* Willoughby Hills, Ohio: Slovenian Research Center of America.

Golec, Boris (1999). Družba v mestih in trgih Dolenjske in Notranjske od poznega srednjega veka do srede 18. stoletja. PhD thesis. Ljubljana: Filozofska fakulteta, Univerza v Ljubljani.

Golouh, Rudlof (1935). Primorska emigracija v Mariboru. *Kronika slovenskih mest* 2/2, 159–160. https://www.dlib.si/stream/URN:NBN:SI:DOC-MYV5JY28/de326bc4-b89d-4127-be24-372df475dea6/PDF, November 11, 2022.

Gombač, Metka; Boris M. Gombač (2013). *Trpljenje otrok v vojni: Sedemdeset let po zaprtju italijanskih taborišč*. Ljubljana: Mladinska knjiga.

Gomiršek, Tanja (2010). Delo in položaj žensk v Goriških brdih 19. stoletja. *Maruščev zbornik: Zbornik prispevkov v počastitev 70-letnice prof. dr. Branka Marušiča* (ed. Petra Kolenc). Nova Gorica: Goriški muzej, 335–369.

Goršič, Fran (1914). O "kasatorjih" in njihovih pogodbah. *Slovenski pravnik* 30, 320–329.

Goršič, France (1911). Naborna dolžnost in Amerika. *Koledar Družbe sv. Mohorja za leto 1911*. Celovec: Družba sv. Mohorja, 122–123.

Gould, John D. (1980). European Inter-Continental Emigration. The Road Home: Return Migration from the USA. *Journal of European Economic History* 9/1, 41–111.

Govorchin, Gerald Gilbert (1961). *Americans from Yugoslavia: A Survey on Yugoslav Immigrants in the United States*. Gainesville: University of Florida Press.

Grafenauer, Danijel (2010). Koroški Slovenci – begunci in njihova družbena integracija v osrednji Sloveniji. *Migracije in slovenski prostor od antike do danes* (eds. Peter Štih, Bojan Balkovec). Ljubljana: Zveza zgodovinskih društev Slovenije, 474–487.

Granda, Stane (1999). Gospodarske krize na slovenskem prostoru v devetnajstem stoletju. *Gospodarske krize in Slovenci* (eds. Neven Borak, Žarko Lazarević). Ljubljana: Inštitut za novejšo zgodovino, 27–45.

Grašič, Mira (1990). Začetki primorskega emigrantskega društva Nanos v predvojnem Mariboru. *Časopis za zgodovino in narodopisje* 1, 23–35.

Green, Nancy L.; Weil, François (2007). Introduction. *Citizenship and Those Who Leave: The Politics of Emigration and Expatriation* (eds. Nancy L. Green, François Weil). Chicago: University of Illinois Press, 1–9.

Grill, Vatroslav (1979). *Med dvema svetovoma*. Ljubljana: Mladinska knjiga.

Grossutti, Javier P. (2013). *Via dall'Istria: L'emigrazione istriana dalla prima metà dell'Ottocento ai primi anni Quaranta del Novecento*. Trieste: Università popolare di Trieste.

Grossutti, Javier P.; Boz, Nadia (2009). Protagoniste o comparse? L'emigrazione femminile dal Friuli. *Il lavoro femminile tra vecchie e nuove migrazioni: Il caso del Friuli Venezia Giulia* (eds. Ariella Verrocchio, Paola Tessitori). Roma: Ediesse, 25–42.

Habjan, Jernej; Koron, Alenka (2014). Periodični tisk: Analiza podatkovne baze. http://pslk.zrc-sazu.si/static/media/analize/sl/Periodika-ANALIZA.pdf, November 11, 2022.

Hahn, Sylvia (2011). Austria. *The Encyclopedia of Migration and Minorities in Europe: From the 17th Century to the Present* (eds. Klaus J. Bade et al.). Cambridge: Cambridge University Press, 83–93.

Hahn, Sylvia (2012). *Historische Migrationsforschung.* Frankfurt, New York: Campus Verlag.

Hahn, Sylvia (2015). Labour Migration and Female Breadwinners. *From Slovenia to Egypt: Aleksandrinke's Trans-Mediterranean Domestic Workers' Migration and National Imagination* (ed. Mirjam Milharčič Hladnik). Göttingen: V&R Unipress, 39–48. https://doi.org/10.14220/9783737004039.39.

Hartman, Bruno (1976). Primorci in slovenska kultura v Mariboru med vojnama. *Jadranski koledar 1977.* Trst: Založništvo tržaškega tiska, 142–150.

Harzig, Christiane (2001). Women Migrants as Global and Local Agents, New research Strategies on Gender and Migration. *Women, Gender and Labour Migration: Historical and Global Perspectives* (ed. Pamela Sharpe). London, New York: Routledge, 15–28.

Hašimbegović, Elma (2008). Novi časi za gospodarstvo. *Četrta stran trikotnika: Znameniti Slovenci in slovenska društva v Bosni in Hercegovini, 1878–2000* (ed. Stanislav Koblar). Ljubljana: Mladinska knjiga, 32–47.

Hecke, Wilhelm (1913). Volksvermehrung, Binnenwanderung und Umgangssprache in den Österreichischen Alpen-Ländern und Südäldern. *Statistiche Monatsschrift XVIII. Jahrgang*, 323–392.

Hecke, Wilhelm (1916). Die österreichisch-ungarische Auswanderung nach Nordamerika. *Statistische Monatsschrift* XVIII. Jahrgang, 483–484.

Herbert, Ulrich (1990). *A History of Foreign Labor in Germany 1880–1980: Seasonal Workers, Forced Laborers, Guest Workers.* Ann Arbor: Michigan University Press.

Herle, Dominik (2020). *Pregled zgodovine slovenske skupnosti v Calumetu, Michigan*. Ljubljana: Združenje Slovenska izseljenska matica.

Hladnik, Miran (1985). Slovene Popular Novels about Emigration in the Nineteenth Century. *Slovene Studies* 1–2, 57–62.

Hočevar, Touissaint (1979). *Slovenski družbeni razvoj: Izbrane razprave.* New Orleans: Prometej.

Hočevar, Toussaint (1977). Češki vpliv pri snovanju slovenskih podpornih "jednot" v ZDA. *Časopis za zgodovino in narodopisje* 48/1–2, 234–240.

Hochstadt, Steven Lawrence (1999). *Mobility and Modernity: Migration in Germany, 1820–1989.* Ann Arbor: The University of Michigan Press.

Hoerder, Dirk (1991). International Labor Markets and Community Building by Migrant Workers in the Atlantic Economies. *A Century of European Migrations*

1830–1930 (eds. Rudolph J. Vecoli, Suzanne M. Sinke). Urbana: University of Illinois Press, 78–107.

Hoerder, Dirk (1996). Migration in the Atlantic Economies. Regional European Origins and Worldwide Expansions. *European Migrants: Global and Local Perspectives* (eds. Dirk Hoerder, Laslie P. Moch). Boston: Northeastern University Press, 21–50.

Hoerder, Dirk (2002). *Cultures in Contact: World Migrations in the Second Millennium*. Durham: Duke University Press.

Hoerder, Dirk (2015). Re-Remembering Women Who Chose Caregiving Careers in a Global Perspective: Mothers of the Nation or Agents in Their Own Lives? *From Slovenia to Egypt: Aleksandrinke's Trans-Mediterranean Domestic Workers' Migration and National Imagination* (ed. Mirjam Milharčič Hladnik). Göttingen: V&R Unipress, 117–130. https://doi.org/10.14220/9783737004039.117.

Hoerder, Dirk; van Nederveen Meerkerk, Elise; Neunsinger, Silke (eds.) (2015). *Towards a Global History of Domestic and Caregiving Workers*. Leiden: Brill. https://doi.org/10.1017/S0020859016000134.

Horvat, Avgust (1997). Prizadevanja za izobrazbo med slovenskimi izseljenci v Argentini. *Dve domovini / Two Homelands* 8, 37–48.

Horvath, Ferenc (1989). Zgodovina gospodarstva v delu Prekmurja, ki je pripadalo Železni županiji (1828–1910). *Časopis za zgodovino in narodopisje* 2, 177–184.

Hranilović, Nada (1987). Međuratna iseljenička politika. *Migracijske teme* 3/4, 325–334.

Hranilović, Nada (1988). Radnička straža kao izvor za izučavanje položaja naših iseljenika u radničkom i socijalističkom pokretu SAD (1907–1918). *Migracijske i etničke teme* 4/1–2, 89–102.

Hubbard, William H. (1984). *Auf dem Weg zur Grossstadt: Eine Sozialgeschichte der Stadt Graz 1850–1914*. Munchen: R. Oldenbourg Verlag.

Humar, Vesna (2019). Aleksandrinke v virih avstroogrske konzularne službe v Egiptu. Seminar paper. Koper: Univerza na Primorskem, Fakulteta za humanistične študije.

Hutter, Martha (1999). Die ersten Amerika-Gotscheer. *Festschrift: 80 Jahre Gottscheer Landsmannschaft in Graz:... den Landsleuten zu helfen*. Graz: Gottscheer Landsmannschaft, 89–96.

Iancis, Paolo (2006). La mobilità del lavoro manifatturiero a Gorizia nel Settecento. *Barok na Goriškem – Il barocco nel goriziano* (ed. Ferdinand Šerbelj). Nova Gorica: Goriški muzej, 173–177.

Irwin, Douglas A. (2003). Explaining America's Surge in Manufactured Exports, 1880–1913. *The Review of Economics and Statistics* 2, 364–376. https://doi.org/10.1162/003465303765299873.

Ismail, Annelies, Gabriel, Mona (2011). *Alexandria Was Our Destiny: Based on the Memories of Marie-Luise Nagel*. Aleksandrija: Bibliotheca Alexandrina.

Ivanush, Mary (1964). Ob 30-letnici Progresivnih Slovenk Amerike. *Slovenski izseljenski koledar 1965*, 216–220.

Izseljenci. (1935). *Koledar Družbe sv. Mohorja 1935*, 54–65.

Izseljeniška spomenica Jugoslovanske Izseljeniške katoliške akcije v državah zapadne Evrope. (1933). *Izseljenski vestnik* 3/2, 2–4. https://www.dlib.si/stream/URN:NBN:SI:DOC-YOV7K58Q/75e31862-4d4e-4924-8633-66df768050b6/PDF, November 11, 2022.

Izseljenski zbornik Izseljenske zbornice v Ljubljani. (1938). Ljubljana: Družba sv. Rafaela.

Izseljevanje v Zedinjene države ameriške. (1922). *Slovenec*, October 8, 1922, 3.

Izseljevanje. (1905). *Slovenec*, October 14, 1905, 1. https://www.dlib.si/details/URN:NBN:SI:DOC-NSV7A52T/?euapi=1&query=%27keywords%3dslovenec+(1873)%27&sortDir=ASC&sort=date&fyear=1905&pageSize=100&page=3, November 11, 2022.

Izveštaj narodnoj skupštini za 1926/27. godinu. (1928). Beograd: Ministarstvo socijalne politike Kraljevine Srba, Hrvata i Slovenaca.

Janjetović, Zoran (2012). National Minorities in Yugoslavia 1918–1941. *Revue Für Kroatische Geschichte / Revue d'histoire Croate* 8, 1, 61–75.

Janša Zorn, Olga (1962). Agitacija "Novic" za preseljevanje Slovencev v Srbijo v letu 1865 in 1866. *Kronika* 10/3, 185–187. https://www.dlib.si/stream/URN:NBN:SI:DOC-PD1G5MQB/2a5ad948-f027-4a90-aba6-c5812b59f74a/PDF, November 11, 2022.

Jergović, Miljenko (2015). Gortan-Bazovica. https://www.jergovic.com/ajfelov-most/gortan-bazovica/, November 11, 2022.

Jerman, Silvin; Todorovski, Ilinka (1999). *Slovenski dom v Zagrebu 1929–1999*. Zagreb: KPD Slovenski dom Zagreb.

Jevnikar, Martin; Mislej, Irene (1999). Slovenci v Južni Ameriki. *Slovenska izseljenska književnost 3: Južna Amerika* (eds. Janja Žitnik, Helga Glušič). Ljubljana: Založba ZRC, 9–21.

Jezeršek, Ciril (1929–1930). Naša povojna prekomorska emigracija. *Geografski vestnik* 1/4, 176–179.

Josipovič, Damir (2014). Slovenci na Balkanu skozi moderne popise prebivalstva 1880–2012. *Priseljevanje in družbeno delovanje Slovencev v drugih delih jugoslovanskega prostora: Zgodovinski oris in sedanjost* (ed. Janja Žitnik Serafin). Ljubljana: Založba ZRC, 67–89. https://doi.org/10.3986/9789612546687.

Jovanović, Amalija (2001). Slovenci v Makedoniji. *Slovensko izseljenstvo: Zbornik ob 50-letnici Slovenske izseljenske matice* (ed. Milica Trebše Štolfa). Ljubljana: Združenje Slovenska izseljenska matica, 319–327.

Jurak, Mirko (1999). Kanada. *Slovenska izseljenska književnost 2: Severna Amerika* (eds. Janja Žitnik, Helga Glušič). Ljubljana: Založba ZRC, Založba Rokus, 305–383.

Jurkovič, Filip (2018). *Pa zbogom, junaki ... Vojni dnevnik Filipa Jurkoviča (1914–1918)*. Ljubljana: ZKP RTV.

Kacin Wohinz, Milica (1972). *Primorski Slovenci pod italijansko zasedbo, 1918–1921*. Maribor: Založba Obzorja.

Kacin Wohinz, Milica (1990). *Prvi antifašizem v Evropi: Primorska 1925–1935*. Koper: Lipa.

Kacin Wohinz, Milica (1995). Raznarodovanje primorskih Slovencev – dejavnik za izseljevanje. *Kulturno ustvarjanje Slovencev v Južni Ameriki* (ed. Irene Mislej). Ljubljana: Znanstveni inštitut Filozofske fakultete.

Kalan, Janez (1936). Notranja kolonizacija. *Izseljenski vestnik* 6/8, 84–85. https://www.dlib.si/stream/URN:NBN:SI:DOC-XNLT1S0U/5c3ab55a-53d6-42ab-8cee-f2edf9df5c8d/PDF, November 11, 2022.

Kalc, Aleksej (1992). Prekooceansko izseljevanje skozi Trst 1903–14. *Zgodovinski časopis* 46/4, 479–496. https://www.dlib.si/stream/URN:NBN:SI:DOC-PKFOGKU1/372065c1-20b1-45a6-bc67-64fb88ffedce/PDF, November 11, 2022.

Kalc, Aleksej (1995). Nekateri vidiki primorskega izseljevanja v Južno Ameriko do prve svetovne vojne. *Kulturno ustvarjanje Slovencev v Južni Ameriki* (ed. Mirko Jurak). Ljubljana: Filozofska fakulteta, 33–55.

Kalc, Aleksej (1996a). L'emigrazione slovena e croata dalla Venezia Giulia e il suo ruolo politico. *Annales: Series historia et sociologia* 8, 23–60.

Kalc, Aleksej (1996b). Ladijske potniške evidence kot vir za zgodovino izseljenstva, s posebnim ozirom na izseljenske sezname tržaškega pristanišča. *Dve domovini / Two Homelands* 7, 51–69.

Kalc, Aleksej (1997). Razvojni obrisi izseljenske zakonodaje v 19. stoletju in do prve svetovne vojne. *Dve domovini / Two Homelands* 8, 9–35.

Kalc, Aleksej (2001). Brazilija je vabila. *Izseljenec: Življenjske zgodbe Slovencev po svetu* (ed. Monika Kokalj). Ljubljana: Muzej novejše zgodovine, 57–60.

Kalc, Aleksej (2002). *Poti in usode: Selitvene izkušnje Slovencev z zahodne meje*. Koper: Zgodovinsko društvo za južno Primorsko, Znanstveno raziskovalno središče.

Kalc, Aleksej (2004). Žensko prebivalstvo v Trstu leta 1775: Nekaj socialno-demografskih in gospodarskih vidikov ter metodoloških vprašanj. *Zgodovinski časopis* 58/3–4, 337–376. https://www.dlib.si/stream/URN:NBN:SI:DOC-FWZA00NM/4cfb081c-a815-479b-bc4b-d1e19d075d49/PDF, November 11, 2022.

Kalc, Aleksej (2007). *Tržaško prebivalstvo v 18. stoletju: Priseljevanje kot gibalo demografske rasti in Družbenih sprememb*. Koper: Annales.

Kalc, Aleksej (2009). Tržaški teritorij v luči konskripcije iz let 1777/1778: Prvi izsledki. *Annales: Series historia et sociologia* 19/1, 59–88.

Kalc, Aleksej (2010). Migracijska gibanja med 18. in 19. stoletjem: Primeri iz zahodnega slovenskega prostora. *Migracije in slovenski prostor od antike do danes* (eds. Bojan Štih, Bojan Balkovec). Ljubljana: Zveza zgodovinskih društev Slovenije, 101–113.

Kalc, Aleksej (2013a). Vidiki razvoja prebivalstva Goriške-Gradiške v 19. stoletju. *Acta Histriae* 4, 683–706.

Kalc, Aleksej (2013b). Istrani v Trstu v 18. stoletju. *Povijesni prilozi* 45, 333–359.

Kalc, Aleksej (2013c). Trieste as a Port of Emigration from East and Southeast Europe. *East Central Europe in Exile. Volume 1: Transatlantic Migrations* (ed. Anna Mazurkiewicz). Newcastle upon Tyne: Cambridge Scholars Publishing, 127–40.

Kalc, Aleksej (2014a). Žensko priseljevanje in zaposlovanje v Trstu na prelomu 19. in 20. stoletja. *Dve domovini / Two Homelands* 40, 11–22.

Kalc, Aleksej (2014b). Vprašanje nezakonitega izseljevanja vojaških obveznikov v Avstriji na predvečer prve svetovne vojne. *Prispevki za novejšo zgodovino* 2, 30–44.

Kalc, Aleksej (2015a). Migration Movements in Goriška in the Time of Aleksandrinke. *From Slovenia to Egypt: Aleksandrinke's Trans-Mediterranean Domestic Workers' Migration and National Imagination* (ed. Mirjam Milharčič Hladnik). Göttingen: V&R Unipress, 49–72.

Kalc, Aleksej (2015b). Trst kot stičišče prekomorskega izseljevanja pred prvo svetovno vojno. *Dve domovini / Two Homelands* 41, 75–88.

Kalc, Aleksej (2016a). Nadzor migracijskih gibanj in migrantov v cesarski Avstriji od 18. stoletja do prve svetovne vojne. *Dve domovini / Two Homelands* 44, 111–123. https://doi.org/10.3986/dd.2016.2.09.

Kalc, Aleksej (2016b). Antifascisti sloveni della Venezia Giulia in Argentina tra le due guerre. *Storia contemporanea in Friuli* 46, 31–55.

Kalc, Aleksej (2016c). Nadzor migracijskih gibanj in migrantov: Pogled v zgodovino. *Dve domovini / Two Homelands* 43, 23–34.

Kalc, Aleksej (2018a). Izseljevanje iz Beneške Slovenije v Severno Ameriko na začetku 20. stoletja na podlagi ladijskih potniških seznamov. *Dve domovini / Two Homelands* 47, 107–126. https://doi.org/10.3986/dd.2018.1.07.

Kalc, Aleksej (2018b). Prebivalstvo in gospodarski viri v tržaškem kraškem zaledju (16.–20. stoletje). *Preživetje in podjetnost: Integrirana kmečka ekonomija na Slovenskem od srednjega veka do danes* (eds. Aleksander Panjek, Žarko Lazarević). Koper: Založba Univerze na Primorskem, 155–188. https://doi.org/10.26493/978-961-7023-81-7.155-188.

Kalc, Aleksej (2021a). O vzpostavitvi izseljenske zakonodaje v Kraljevini Srbov, Hrvatov in Slovencev. *Annales: Series Historia et Sociologia* 3, 507–518.

Kalc, Aleksej (2021b). Ravni, akterji in dinamike nadzora ter politike izseljevanja v desetletjih pred prvo svetovno vojno. *Nadzor migracij na Slovenskem od liberalizma do socializma* (ed. Aleksej Kalc). Ljubljana: Založba ZRC, 27–62.

Kalc, Aleksej (2023). Prihranki izseljencev in njihove poti v domovino v desetletjih pred prvo svetovno vojno. *Dve domovini / Two Homelands* 58, 145–166. https://doi.org/10.3986/dd.2023.2.08.

Kalc, Aleksej; Kodrič, Majda (1992). Izseljevanje iz Beneške Slovenije v kontekstu furlanske emigracije s posebnim ozirom na obdobje 19. stoletja in do prve svetovne vojne. *Zgodovinski časopis* 46/2, 197–209. https://www.dlib.si/stream/URN:NBN:SI:DOC-I4U8VHK6/6004f80d-13ff-4f42-b92e-da0446210fb9/PDF, November 11, 2022.

Kalc, Aleksej; Zobec, Miha (2021). Med teritorialno in nacionalno suverenostjo. Migracijski režim in odnosi med "domovino" in diasporo v prvi Jugoslaviji. *Nadzor migracij na Slovenskem od liberalizma do socializma* (ed. Aleksej Kalc). Ljubljana: Založba ZRC, 123–159.

Kalc, Andrej (1889). Iz Brazilije - nazaj v Trst. *Edinost*, 16, 19, 23, 26, 30 January 1889.

Kante, Viktorija (2000). Ribniški krošnjar. *Izseljenec: Življenjske zgodbe Slovencev po svetu* (ed. Monika Kokalj). Ljubljana: Muzej novejše zgodovine, 69–72.

Karanović, Milenko (1978). Povratak jugoslovenskih iseljenika iz SAD neposredno posle Prvog svetskog rata. *Iseljeništvo naroda in narodnosti Jugoslavije i njegove uzajamne veze sa domovinom.* Zagreb: Zavod za migracije i narodnosti, 327–338.

Kavrečič, Petra (2015). *Turizem v Avstrijskem Primorju: Zdravilišča, kopališča in kraške jame (1819–1914).* Koper: Založba Univerze na Primorskem. https://doi.org/10.14712/23363525.2017.45.

Kavrečič, Petra; Koderman, Miha (2014). Slovenski izseljenci in turistični obisk Dravske Banovine s posebnim poudarkom na poročanju Izseljeniškega vestnika. *Annales: Series historia et sociologia* 24/1, 77–94.

Keeling, Drew (2012). *The Business of Transatlantic Migration between Europe and the United States, 1900–1914*. Zürich: Chronos Verlag.

Klemenčič, Matjaž (1978). Metodološki problemi v zvezi z ugotavljanjem realnosti ameriških statistik materinega jezika izseljencev iz Habsburške monarhije. *Časopis za zgodovino in narodopisje* 1, 42–53.

Klemenčič, Matjaž (1979). Germanizacijski procesi na Štajerskem od srede 19. stoletja do prve svetovne vojne. *Časopis za zgodovino in narodopisje* 1/2, 350–369.

Klemenčič, Matjaž (1984). Slovenci v ameriški statistiki v 19. in 20. stoletju. *Celovški zvon* 4, 23–31.

Klemenčič, Matjaž (1985). Proletarec and the Acculturation of Slovene Workers in the United States: Labor Migration Project. *Labor Newspaper Preservation Project: The Press of Labor Migrants in Europe and North America 1880s to 1930s* (eds. Christiane Harzig, Dirk Hoerder). Bremen: Universität, 475–486.

Klemenčič, Matjaž (1987). *Ameriški Slovenci in NOB v Jugoslaviji*. Maribor: Založba Obzorja.

Klemenčič, Matjaž (1991). Slovenski izseljenski tisk / Slovenian Emigrant Newspapers. *Humanistika* 3/2, 299–317.

Klemenčič, Matjaž (1992). Elections for Councilmen in the City of Cleveland and Slovenes of Cleveland. *Dve domovini / Two Homelands* 2/3, 353–369.

Klemenčič, Matjaž (1995). *Slovenes of Cleveland: The Creation of a New Nation and a New World Community Slovenia and the Slovenes of Cleveland, Ohio*. Novo mesto: Dolenjska založba.

Klemenčič, Matjaž (1999). *Jurij Trunk med Koroško in Združenimi državami Amerike ter zgodovina slovenskih naselbin v Leadvillu, Kolorado, in v San Franciscu, Kalifornija*. Celovec: Mohorjeva založba.

Klemenčič, Matjaž (2001a). Slovenska tvarna kulturna dediščina v ZDA. *Dve domovini / Two Homelands* 14, 169–180.

Klemenčič, Matjaž (2001b). Slovenske naselbine v Združenih državah Amerike. *Slovensko izseljenstvo: Zbornik ob 50-letnici Slovenske izseljenske matice* (ed. Milica Trebše Štolfa). Ljubljana: Združenje Slovenska izseljenska matica, 179–186.

Klemenčič, Matjaž (2001c). Odnos slovenskih izseljencev do stare domovine. *Slovensko izseljenstvo: Zbornik ob 50-letnici Slovenske izseljenske matice* (ed. Milica Trebše Štolfa). Ljubljana: Združenje Slovenska izseljenska matica, 53–63.

Klemenčič, Matjaž (2002). Slovenes in Cleveland. *Identity, Conflict and Cooperation: Central Europeans in Cleveland 1850–1930* (eds. David C. Hammack, Diane L. Grabowsky, John J. Grabowski). Cleveland: The Western Reserve Historical Society, 307–344.

Klemenčič, Matjaž (2011). *Zgodovina skupnosti slovenskih Američanov v Pueblu, Kolorado*. Maribor: Mednarodna založba Oddelka za slovanske jezike in književnosti. https://doi.org/10.17161/SCN.1808.9958.

Klemenčič, Matjaž (2013). Slovenes and Slovene Americans. *Immigrants in American History: Arrival, Adaptation, and Integration* (ed. Elliot Robert Barkan). Santa Barbara: ABC CLIO, 613–622.

Klemenčič, Matjaž, Emeršič, Darja (1990). Ideja in poskusi izgradnje Slovenskega narodnega doma na St. Clairu v Clevelandu. *Dve domovini / Two Homelands* 1, 137–159.

Klemenčič, Matjaž; Javernik, Igor (1996). Ustanovitev in razvoj Kranjsko slovenske katoliške jednote – prve slovenske centralizirane podporne organizacije v Združenih državah Amerike. *Etnični fraternalizem v priseljenskih državah / Ethnic Fraternalism in Immigrant Countries* (ed. Matjaž Klemenčič). Maribor: Pedagoška fakulteta, 345–356.

Klemenčič, Matjaž; Maver, Aleš (2017). *Izbrana poglavja iz zgodovine selitev od začetkov do danes*. Maribor: Univerzitetna založba Univerze. https://doi.org/10.18690/978-961-286-116-2.

Klemenčič, Matjaž; Šeruga, Tadej (2019). *Pregled zgodovine slovenske skupnosti v Elyju, Minnesota*. Maribor: Univerzitetna založba Univerze. https://doi.org/10.18690/978-961-286-325-8.

Klinec, Rudolf (1978). Füster, Anton (1808–1881). *Primorski slovenski biografski leksikon* 5. Gorica: Goriška Mohorjeva družba, 399–400.

Koblar, Stanislav (2008). 110 let združevanja. *Četrta stran trikotnika: Znameniti Slovenci in slovenska društva v Bosni in Hercegovini: 1878–2000* (ed. Stanislav Koblar). Ljubljana: Mladinska knjiga, 334–455.

Kocjančič, Cvetka (1998). Združevanje kanadskih Slovencev. *Dve domovini / Two Homelands* 9, 149–160.

Kodrič, Majda (1983). Etbin Kristan in socialistično gibanje jugoslovanskih izseljencev v ZDA v letih 1914–1920. *Prispevki za zgodovino delavskega gibanja* 1–2, 63–87.

Kodrič, Majda (1987). "Kranjci" v Stearns County – vezi med etnično zavestjo in zemljepisnimi dejavniki v neki slovenski kmečki naselbini v Minnesoti. *Geografski obzornik* 2, 108–112. https://www.dlib.si/stream/URN:NBN:SI:DOC-HUCHE71X/93715f11-451c-4885-bfde-3a58b2cb92d3/PDF, November 11, 2022.

Kodrič, Majda (2003). Franc Pirc in nastanek prvih naselbin v Združenih državah Amerike. *Dve domovini / Two Homelands* 18, 55–61.

Kokalj, Kočevar Monika (2010). "Mother, Are the Apples at Home Ripe Yet?" Slovenian Forced and Slave Labourers during the Second World War. *Hitler's Slaves: Life Stories of Forced Labourers in Nazi-occupied Europe* (eds. Alexander von Plato, Almut Leh, Christoph Thonfeld). New York: Berghahn Books, 138–150. https://doi.org/10.1515/9781845459901.

Kolar, Bogdan (1991). Periodični tisk slovenskih cerkvenih skupnosti v Severni in Južni Ameriki ter Avstraliji s poudarkom na listu Ave Marija. *Znanstvena revija: Humanistika* 2/3, 393–404.

Kolar, Bogdan (1995). Delo slovenskih duhovnikov med Slovenci v Zagrebu med obema vojnama. *Slovenci v Hrvaški* (ed. Vera Kržišnik-Bukić). Ljubljana: Inštitut za narodnostna vprašanja, 115–131.

Kolar, Bogdan (1999). St. Joseph KSKJ Local Lodge, Bridgeport, Connecticut. KSKJ. *Dve domovini / Two Homelands* 10, 61–78.

Kolar, Bogdan (2001). Mesto cerkve med Slovenci po svetu in njen prispevek za ohranjanje narodnosti. *Slovensko izseljenstvo: Zbornik ob 50-letnici Slovenske izseljenske matice* (ed. Milica Trebše Štolfa). Ljubljana: Združenje Slovenska izseljenska matica, 125–147.

Kolar Dimitrijević, Mira (2012). Artur Benko Grado Bojnički kao evidentičar migracionih i masovnih gospodarsko-socijalnih kretanja u Hrvatskoj u prvoj polovici 20. stoljeća. *Podravina* 22, 108–126. https://hrcak.srce.hr/95369, November 11, 2022.

Komac, Jernej (2017). "Hudič vzemi vojske in politiko!": Begunska izkušnja v kolektivnem spominu bovškega prebivalstva v letih 1914–1941. *Goriški letnik* 41, 217–232.

Komac, Miran (2015). Narodne manjšine v Sloveniji 1920–1941. *Razprave in gradivo. Revija za narodnostna vprašanja* 75, 49–81.

Komlosy, Andrea (2003). Empowerment and Control: Conflicting Central and Regional Interests in Migration Within the Habsburg Monarchy. *Migration Control in the North Atlantic World: The Evolution of State Practices in Europe and the United States from the French Revolution to the Inter-War Period* (eds. Andreas Fahrmeir, Olivier Faron, Patrick Weil). New York: Berghahn Books, 153–166.

Koprivec, Daša (2013). *Dediščina aleksandrink in spomini njihovih potomcev.* Ljubljana: Založba ZRC.

Kordin, Anton (1933). O naših po širnem svetu. Srbija. *Izseljeniški vestnik* 3, 5. https://www.dlib.si/stream/URN:NBN:SI:DOC-VJGKDYS9/f6f82aaf-4f5b-4c2d-b27b-93f7b41d704a/PDF, November 11, 2022.

Korpič-Horvat, Etelka (1992). *Zaposlovanje in deagrarizacija pomurskega prebivalstva*. Murska Sobota: Pomurska založba.

Kosi, Jernej (2018). The Imagined Slovene Nation and Local Categories of Identification: "Slovenes" in the Kingdom of Hungary and Postwar Prekmurje. *Austrian History Yearbook* 49, 87–102. https://doi.org/10.1017/S0067237818000103.

Kosi, Jernej (2020). Summer of 1919: A Radical, Irreversible, Liberating Break in Prekmurje/Muravidék? *The Hungarian Historical Review* 9/1, 51–68.

Kosi, Jernej; Cvirn, Janez (2012). Nemški begunci v Avstriji. *Nemci in Maribor: stoletje preobratov 1846–1946* (ed. Jerneja Ferlež). Maribor: Umetniški kabinet Primož Pemzl, 62–64.

Kosi, Miha (1998). *Potujoči srednji vek: Cesta, popotnik in promet na Slovenskem med antiko in 16. stoletjem*. Ljubljana: Založba ZRC.

Kosnica, Ivan (2018). Odnos državljanstva i nacionalne pripadnosti u Kraljevini SHS/Jugoslaviji. *Zbornik PFZ* 1, 61–83.

Kovács, Attila (2002). Izključitev prekmurskih Madžarov iz agrarne reforme na osnovi podatkov popisa o maternem jeziku iz leta 1921. *Razprave in gradivo* 41, 188–205.

Kovács, Attila (2007). Agrarna reforma in kolonizacija na območju Dolnje Lendave med obema vojnama. *Razprave in gradivo* 53/54, 68–97.

Kozanski (Reya Oskar) (2002). *Nepozabljena Brda: Veselje in žalost briškega ljudstva*. Ljubljana: Vitrum.

Krašovec, Teja (2016). Primorski priseljenci v Ljubljani med obema svetovnima vojnama. PhD thesis. Koper: Univerza na Primorskem, Fakulteta za humanistične študije.

Krek, Janez Evangelist (1913). Izseljevanje in zadružništvo. *Narodni gospodar* 22, November 25, 1913. https://www.dlib.si/stream/URN:NBN:SI:DOC-P81XZF4Q/96f631dc-a9a4-4105-8049-dd3c1c90f912/PDF, November 11, 2022.

Kremenšek, Slavko (1976). Izvir in okolje delavk ljubljanske tobačne tovarne do prve svetovne vojne. *Slovenski etnograf* 29, 91–100. https://www.dlib.si/stream/URN:NBN:SI:DOC-2DYL8X9I/4e26ee99-4021-4d54-8441-6a87043a71ca/PDF, November 11, 2022.

Kresal, France (1976). *Razvoj tekstilne industrije v Sloveniji*. Ljubljana: Borec.

Kresal, France (2005). Socialna politika. *Slovenska novejša zgodovina 1848–1992* (eds. Jasna Fischer et al.). Ljubljana: Mladinska knjiga, 98–101.

Kresal, France; Perovšek, Jurij (2005). Razvoj prebivalstva. *Slovenska novejša zgodovina 1848–1992, Vol. 1* (eds. Jasna Fischer et al.). Lubljana: Mladinska knjiga, 177–178.

Kristan, Anton (1927a). Moja pot čez veliko lužo. *Življenje in svet – tedenska priloga Jutra* 1/27, 700–702. https://www.dlib.si/stream/URN:NBN:SI:DOC-VDZD0JWZ/8754b887-6d68-4bbd-9072-305c1db18183/PDF, November 11, 2022.

Kristan, Anton (1927b). Ameriški vtisi. *Življenje in svet – tedenska priloga Jutra* 1/38, 866–868. https://www.dlib.si/stream/URN:NBN:SI:DOC-V1MZ3FCV/5d382c73-6999-46bb-aff2-662b29ba6c1f/PDF, November 11, 2022.

Kržišnik-Bukić, Vera (1993). Slovenski gostinci in pomen slovenskih gostiln v Zagrebu. *Razprave in gradivo* 28, 136–142.

Kržišnik-Bukić, Vera (1994–1995). Okvirni pregled poteka, vzrokov in pomena izseljevanja Slovencev na Hrvaško. *Razprave in gradivo* 29/30, 85–93.

Kržišnik-Bukić, Vera (1995). O narodnostnem in kulturnem samoorganiziranju Slovencev na Hrvaškem v 20. stoletju. *Slovenci v Hrvaški* (ed. Vera Kržišnik-Bukić). Ljubljana: Inštitut za narodnostna vprašanja, 133–188.

Kržišnik-Bukić, Vera (2003). Slovenci v Hrvaški, Bosni in Hercegovini, Srbiji in Črni gori ter Makedoniji med preteklostjo in sedanjostjo. *Traditiones* 32/2, 117–135.

Kržišnik-Bukić, Vera (2004). Slovenci v Beogradu. *Slovenski etnološki leksikon* (eds. Angelos Baš et al.). Ljubljana: Mladinska knjiga, 546.

Kržišnik-Bukić, Vera (2006). Znameniti Slovenci na Hrvaškem skozi zgodovino. *Migracijske i etničke teme* 4, 421–445.

Kržišnik-Bukić, Vera (2007). *Slovenci v Bosni in Hercegovini skozi pričevanja, spomine in literarne podobe: 1831–2007.* Ljubljana: Inštitut za narodnostna vprašanja.

Kuhar, Alojzij (1939). Naše izseljensko vprašanje. *Spominski zbornik Slovenije: Ob dvajsetletnici Kraljevine Jugoslavije* (eds. Jože Lavrič, Josip Mal, France Stele). Ljubljana: Jubilej, 524–536.

Kulihova, Lucia (2019). Zavajajoči podatki na seznamih ladijskih potnikov kot posledica restriktivne ameriške migracijske zakonodaje. *Dve domovini / Two Homelands* 50, 237–252. https://doi.org/10.3986/dd.v2019i50.7468.

Kuzmič, Mihael (1996). Fraternalistične organizacije prekmurskih Slovencev v ZDA: Prvo slovensko ameriško bratsko podporno društvo 1912–1942. *Etnični fraternalizem v priseljenskih državah / Ethnic Fraternalism in Immigrant Countries* (ed. Matjaž Klemenčič). Maribor: Pedagoška fakulteta, 357–369.

Kuzmič, Mihael (2001). *Slovenski izseljenci iz Prekmurja v Betlehemu v ZDA 1893–1924: Naselitev in njihove zgodovinske, socialne, politične, literarne in verske dejavnosti.* Ljubljana: Založba ZRC.

La popolazione (1878). *La popolazione di Trieste nel 1875: Resoconto ufficiale del censimento generale della popolazione effettuato secondo lo stato del 31 dicembre 1875*. Trieste: Civico Ufficio Statistico Anagrafico.

Lampe, Eugen (ed.) (1907). *III. slovenski katoliški shod v Ljubljani dne 26., 27. in 28. avgusta 1906: Govori, posveti in sklepi*. Ljubljana: Katoliška bukvarna.

Lanver, Mak (2012). *The British in Egypt: Community, Crime and Crises 1822–1922*. London: Tauris.

Larson, Ethan (2020). The New Old Country. The Kingdom of Yugoslavia and the Creation of a Yugoslav Diaspora 1914–1951. PhD thesis. Urbana, Champaign: University of Illinois Urbana-Champaign.

LaVigne, David (2017). The "Black Fellows" of the Mesabi Iron Range: European Immigrants and Racial Differentiation during the Early Twentieth Century. *Journal of American Ethnic History* 2, 11–39. https://doi.org/10.5406/jamerethnhist.36.2.0011.

Lavrenčič Pahor, Minka (1994). *Primorski učitelji 1914–1941: Prispevek k proučevanju zgodovine slovenskega šolstva na Primorskem*. Trst: Narodna in študijska knjižnica, Odsek za zgodovino.

Lazarević, Žarko (1993). Kmečki dolgovi v avstrijski dobi. *Prispevki za novejšo zgodovino* 33/1–2, 3–15.

Lazarević, Žarko (1994). *Kmečki dolgovi na Slovenskem: socialno-ekonomski vidiki zadolženosti slovenskih kmetov 1848–1948*. Ljubljana: Znanstveno in publicistično središče.

Lazarević, Žarko (2005). Velika gospodarska kriza. *Slovenska novejša zgodovina 1848–1992* (eds. Jasna Fischer et al.). Ljubljana: Mladinska knjiga, 473–479.

Lazarević, Žarko (2015). *Spremembe in zamišljanja: Gospodarsko-zgodovinske refleksije*. Ljubljana: Inštitut za novejšo zgodovino.

Lazarević, Žarko (2017). Migracije in ekonomske interakcije (do prve svetovne vojne). *Raziskovanje slovenskega izseljenstva: Vidiki, pristopi, vsebine* (eds. Janja Žitnik Serafin, Aleksej Kalc). Ljubljana: Založba ZRC, 49–65. https://doi.org/10.3986/9789610500414.

Ledbetter, Eleanor E. (1918). *The Jugoslavs of Cleveland: With a Brief Sketch of their Historical and Political Backgrounds*. Cleveland: Cleveland Americanization Committee.

Lees, Lorraine L. (2007). *Yugoslav-Americans and National Security during World War II*. Urbana: University of Illinois Press.

Lesjak, Hans-Peter (1996). Kako so se v poznem 19. stoletju izseljevali iz zgornje Baške doline na Koroško. *Kronika* 44/1, 30–31. https://www.dlib.si/stream/URN:NBN:SI:DOC-MMK0KZKQ/e242ffb1-bb5e-4c71-878c-d1030af6c87a/PDF, November 11, 2022.

Levitt, Peggy; Glick Schiller, Nina (2004). Conceptualizing Simultaneity: A Transnational Social Field Perspective on Society. *International Migration Review* 38/3: 1002–1039. https://doi.org/10.1111/j.1747-7379.2004.tb00227.x.

Light, Ivan (2013). The Migration Industry in the United States, 1882–1924. *Migration Studies* 1/3, 258–275. https://doi.org/10.1093/migration/mnt021.

Lilek, Emilijan (1933). *Slovenski v tujini službujoči šolniki*. Celje: self-published.

Lipoglavšek-Rakovec, Slava (1950). Slovenski izseljenci. *Geografski vestnik* 22, 3–60. https://www.dlib.si/stream/URN:NBN:SI:DOC-Z8BXU1RB/f718a112-dbea-40c5-9d87-57cc89943659/PDF, November 11, 2022.

Lipovec Čebron, Uršula (2017). Benigar, Juan Ivan (1883–1950). *Novi Slovenski biografski leksikon: 2. zv. B-Bla* (eds. Barbara Šterbenc Svetina et al.). Ljubljana: Založba ZRC, 273–276.

Lokar, Janko (1911). Iz Bele krajine. *Carniola: Izvestja Muzejskega društva za Kranjsko*. Ljubljana: Muzejsko društvo za Kranjsko, 1–31.

Lorenzetti, Luigi (2000). Immigration et mariage à Gèneve: Aspects de l'insertion urbaine à la fin du XIXe siècle. *Migration in die Städte. Ausschluss – Assimilierung – Integration – Multikulturalität / Migrations vers les villes. Exclusion – assimilation – intégration – multiculturalité* (eds. Hans-Jörg Gilomen, Anne-Lise Head-König, Anne Radeff). Zürich: Chronos Verlag, 225–238. https://doi.org/10.5169/seals-81245.

Lorenzetti, Luigi; Schumacher, Reto (2005). L'endogamie matrimoniale dans les villes suisses, 1880–1930. Effets structurels et préférences individuelles. *Histoire & mesure* XX, 1–23. https://doi.org/10.4000/histoiremesure.783.

Lubotina, Paul (2015). Corporate Supported Ethnic Conflict on the Mesabi Range, 1890–1930. *Upper Country: A Journal of the Lake Superior Region* 3, 29–52.

Lucassen, Jan; Lucassen, Leo (2009). The Mobility Transition Revisited, 1500–1900: What the Case of Europe can Offer to Global History. *Journal of Global History* 3/4, 347–377. https://doi.org/10.1017/S174002280999012X.

Lucassen, Leo; Lucassen, Jan (2015). Quantifying and Qualifying Cross-cultural Migrations in Europe since 1500: A Plea for Broader View. *The History of Migration in Europe: Perspectives from Economics, Politics and Sociology* (ed. Francesca Fauri). London: Routledge, 13–38. https://doi.org/10.4324/9781315772400.

Lukić, Dijana (2017). "Drugega skoraj ne ostaja, ko zapustiti milo domovino in si drugje domovja iskati" ali kratek pregled migracijskih tokov Slovencev v Vojvodino. *Slovenika* 3, 35–57. https://doi.org/10.18485/slovenika.2017.3.1.2.

Luthar, Oto (2000). *O žalosti niti besede: Uvod v kulturno zgodovino velike vojne*. Ljubljana: Založba ZRC. https://doi.org/10.3986/9789612544959.

Luthar, Oto (2015). Take It Personally (!) *On the Eve of the Great War: Political Concepts, Conspiracies and Theories.* (eds. Gregor Antoličič et al.). Ljubljana: Založba ZRC, 7–11.

Luthar, Oto (ed.) (2013). *The Land Between: A History of Slovenia*. Frankfurt am Main: Peter Lang.

Mackaman, Tom (1998). Life and Labor in Ely, Minnesota, 1886–1919: Community and Worker Identity on the American Industrial Frontier. BA thesis. Minneapolis: University of Minnesota.

Mačkovšek, Janko (1911). Statistika Slovencev. *Slovanski svet: Zemljepisna in statistična slika današnjega slovanstva* (ed. Lubor Niederle). Ljubljana: Omladina, 245–260.

Mädchenschutz. Mädchenhandel von Krain nach Nordamerika. (1911). *Der Auswanderer* II/5, 67–68.

Maister, Hrvoj; Uratnik, Filip (1938). *Socialni problemi slovenske vasi*. Ljubljana: Socialno ekonomski institut v Ljubljani.

Mak (1937). Slovenci na Kosovu. *Izseljenski vestnik* 7/8, 81–82. https://www.dlib.si/stream/URN:NBN:SI:DOC-ATDPX8FP/96bb0ee7-96d9-4256-ba5b-e1925c71e62e/PDF, November 11, 2022.

Makuc, Dorica (1985). Prišleki kak kastiga kak boži bič. *Jadranski koledar 1986*, 49–62.

Makuc, Dorica (1987). Primorci v severni slovenski metropoli. *Jadranski koledar 1988*, 167–173.

Makuc, Dorica (1994). *Aleksandrinke*. Gorica: Goriška Mohorjeva družba.

Malni, Paolo (2001). Evaquati e fuggiaschi dal fronte dell'Isonzo: I profughi della Grande Guerra in Austria e Italia. *"Un esodo che non ha pari." 1914–1919. Profughi, internati ed emigrati di Trieste, dell'Isontino e dell'Istria* (ed. Franco Cecotti). Gorica: LEG, 99–153.

Mamić, Tino (2008). Anton Laščak Beg – življenjepis kraljevega arhitekta. *Kronika* 56/1, 71–84. https://www.dlib.si/stream/URN:NBN:SI:DOC-8CAV9252/c8fc98ca-93e2-4406-899d-914afacd5b07/PDF, November 11, 2022.

Marentič, Janez (1957). *Slovenska vas pod kapitalističnim jarmom*. Ljubljana: Kmečka knjiga.

Markitan, Franz (1912). *Auswandererverkehrswege in Österreich. Vortrag gehalten von Herrn k.k. Polizeikommissar Markitan aus der Generalversammlung des österreichischen St.-Raphael Vereins*. Wien: Opitz.

Markitan, Franz (1913). *Die österreichische Saisonwanderung: Vortrag gehalten in Wien am 21. November 1912 von Franz Markitan*. Wien: Verlag des österreichischen St. Raphael Vereines.

Marković, Irena (2017). *Irenej Friderik Baraga – misijonar in škof med Otavci in Očipvejci.* Celovec: Mohorjeva družba, Slovenski narodopisni inštitut Urban Jarnik.

Martellone, Anna Maria (1980). Introduzione. *La "questione" dell'immigrazione degli Stati Uniti* (ed. Anna Maria Martellone). Bologna: Il Mulino, 7–81.

Martinčič, Andrej (1960). Borba za srečo in kruh. *Slovenski izseljenski koledar 1960*, 231–234.

Marušič, Branko (1995). Prispevka o slovenskem izseljevanju v Južno Ameriko. *Kulturno ustvarjanje Slovencev v Južni Ameriki* (ed. Mirko Jurak). Ljubljana: Znanstveni inštitut Filozofske fakultete, 57–66.

Marušič, Branko (2012). Ljudsko štetje 31. decembra leta 1910 na Goriškem. *Zgodovinski pogledi na zadnje državno ljudsko štetje v Avstrijskem primorju 1910: Jezik, narodnost, meja* (eds. Barbara Šterbenc Svetina, Petra Kolenc, Matija Godeša). Ljubljana: Založba ZRC, 77–86.

Matt, Susan J. (2011). Current Emotion Research in History: Or, Doing History from the Inside Out. *Emotion Review* 3/1, 117–124. https://doi.org/10.1177/1754073910384416.

Maučec, Matija (1933). Prenaseljenost in sezonsko izseljevanje v Prekmurju. *Geografski vestnik* 1–4, 107–118. https://www.dlib.si/stream/URN:NBN:SI:DOC-SGWZ42GN/969412fb-9d79-4ddb-9f7e-5aa38d9b66b2/PDF, November 11, 2022.

Meinzinger, Franz von (1903). Die Wanderbewegung auf Grund von Gebürtigkeitsdaten der Volszählung vom 31. Dezember 1900. *Statistische Monatsschrift* VIII, 133–161.

Melik, Anton (1938). Populacijski problemi Jugoslavije v dobi 1918–1938. *Misel in delo* 4/12, 247–262. https://www.dlib.si/stream/URN:NBN:SI:DOC-SOWRB7NL/cfa6021d-3fa8-439d-b05a-377698446485/PDF, November 11, 2022.

Melik, Anton (1956). *Amerika in Ameriška Slovenija.* Ljubljana: Državna založba Slovenije.

Melik, Anton (1963). *Slovenija I. Geografski opis.* Ljubljana: Slovenska matica.

Melik, Vasilij (1956). Rast mestnega prebivalstva na Slovenskem pred prvo svetovno vojno. *Ekonomska revija*, 193–210.

Mesesnel, France (2013). Jager, Ivan (1871–1959). *Slovenska biografija.* Slovenska akademija znanosti in umetnosti, Znanstvenoraziskovalni center SAZU. http://www.slovenska-biografija.si/oseba/sbi245658/#slovenski-biografski-leksikon, November 11, 2022.

Michigan Copper District Strike. (1914). Michigan Copper District Strike. *Bulletin of the U. S. Bureau of Labor Statistics* 139, September 7, 1914. https://fraser.stlouisfed.org/title/3806, November 11, 2022.

Miklavčič, France (1927). Kulturne zveze naših izseljencev s staro domovino, s posebnim ozirom na ameriške razmere. *Socialna misel* 11, 221–225. https://www.dlib.si/stream/URN:NBN:SI:DOC-GXQZGBI2/a3a3a65a-f295-4da1-a456-666f60e76307/PDF, November 11, 2022.

Miklič, Radon (1939). Sezonsko izseljevanje. *Tehnika in gospodarstvo* 3–4, 87–96.

Milanič, Irena (2001). Skrb za drugo generacijo: Kulturna investicija za ohranitev slovenske pripadnosti v ZDA (primer SNJP). *Dve domovini / Two Homelands* 14, 19–28.

Miletić, Aleksandar (2009). *Journey under Surveillance: The Overseas Emigration Policy of the Kingdom of Serbs, Croats and Slovenes in Global Context 1918–1928*. Beograd: Institut za noviju istoriju Srbije.

Milharčič Hladnik, Mirjam (2004). Ohranjanje etnične identitete in tradicije med slovenskimi izseljenci in njihovimi potomci v združenih državah Amerike. *Dve domovini / Two Homelands* 19, 121–140.

Milharčič Hladnik, Mirjam (2007). Marie Prisland – her Role in Preserving Slovenian Culture and Tradition among Slovenian Migrants in the United States. *Dve domovini / Two Homelands* 25, 229–247.

Milharčič Hladnik, Mirjam (2011). Stopinje skozi čas v avto/biografskih zapisih slovenskih migrantk. *Avtobiografski diskurz: Teorija in praksa avtobiografije v literarni vedi, humanistiki in družboslovju* (eds. Alenka Koron, Andrej Leben). Ljubljana: Založba ZRC, 189–206.

Milharčič Hladnik, Mirjam (2015a). Otroci v migracijskih kontekstih. *Slovensko izseljenstvo v luči otroške izkušnje* (ed. Janja Žitnik Serafin). Ljubljana: Založba ZRC, 13–34. https://doi.org/10.3986/9789612548469.

Milharčič Hladnik, Mirjam (2015b). Trans-Mediterranean Women Domestic Workers: Historical and Contemporary Perspectives. *From Slovenia to Egypt: Aleksandrinke's Trans-Mediterranean Domestic Workers' Migration and National Imagination* (ed. Mirjam Milharčič Hladnik). Göttingen: V&R Unipress, 11–38. https://doi.org/10.14220/9783737004039.11.

Milharčič Hladnik, Mirjam (2015c). Žensko delo in gospodarstvo v migracijskem kontekstu – zgodovinski primeri v luči sedanjosti. *Žensko delo: Delo žensk v zgodovinski perspektivi* (eds. Mojca Šorn, Nina Vodopivec, Žarko Lazarević). Ljubljana: Inštitut za novejšo zgodovino, 39–55.

Milharčič Hladnik, Mirjam (2021). "People Cannot Live on Love Alone": Negotiating Love, Gender Roles, and Family Care between Slovenia and Egypt. *Emotional Landscapes: Love, Gender, and Migration* (eds. Marcelo J. Borges,

Sonia Cancian, Linda Reeder). Urbana: University of Illinois Press, 57–74. https://doi.org/10.5406/j.ctv1f45rkt.

Milharčič Hladnik, Mirjam; Mlekuž, Jernej (eds.) (2014). *Going Places: Slovenian Women's Stories on Migration*. Akron: Akron University Press.

Mislej, Irene (ed.) (1989a). *Arhitekt Viktor Sulčič: Trst 1895, Buenos Aires 1973*. Ajdovščina: Pilonova galerija.

Mislej, Irene (1989b). Odsek Slovenija v Buenos Airesu: Ob 70. letnici prve slovenske organizacije v Južni Ameriki. *Slovenski koledar 1990*. Ljubljana: Slovenska izseljenska matica, 143–145.

Mislej, Irene (1990). Slovensko šolstvo v Argentini. *Slovenski koledar '91*. Ljubljana: Slovenska izseljenska matica, 187–190.

Mislej, Irene (1992). Slovenski tisk v Argentini po drugi svetovni vojni. *Dve domovini / Two Homelands* 2–3, 185–194.

Mislej, Irene (1995a). Kulturna zgodovina Slovencev v Južni Ameriki. *Kulturno ustvarjanje Slovencev v Južni Ameriki* (ed. Mirko Jurak). Ljubljana: Filozofska fakulteta, 15–21.

Mislej, Irene (1995b). Primorski odbor: Politično delovanje primorskih Slovencev med drugo svetovno vojno v Argentini. *Dve domovini / Two Homelands* 5, 85–113.

Mislej, Irene (1999). Literarno delo južnoameriških Slovencev pred drugo svetovno vojno: Proza. *Slovenska izseljenska književnost 3: Južna Amerika* (eds. Janja Žitnik, Helga Glušič). Ljubljana: Založba ZRC, 23–79.

Mislej, Irene (2001). Slovenski izseljenci v Južni Ameriki. *Slovensko izseljenstvo: Zbornik ob 50-letnici Slovenske izseljenske matice* (ed. Milica Trebše Štolfa). Ljubljana: Združenje Slovenska izseljenska matica, 237–247.Mislej, Irene; Jurak, Mirko (eds.) (1995). *Kulturno ustvarjanje Slovencev v Južni Ameriki*. Ljubljana: Znanstveni inštitut Filozofske fakultete: Slovenska izseljenska matica.

"Mišljenje eksperta Iseljeničkog komesarijata Artura Benka Bojničkoga: Imigraciona politika prema narodnosnim manjinama, Zagreb, April 17, 1925." Hrvatski državni arhiv, 790 *Benko Grado Artur*, Grada nastala za vrijeme vršenja funkcije eksperta Iseljeničkog komesarijata Ministarstva socialne politike u Zagrebu (1923–1931).

Mlakar, Frank (1950). *He, the Father*. New York: Harper & Brothers.

Mlekuž, Jernej (2004). Izbrani vidiki zaposlovanja beneških deklet v gospodinjstvih italijanskih mest: Tiha, grenko-sladka, nikoli povsem izrečena in slišana zgodba. *Dve domovini / Two Homelands* 19, 141–164.

Mlekuž, Jernej (2016). Aleksandrinke kot nosilke časti narodne skupnosti v dopisu Karola Pečnika iz Egipta (1897). *Dve domovini / Two Homelands* 43, 143–156. https://doi.org/10.3986/dd.2016.1.12.

Moch, Leslie Page (1992). *Moving Europeans: Migration in Western Europe since 1650*. Bloomington: Indiana University Press.

Molek, Ivan (1934). *Veliko mravljišče*. Ljubljana: Cankarjeva družba.

Molek, Ivan (1979). *Slovene Immigrant History, 1900–1950: Autobiographical Sketches*. Dover, Delaware: Mary Molek.

Molek, Mary (1976). *Immigrant Woman*. Dover, Delaware: Mary Molek.

Molek, Nadia (2019). *Biti Slovenec v Argentini: Kompleksnost identitetnih procesov argentinskih Slovencev*. Ljubljana: Založba ZRC.

Morassi, Luciana, Panjek, Giovanni (1984). Strategie familiari in Val di Resia (Sec. XIX). *Economia e Storia* 4, 439–456.

Moric, Anja (2016). Slovenski Nemci v diaspori. PhD thesis. Ljubljana: Univerza v Ljubljani, Filozofska fakulteta.

Moric, Anja (2020). Gottscheer Folk Song – From a Means of Fostering National Differentiation to a Memory Keeper in the Diaspora. *Traditiones* 49/2, 141–166. https://doi.org/10.3986/Traditio2020490208.

Morys, Matthias (2010). The Original Sin that Started Only Later: How Austria-Hungary's Paper Debt Turned Golden, 1870s–1913. CHERRY Discussion Paper Series: Centre for Historical Economics and Related Research at York.

Mrkun, Anton (1932). Naš narodni izseljeniški program. *Izseljenski vestnik* 2/4, 2–3. https://www.dlib.si/stream/URN:NBN:SI:DOC-MJQT8M0Z/1c44f513-f338-4cfe-81f9-15ca957a92bb/PDF, November 11, 2022.

Mrkun, Anton (1940). Kmetsko delavstvo v Dobrépoljah in okolici. *Etnolog* 13, 116–124.

Mulaček, Ivan (1913). Naše izseljevanje v številkah. *Čas* 4, 256–266.

Mulders, Milena (2010). *S trebuhom za kruhom: Moja slovenska zgodovina*. Ljubljana: Slovenska izseljenska matica.

Munda Hirnök, Katalin (1992). Vzroki in razvoj poljedelskega zaposlovanja med porabskimi Slovenci na Madžarskem. *Traditiones* 21, 199–208.

Munda, Jože (1969). *Bibliografija slovenskega marksističnega tiska: 11. april 1920–26. marec 1941*. Ljubljana: Inštitut za zgodovino delavskega gibanja.

Nećak, Dušan (1973). Prispevek k vprašanju primorskih beguncev v letih 1918–1920. *Kronika* 21/2, 120–126. https://www.dlib.si/stream/URN:NBN:SI:DOC-DTQ55BCM/0148a588-eafe-4898-9777-f9aa83d9e9e7/PDF, November 11, 2022.

Nećak, Dušan (ed.) (2004). *Slovensko-avstrijski odnosi v 20. stoletju*. Ljubljana: Oddelek za zgodovino Filozofske fakultete.

Nemanic, Mary Lou (2007). *One Day for Democracy: Independence Day and the Americanization of Iron Range Immigrants*. Athens: Ohio University Press.

Nobile, Annunziata (1974). Politica migratoria e vicende dell'emigrazione durante il fascismo. *Il Ponte*, November/December, 1322–1341.

Noiriel, Gérard (2007). *Immigration, antisémitisme et racisme en France (XIXe–XXe siècle): Discurs publics, humiliations privées*. Paris: Fayard. https://doi.org/10.4000/questionsdecommunication.1941.

Novak Popov, Irena (2010). Slovenski dnevniki iz prve svetovne vojne. *Kultura spominjanja* (ed. Vincenc Rajšp). Dunaj: Slovenski znanstveni inštitut; Ljubljana: Založba ZRC, 105–120.

Novak, Valerija (2016). Nemška okupacija ozemlja današnje Slovenije med drugo svetovno vojno in izgon prebivalstva v Nemčijo: Na podlagi ustnih pričevanj izgnancev v Nemčijo. MA thesis. Koper: Univerza na Primorskem, Fakulteta za humanistične študije.

Novak, Vladimir; Zwitter, Fran (1945). *Oko Trsta*. Beograd: Državni izdavački zavod Jugoslavije.

Obravnave deželnega zbora kranjskega v Ljubljani. (1914). Obravnave deželnega zbora kranjskega v Ljubljani. Stenografski zapisniki, 49. zvezek, zasedanje od 5. februarja do 3. marca 1914.

Odorizzi Planinsek, Irene (1978). *Footsteps Through Time*. Washington: Landmark Tours.

Ogris, Tomaž (2011). *Anisja, prisilna delavka na Koroškem*. Celovec: Drava.

Olas, Ludvik (1957). Razvoj in problemi sezonskega zaposlovanja prekmurskega prebivalstva. *Geografski vestnik* 27/28, 176–208. https://www.dlib.si/stream/URN:NBN:SI:DOC-PNYIHWOL/ba7b5343-3b4d-4592-a9e0-253df7e008bd/PDF, November 11, 2022.

Olas, Ludvik (1990). Izseljevanje iz Prekmurja. *Zgodovinski časopis* 44/4, 589–594. https://www.dlib.si/stream/URN:NBN:SI:DOC-0MO0DU74/2db06ed7-abd5-41bd-a417-1ff85692bab0/PDF, November 11, 2022.

Österreichische Statistik (1902). *Die Ergebnisse der Volkszählung vom 31. December 1900 in den im Reichsrathe vertretenen Königreichen und Ländern. 64. Band., 1. Heft, Tafel A: Die Anwesende Bevölkerung nach ihrer heimatsberechtigung*. Wien: K. k. Statistische Zentral-Commission.

Österreichische Statistik (1903). *Die Ergebnisse der Volkszählung vom 31. December 1900 in den im Reichsrathe vertretenen Königreichen und Ländern. 63. Band, 2. Heft: Die Bevölkerung nach den Grössenkategorien der Ortschaften, nach der Gebürtigkeit, nach der Confession und Umgangssprache in Verbindung mit dem Geschlechte, nach dem Bildungsgrade innerhalb der Grössenkategorien der Ortschaften und dem Familienstande.* Wien: K. k. Statistische Zentral-Commission.

Österreichische Statistik (1905). *Die Ergebnisse der Volkszählung vom 31. December 1900 in den im Reichsrathe vertretenen Königreichen und Ländern. 64. Band, 2. Heft: Die Ausländer in den im Reichsrathe vertretenen Königreichen und Ländern sowie die Angehörigen dieser letzteren im Auslande.* Wien: K. k. Statistische Zentral-Commission.

Österreichische Statistik (1914). *Die Ergebnisse der Volkszählung vom 31. Dezember 1910 in den im Reichsrate vertretenen Königreichen und Ländern. Neue Folge, 1. Band, 2. Heft: Die Bevölkerung nach der Gebürtigkeit, Religion und Umgangssprache in Verbindung mit dem Geschlechte, nach dem Bildungsgrade und Familienstande; die körperlichen Gebrechen; die soziale Gliederung der Haushaltungen.* Wien: K. k. Statistische Zentral-Commission.

Österreichische Statistik (1919). *Ergebnisse der Volkszählung vom 31. Dezember 1910 in Österreich. Neue Folge., Band 2, Heft 3: Die Österreicher im Auslande.* Wien: K. k. Statistische Zentral-Commission.

Ovsenar, Tita (1993). Meksikajnarji. *Kronika* 41/2, 14–17. https://www.dlib.si/stream/URN:NBN:SI:DOC-3L579NXE/a311ec69-0ae5-41a2-8385-fa2fe305b24c/PDF, November 11, 2022.

Pahor, Milan (1987). Opcije v Kanalski dolini (1939–1942). *Zgodovinski časopis* 41/2, 321–328. https://www.dlib.si/stream/URN:NBN:SI:DOC-5J8JGO1N/472a2503-365c-4768-8327-ca310725cbbe/PDF, November 11, 2022.

Pahor, Milan (ed.) (1990). *Tabor "Rezija 89"*. Trst: SLORI.

Pajsar, Breda (2005). Muzejska zbirka arhitekta Ivana Jagra. *Etnolog* 15, 395–404.

Palko, Olena; Foster, Samuel (2021). Contested minorities in the 'New Europe': national identities in interwar Eastern and Southeastern Europe. *National Identities*, 23/4, 303–323. https://doi.org/10.1080/14608944.2020.1749837.

Panjek, Aleksander (2018). Od kmečke trgovine do integrirane kmečke ekonomije. *Preživetje in podjetnost: Integrirana kmečka ekonomija na Slovenskem od srednjega veka do danes* (eds. Aleksander Panjek, Žarko Lazarević). Koper: Založba Univerze na Primorskem, 15–50. https://doi.org/10.26493/978-961-7023-81-7.15-50.

Panjek, Aleksander (2022). Land Will Tear Us Apart. Farm-Division and Real-Estate Market in Rural Slovenia (Sixteenth to Eighteenth Centuries). *The History of the Family* 27:1, 54–81. https://doi.org/10.1080/1081602X.2021.2005654.

Panjek, Giovanni (2011). Gospodarski odnosi med Trstom in Aleksandrijo v XIX. stoletju. *Le rotte di Alexandria / Po aleksandrijskih poteh* (eds. Franco Però, Patrizia Vascotto). Trst: EUT, 53–64.

Papež Adamič, Vera (2009). *Od prednikov do potomcev: Slovenci v Slatini in Banjaluki, 1923–2008*. Banjaluka: Društvo Slovencev Triglav.

Peck, Gunther (2000). *Reinventing Free Labor: Padrones and Immigrant Workers in North American West*. Cambridge: Cambridge University Press.

Pečnik, Karol (1897). Egiptovsko iverje. *Soča*, June 25, 1897, 1. https://www.dlib.si/stream/URN:NBN:SI:DOC-D5BSODX3/a20afb6f-46c5-408f-b8dd-87feb6f87d50/PDF, November 11, 2022.

Penn, Allie (2018). Female Agitators: The Women of the 1913–1914 Keweenaw Copper Strike. *Upper Country: A Journal of the Lake Superior Region* 6, Article 5.

Però, Franco; Vascotto, Patrizia (eds.) (2011). *Le rotte di Alexandria / Po aleksandrijskih poteh*. Trst: EUT.

Perusini, Gaetano (1948). Rezijanski izseljenci v 16. Stoletju. *Slovenski etnograf* 1, 57–65. https://www.dlib.si/stream/URN:NBN:SI:DOC-WRWIZ83N/e37b0fd0-979b-4a4b-aa98-ba1ae5181090/PDF, November 11, 2022.

Peruško, Tone (1953). *Slovensko primorje in Istra*. Beograd: Rad.

Petrič, Jerneja (ed.) (1982a). *Naši na tujih tleh*. Ljubljana: Cankarjeva založba.

Petrič, Jerneja (1982b). Nekatere pomembnejše smeri in tipi ameriško-slovenske književnosti. *Naši na tujih tleh* (ed. Jerneja Petrič). Ljubljana: Cankarjeva založba, 403–406.

Petrič, Jerneja (1999). Književnost slovenskih priseljencev in njihovih potomcev v ZDA od leta 1891. *Slovenska izseljenska književnost 2: Severna Amerika* (eds. Janja Žitnik, Helga Glušič). Ljubljana: Založba ZRC, 89–263.Petricioli, Marta (2007). *Oltre il mito: L'Egitto degli italiani (1917–1947)*. Milano: Bruno Mondadori.

Pezo, Edvin (2009). "Re-conquering" Space. Yugoslav Migration Policies and the Emigration of Non-Slavic Muslims to Turkey (1918–1941). *Transnational Societies, Transnational Politics. Migrations in the (Post-)Yugoslav Region, 19th–21st Century* (ed. Ulf Brunnbauer). München: Oldenburg Verlag, 73–94.

Pfaundler, Richard (1907). Die Grundlagen der nationalen Bevölkerungsentwicklung Steiermarks. *Statistische Monatsschrift*, N. F XII, 557–592.

Phelps, Nicole M. (2013). *U.S.-Habsburg Relations from 1815 to the Paris Peace Conference: Sovereignty Transformed*. New York: Cambridge University Press. https://doi.org/10.2307/j.ctt1n2txft.20.

Picotti, Guido (1909). L'intensità del fenomeno migratorio nel circondario di Udine. *La Patria del Friuli*, October 31, 1909. http://periodicifriulani.sbhu.it/easyne2/Download.aspx?Code=PERF&filename=Archivi/PERF/PDF/0055/55904.PDF, November 11, 2022.

Piljušić, Miroslav Miha (2013). *Naših prvih deset let: Kronika Slovenske kulturne skupnosti Timoške krajine Ivan Cankar*. Bor: Qwerty.

Pislar Fernandez, Marie (2004). *Slovenci v železni Loreni (1919–1939) skozi družinske pripovedi / Slovènes en Lorraine du fer (1919–1939) à travers des récits de familles*. Ljubljana: Založba ZRC.

Plato, Alexander von; Leh, Almut; Thonfeld, Christoph (eds.) (2010). *Hitler's Slaves: Life Stories of Forced Labourers in Nazi-occupied Europe*. New York: Berghahn Books.

Plut Pregelj, Leopoldina (1992). Katoliški šolski sistem v Združenih državah Amerike in ustanavljanje osnovnih šol pri slovenskih župnijah (1895–1941). *Dve domovini / Two Homelands* 2–3, 391–407.

Podbersič, Renato ml. (2018). Dr. Renato Podbersič ob 110-letnici Rafaelove družbe. http://www.rafaelova-druzba.si/rafaelova_druzba/o_rafaelovi_druzbi, November 11, 2022.

Pogačar, Timothy (2014). Slovenska proza v dveh slovensko-ameriških publikacijah: Recepcija zgodnjega 20. stoletja. *Recepcija slovenske književnosti* (ed. Alenka Žbogar). Ljubljana: Znanstvena založba Filozofske fakultete, 325–332.

Pogačnik, Teodora (2000). Težave slovenskih izseljencev na poti v Ameriko v letih od 1900 do 1914. BA thesis. Ljubljana: Univerza v Ljubljani, Filozofska fakulteta.

Pörš, Sidika (1940). Sezonsko izseljevanje v Prekmurju. *Mladi Prekmurec* 4/5–6, 65–72. https://www.dlib.si/stream/URN:NBN:SI:DOC-V2HF3A9Q/8803dfe3-5965-40d0-b75c-e2848c37cca5/PDF, November 11, 2022.

Prečastiti duhovščini in vernikom ilirske cerkvene pokrajine pozdrav in blagoslov v Gospodu! (1913). *Ljubljanski škofijski list* XI/123, 137–142. https://www.dlib.si/stream/URN:NBN:SI:DOC-F7VZN2QH/a03479fb-034e-4d54-ab64-e938bb1db059/PDF, November 11, 2022.

Prešeren, Jože (2001). Slovenske sledi v svetu. *Slovensko izseljenstvo: Zbornik ob 50-letnici Slovenske izseljenske matice* (ed. Milica Trebše Štolfa). Ljubljana: Združenje Slovenska izseljenska matica, 115–124.

Prisland, Marie (1950). The Slovenians, Most Recent American Immigrants. *The Wisconsin Magazine of History* 33/3, 265–280.

Prisland, Marie (1960). Ob 30-letnici Zarje. *Slovenski izseljenski koledar 1960*, 182–184.

Prisland, Marie (1964). Zgodovina slovenske ženske zveze. *Slovenski izseljenski koledar 1965*, 208–215.

Prisland, Marie (1966). Zgodovina slovenske naselbine v Sheboyganu. *Slovenski izseljenski koledar 1966*, 182–190.

Prisland, Marie (1968). *From Slovenia to America: Recollections and Collections*. Chicago: Slovenian Women's Union of America.

Protokoll der im k. k. Handelsministerium durchgeführten vernehmung von Auskunftspersonen über die Auswanderung aus Österreich. (1912b). Wien: K. k. Hof- und Staatsdruckerei.

Prvi slovenski izseljenski kongres v Ljubljani dne 1. julija 1935. (1936). Ljubljana: Družba sv. Rafaela.

Purini, Piero (2010). *Metamorfosi etniche: I cambiamenti di popolazione a Trieste, Gorizia, Fiume e in Istria 1914–1975*. Udine: Kappa Vu.

Puskás, Julianna (1982). *From Hungary to the United States (1880–1914)*. Budimpešta: Akadémiai Kiadó.

Rahten, Andrej (2018). *Med Kakanijo in Wilsonio: Poklicne in politične preizkušnje Hansa Schwegla alias Ivana Švegla*. Ljubljana: Mohorjeva založba.

Ramella, Franco (2009). Reti sociali, famiglie e strategie migratorie. *Storia dell'emigrazione italiana. Partenze* (eds. Piero Bevilacqua, Andreina De Clementi, Emilio Franzina). Roma: Donzelli, 143–160.

Ravnik, Mojca (1981). Način življenja in izseljevanje prebivalcev Grosupljega in okolice do prve svetovne vojne. *Louis Adamič: Simpozij: Ljubljana 16.–18. september 1981*, 41–48.

Ravnik, Mojca (1996). *Bratje, sestre, strniči, zermani: Družina in sorodstvo v vaseh v Slovenski Istri*. Ljubljana: Založba ZRC.

Ravnik, Mojca (2002). Z meje na mejo. *Poti in usode: Selitvene izkušnje Slovencev z zahodne meje* (ed. Aleksej Kalc). Koper: Zgodovinsko društvo za južno Primorsko, 54–61.

Ravnik, Mojca (2010). Nekaj pogledov v preteklost in sedanjost beograjskih Slovencev. *Traditiones* 1, 167–193. https://doi.org/10.3986/Traditio2010390114.

Repe, Božo (2010). Izgon: Usoda slovenskih izgnancev med leti 1941–1945. *Migracije in slovenski prostor od antike do danes* (eds. Peter Štih, Bojan Balkovec). Ljubljana: Zveza zgodovinskih društev Slovenije, 245–256.

Reports of the Immigrant Commission (1911a). *Immigrants in Industries, Part 2: Iron and Steel Manufacturing in the Middle West* Vol. 2. Washington: Government Printing Office.

Reports of the Immigrant Commission (1911b). *Immigrants in Industries, Part 23: Summary Report on Immigrants in Manufacturing and Mining* Vol. 1. Washington: Government Printing Office.

Reports of the Immigrant Commission (1911c). *Emigration Conditions in Europe*. Washington: Government Printing Office.

Reports of the Immigrant Commission (1911d). *Children of Immigrants in Schools*. Washington: Government Printing Office.

Reports of the Immigrant Commission (1911e). *Abstracts of the Reports of the Immigration Commission* Vol. I. Washington: Government Printing Office.

Ribnikar, Vojteh (ed.) (1889). *Logaško okrajno glavarstvo: Zemljepisni in zgodovinski opis*. Logatec: Društvo učiteljev in šolskih prijateljev okraja Logaškega.

Roš, Fran (1967). *Slovenski izgnanci v Srbiji 1941–1945*. Maribor: Obzorja.

Rosental, Paul-André (1999). *Les sentiers invisibles: Espaces, familles et migrations dans la France du XIXe siècle*. Paris: Éditions de l'École des hautes études en sciences sociales.

Rosental, Paul-André (2006). Between Micro and Macro: Theorizing Agency in Nineteenth-Century French Migrations. *French Historical Studies* 29/3, 457–481. https://doi.org/10.1215/00161071-2006-007.

Rosenwein, Barbara H. (2010). Problems and Methods in the History of Emotions. *Passions in Context: Journal of the History and Philosophy of the Emotions* 1/1: 11.

Roškar, Saša (2014). Slamnikarice Abroad and at Home: Ladies and Entrepreneurs. *Going Places: Slovenian Women's Stories on Migration* (eds. Mirjam Milharčič Hladnik, Jernej Mlekuž). Akron: Akron University Press, 171–212.

Rostohar, Mihajlo (1912). Naš narodni problem. *Napredna misel* 6, 245–253. https://www.dlib.si/stream/URN:NBN:SI:DOC-KLXGT5Z5/1b54c809-85fe-42f9-a58a-76cf75142b18/PDF, November 11, 2022.

Rotim Malvić, Jasna (2015). Preventivna zdravstvena provjera emigranata u SAD početkom 20. stoljeća na primjeru Hotela "Emigranti" u Rijeci. *Acta medico-historico adriatica* 13/2, 237–250.

Rowbotham, Sheila (2001). Foreword. *Women, Gender and Labour Migration: Historical and Global Perspectives* (ed. Pamela Sharpe). London: Routledge, XVI–XVII.

Rozman, Franc (1979). *Socialistično delavsko gibanje na slovenskem Štajerskem*. Ljubljana: Založba Borec.

Rupel, Lidia (2018). Srednješolsko izobraževanje tržaških Slovencev za časa Avstro-Ogrske in ustanovitev slovenske gimnazije leta 1945. *Šolska kronika*

27/3, 81–94. http://www.dlib.si/listalnik/URN_NBN_SI_doc-OQXJNH20/index.html, November 11, 2022.

Rusić, Branislav; Novak, Vilko (1973). Slovenci v Birtrenici v Makedoniji. *Traditiones* 2, 177–201.

Ruttar, Riccardo (2002). 'Buoh je dau, Buoh je uzeu'. *Poti in usode: Selitvene izkušnje Slovencev z zahodne meje* (ed. Aleksej Kalc). Koper: Zgodovinsko društvo za Južno Primorsko, 28–34.

S. P. B. (1959). Brazilija, dobra zemlja … Ob stoletnici prihoda prvih Slovencev v Brazilijo. *Zbornik koledar Svobodne Slovenije 1960*. Buenos Aires, 185–186.

Sala, Teodoro (1974). Programmi di snazionalizzazione del "fascismo di frontiera" (1938–1939). *Bollettino dell'Istituto regionale per la storia delmovimento di liberazione del Friuli-Venezia giulia* 2, 24–30.

Salazar Parreñas, Rhacel (2001). *Servants of Globalization: Women, Migration, and Domestic Work*. Stanford, CA: Stanford University Press.

Salazar Parreñas, Rhacel (2005). *Children of Global Migration: Transnational Families and Gendered Woes*. Stanford, CA: Stanford University Press.

Sarić, Samija; Štimac, Vera (2010). Kolonizacija Slovencev v Bosni in Hercegovini od 1878 do 1918. *Kronika* 2, 397–406.

Scartabellati, Andrea (2009). Un'altra Trieste asburgica tra consapevolezza dei poveri e coscienza operaia. *InStoria* 9, http://www.instoria.it/home/trieste_movimento_operaio.htm, November 11, 2022.

Schwartz Seller, Maxine (1988). *To Seek America: A History of Ethnic Life in the United States*. Englewood: Jerome S. Ozer.

Schwartz Seller, Maxine (ed.) (1994). *Immigrant Women*. Albany: State University of New York Press.

Schwegel, Hans (1904). *Die Einwanderung in die Vereinigten Staaten von Amerika: Mit besonderer Rücksicht auf die österreichisch-ungarische Auswanderung*. Wien: Braumüller.

Sedmak, Drago (1994). Ekonomska in politična migracija prebivalstva na Bovškem v letih 1850 do 1940. *Soški protokol* (eds. Andreas Moritsch, Gudmund Tributsch). Celovec: Mohorjeva založba, 79–125.

Sedmak, Drago; Bajt, Anton (eds.) (2006). *Kanal ob Soči v dnevniku občinskega tajnika Antona Bajta*. Nova Gorica: Goriški muzej.

Seymour, Mark (2010). Epistolary Emotions: Exploring Amorous Hinterlands in 1870s Southern Italy. *Social History* 35/2, 148. https://doi.org/10.1080/03071021003719139.

Shiffman, Dan (2003). *Rooting Multiculturalism: The Work of Louis Adamic*. Madison (NJ): Fairleigh Dickinson University Press.

Sielemann, Jürgen (1994). "Haben alle Passagiere auch Geld?": Zur Geschichte der Auswanderung über den Hamburger Hafen 1892–1954. *Hoffnung Amerika: Europäische Auswanderung in die Neue Welt* (ed. Karin Schulz). Breerhaven: Nordwestdeutsche Verlagsgesellschaft, 91–112.

Silvestri, Marina (2017). *Lassù nella Trieste asburgica: La questione dei regnicoli e l'identità rimossa*. Gorizia: LEG.

Sinke, Suzanne M. (2002). *Dutch Immigrant Women in the United States, 1880–1920*. Urbana: University of Illinois Press.

Sirjamaki, John (1946). The People of Mesabi Range. *Minnesota History* 28, 202–215.

Slokan, Ina (1983). Zlato leto Progresivnih Slovenk Amerike. *Slovenski koledar '84*, 156–161.

Slovani v Egiptu. (1897). *Soča*, April 9, 1897, 1. https://www.dlib.si/stream/URN:NBN:SI:DOC-VZIBXKZR/c7d224a3-03ff-4c19-9d99-4e6dc5c757d0/PDF, November 11, 2022.

Slovenska kri v Južni Srbiji. (1937). *Ponedeljski Slovenec*, January 11, 1937, 2. https://www.dlib.si/stream/URN:NBN:SI:DOC-ESAYX45N/ecb8732c-39e0-4ff6-87f3-5413eee254e5/PDF, November 11, 2022.

Slovensko Republičansko Združenje in Slovenska N. Zveza za skupni shod. (1918). *Clevelandska Amerika*, November 27, 1918, 1. https://www.dlib.si/stream/URN:NBN:SI:DOC-DXZDPDZ4/4072c0a4-13a5-499b-90bc-74b30 2bebdd7/PDF, November 11, 2022.

Smolik, Marijan (2013). Zakrajšek, Kazimir (1878–1958). Slovenska biografija. http://www.slovenska-biografija.si/oseba/sbi854497/#slovenski-biografski-leksikon, November 11, 2022.

Sori, Ercole (1979). *L'emigrazione italiana dall'Unità alla seconda guerra mondiale*. Bologna: Il Mulino.

Spominska zgodovina - Historical Memories. (1982). Willard, Wisconsin: Slovenska družba.

Srebernič, Josef (1914). Die Wanderbewegung in der Umgebung von Görz. *Jahresbericht des Mädchenlyzeums der Armen Schulschwestern de Notre Dame in Görz*, 5–17.

Stanič, Stane (1987). Beograd in Slovenci. *Enciklopedija Slovenije*, zv. 1. Ljubljana: Mladinska knjiga, 243–244.

Stanley, Jerry (1996). *Big Annie of Calumet: A True Story of the Industrial Revolution*. New York: Crown Publishers.

Stanonik, Janez (1979). Ameriška leta slovenskega izobraženca Antona Fistra. *Slovenski koledar 1980*, 197–204.

Stanonik, Janez (1997). Slovenci v Clevelandu. *Zgodovinski časopis* 51/1, 21–32. https://www.dlib.si/stream/URN:NBN:SI:DOC-1HZDIFSL/7d6a13b4-2553-4127-b72e-458c1ffa3803/PDF, November 11, 2022.

Stanonik, Janez (1999). Književnost Slovencev v ZDA pred letom 1981. *Slovenska izseljenska književnost 2: Severna Amerika* (eds. Janja Žitnik, Helga Glušič). Ljubljana: Založba ZRC, 385–456.

Steidl, Annemarie (2007). The "Relatives and Friends Effect": Migration Networks of Transatlantic Migrants from the Late Habsburg Monarchy. *Maritime Transport and Migration: The Connections between Maritime and Migration Networks* (eds. Torsten Feys et al.). St. John's Newfoundland: International Maritime Economic History Association – IMEHA, 75–96. https://doi.org/10.5949/liverpool/9780973893434.003.0006.

Steidl, Annemarie (2009). Transatlantic Migration from the Late Austrian Empire and its Relation to Rural-Urban Stage Migration. *European Mobility: Internal, International, and Transatlantic Moves in the 19th and 20th Centuries* (eds. Annemarie Steidl et al.). Göttingen: V&R unipress, 69–88. https://doi.org/10.2307/j.ctt1t89kvv.6.

Steidl, Annemarie (2021). *On Many Routes. Internal, European, and Transatlantic Migration in the Late Habsburg Empire.* West Lafayette, Indiana: Purdue University Press. https://doi.org/10.2307/j.ctv15pjz1z.

Steidl, Annemarie; Fischer-Nebmaier, Wladimir; Oberly, James W. (2017). *From a Multiethnic Empire to a Nation of Nations: Austro-Hungarian Migrants in the US, 1870–1940.* Innsbruck: Studien Ferlag. https://doi.org/10.1086/704407.

Steinicke, Ernst (1984). *Das Kanaltal – Val Canale: Sozialgeographie einer alpinen Minderheitenregion.* Innsbruck: Geographische Studien.

Stek, Pamela (2015). Militant Mothers: Women and the Minnesota Iron Range Strike of 1916. *Mining History Journal*, 29–50. https://www.mininghistoryassociation.org/Journal/MHJ-v22-2015-Stek.pdf.

Stelè, France (2013). Prusheck, Harvey Gregory (1887–1940). Slovenska biografija. Slovenska akademija znanosti in umetnosti, Znanstvenoraziskovalni center SAZU. http://www.slovenska-biografija.si/oseba/sbi472745/#slovenski-biografski-leksikon, November 11, 2022.

Stergar, Rok (2004). *Slovenci in vojska, 1867–1914: Slovenski odnos do vojaških vprašanj od uvedbe dualizma do začetka 1. svetovne vojne.* Ljubljana: Oddelek za zgodovino Filozofske fakultete.

Strašek, Franc (2006). Slovenci na Hrvaškem – samoorganiziranost nekdaj in sedaj. *Slovenci na Hrvaškem – dediščina in sedanjost* (eds. Mojca Ravnik, Katalin Munda Hirnök). Ljubljana: Slovensko etnološko društvo, 97–104.

Strauss, Johann (1897). Die Oesterreichischen und Ungarischen Staatsangehörigen in Bosnien und der Herzrgovina. *Statistische Monatsschrift*, N. F. 2, 385–420.

Strikwerda, Carl (1993). The Troubled Origins of European Economic Integration: International Iron and Steel and Labor Migration in the Era of World War I. *The American Historical Review* 4, 1106–1129. https://doi.org/10.1086/ahr/98.4.1106

Strle, Urška (2015a). K razumevanju ženskega dela v veliki vojni. *Prispevki za novejšo zgodovino* LV 2, 103–125. https://doi.org/10.51663/pnz.55.2.06.

Strle, Urška (2015b). Otroštvo, vojna in prisilne migracije. *Slovensko izseljenstvo v luči otroške izkušnje* (ed. Janja Žitnik Serafin). Ljubljana: Založba ZRC, 91–108.

Strle, Urška (2016). Begunke in vélika vojna na Slovenskem. *Begunci, slovenski begunci s soške fronte* (eds. Ines Beguš, Marko Klavora, Katarina Brešan). Nova Gorica: Goriški muzej, 44–51.

Studen, Andrej (2005a). Revolucija v prometu. *Slovenska novejša zgodovina 1848–1992* (eds. Jasna Fischer et al.). Ljubljana: Mladinska knjiga, 86–90.

Studen, Andrej (2005b). Modernizacija načina življenja. *Slovenska novejša zgodovina 1848–1992* (eds. Jasna Fischer et al.). Ljubljana: Mladinska knjiga, 104–122.

Suhadolnik, Violeta (2006). Poravnava vojnih krivic civilnim žrtvam vojnega nasilja 2. svetovne vojne v Sloveniji. MA thesis. Ljubljana: Univerza v Ljubljani, Fakulteta za družbene vede.

Sulič, Nives (1996). A Home Away from Home: Brothers in Distress. *Etnični fraternalizem v priseljenskih državah / Ethnic Fraternalism in Immigrant Countries* (ed. Matjaž Klemenčič). Maribor: Pedagoška fakulteta, 45–52.

Susel, Rudolf M. (1992). Poslanstvo in vloga časopisa Ameriška Domovina v zgodovini ameriških Slovencev, 1919–1991. *Dve domovini / Two Homelands* 2–3, 237–251.

Susel, Rudolf M. (1996). The American Mutual Life Association 1910–1994. *Etnični fraternalizem v priseljenskih državah / Ethnic Fraternalism in Immigrant Countries* (ed. Matjaž Klemenčič). Maribor: Pedagoška fakulteta, 101–113.

Susel, Rudolf M. (n.d.). Slovenes. *Encyclopedia of Cleveland History*. http://case.edu/ech/articles/s/slovenes/, November 11, 2022.

Svoljšak, Petra (1991). *Slovenski begunci v Italiji med prvo svetovno vojno.* Ljubljana: Zveza zgodovinskih društev Slovenije, Založba ZRC.

Svoljšak, Petra (2004). "Tudi jaz sem pomagala do velike zmage!": Položaj in vloge žensk na Slovenskem med 1. svetovno vojno. *Ženske skozi zgodovino: Zbornik referatov 32. zborovanja slovenskih zgodovinarjev* (ed. Aleksander Žižek). Celje: ZZDS, 153–158.

Svoljšak, Petra (2009). The Social History of the Soča/Isonzo Region in the First World War. *Mitteilungsblatt des Instituts fur soziale Bewegungen* 41, 89–110. https://doi.org/10.13154/mts.41.2009.89-109.

Svoljšak, Petra; Antoličič, Gregor (2018). *Leta strahote. Slovenci in prva svetovna vojna*. Ljubljana: Cankarjeva založba.

Szabo, Agneza (1995). Socijalna struktura Slovenaca u Kraljevini Hrvatskoj i Slavoniji između 1880. i 1910. godine. *Slovenci v Hrvaški* (ed. Vera Kržišnik-Bukić). Ljubljana: Inštitut za narodnostna vprašanja, 85–105.

Šašek, A. (1912). Izseljevalno vprašanje. *Napredna misel* 6, 253–257. https://www.dlib.si/stream/URN:NBN:SI:DOC-KLXGT5Z5/1b54c809-85fe-42f9-a58a-76cf75142b18/PDF, November 11, 2022.

Šega, Polona (1997). *Slovenski kostanjarji na Dunaju: Prebivalci nekdanjega velikolaškega okraja kot kostanjarji v cesarskem mestu*. Novo mesto: Dolenjska založba.

Šega, Polona (1998). Ribniški krošnjarji na Dunaju in širši okolici in njihovi medsebojni odnosi. *Traditiones* 27, 145–157.

Šifrer, Živko (1962). *Izseljevanje iz slovenskega ozemlja*. Ljubljana: Zavod SRS za statistiko.

Šifrer, Živko (1974). *Izseljevanje s slovenskega ozemlja. Prvi slovenski demografski simpozij*. Ljubljana: Raziskovalni center Ekonomske fakultete Univerze v Ljubljani.

Škafar, Ivan (1939a). Prekmurski sezonski delavci v Slavoniji. *Izseljenski vestnik* 3, 34–36. https://www.dlib.si/stream/URN:NBN:SI:DOC-0SUURT0S/c10a611d-585a-4d80-a353-59c6cb118950/PDF, November 11, 2022.

Škafar, Ivan (1939b). Socialni in gospodarski položaj prekmurskih in medjimurskih sezonskih delavcev v Baranji, Bački in Banatu v letu 1939. *Izseljenski vestnik* 8/9, 121–143. https://www.dlib.si/stream/URN:NBN:SI:DOC-FO7AENLT/b6760dab-3178-4d85-8cd5-49502f96819b/PDF, November 11, 2022.

Škrlj Počkaj, Sonja (2011). Prebivalstvo župnije Trnovo v 19. stoletju skozi matične knjige in zapisnike duš. MA thesis. Koper: Univerza na Primorskem, Fakulteta za humanistične študije.

Škrlj, Katja (2009). Komaj sem čakala, da zrastem in postanem aleksandrinka: Demitizacija aleksandrink. *Krila migracij: Po meri življenjskih zgodb* (eds. Mirjam Milharčič Hladnik, Jernej Mlekuž). Ljubljana: Založba ZRC, 143–189.

Šmitek, Zmago (1981). *Občina Kočevje: Etnološka topografija slovenskega etničnega ozemlja —20. stoletje*. Ljubljana: Znanstveni inštitut Filozofske fakultete Univerze v Ljubljani.

Šteiner, Martin (1994). Akulturacija prekmurskih sezoncev. *Kronika* 42/2, 72–76. https://www.dlib.si/stream/URN:NBN:SI:DOC-3HTD4GSV/84053aca-e050-4127-b007-b0a4338c9c96/PDF, November 11, 2022.

Štih, Peter (2010). Migracije in oblikovanje kompleksnega kulturnega prostora med vzhodnimi Alpami in severnim Jadranom v srednjem veku. *Migracije in slovenski prostor od antike do danes* (eds. Bojan Štih, Bojan Balkovec). Ljubljana: Zveza zgodovinskih društev Slovenije, 59–77.

Štrekelj, Anton (1919). Naselitev Goričanov v Makedoniji. *Slovenski narod*, September 10, 1919. https://www.dlib.si/stream/URN:NBN:SI:DOC-YSQHL0WU/f7bbe0a7-7af2-40cb-aea0-2c7d0f538d4c/PDF, November 11, 2022.

Štrucelj, Dragotin (1909). Iz S. Paula v Braziliji. *Soča* 68, July 16, 1909, 1. https://www.dlib.si/stream/URN:NBN:SI:DOC-YXSL4VFA/2326d246-3782-4332-b840-acb095a9dd58/PDF, November 11, 2022.

Šuklje, Franc (1905). Izseljevanje. Govor poslanca Šukljeta v državnem zboru, dne 5. 10. 1905. *Slovenec*, October 13, 1905, 1. https://www.dlib.si/stream/URN:NBN:SI:DOC-YAWIARWF/27465a4b-0f9d-4df0-8748-e2eb38e7bfd4/PDF, November 11, 2022.

Šuklje, Marko (1936/1937). Izseljevanje in doseljevanje v Žumberku. *Geografski vestnik* XII.–XIII., 170–183. https://www.dlib.si/stream/URN:NBN:SI:DOC-7SQSIYMM/bd091b8d-e7d5-4644-9e58-b22bfdc2ef53/PDF, November 11, 2022.

Šušteršič, Franc (1891). Slovenci v Ameriki. *Koledar Družbe sv. Mohorja*, 15–27.

Švent, Rozina (2009). Prizadevanja slovenskih izseljencev za novo jugoslovansko državo. *Jugoslavija v času / Yugoslavia through Time* (ed. Bojan Balkovec). Ljubljana: Znanstveni inštitut Filozofske fakultete, 267–276.

Tavčar, Zora (1999). Literarno delo južnoameriških Slovencev pred drugo svetovno vojno: Poezija. *Slovenska izseljenska književnost 3: Južna Amerika* (eds. Janja Žitnik, Helga Glušič). Ljubljana: Založba ZRC, 46–79.

Terpin, Marija (2008). *Idrija Lace: A History Written in Thread*. Idrija: Idrija Municipal Museum.

Testen, Petra (2013). Služkinje kot del slovenske narodne skupnosti na Goriškem ob koncu 19. in na začetku 20. stoletja. *Eliminacionizem in emancipacija: Zbornik razprav* (ed. Igor Grdina). Ljubljana: Založba ZRC, 45–61.

Ther, Philipp (2016). *The Dark Side of Nation-States: Ethnic Cleansing in Modern Europe*. New York: Berghahn Books. https://doi.org/10.2307/j.ctt9qd3ng.

Thomson, Alistair (2011). *Moving Stories: An Intimate History of Four Women Across Two Countries*. Manchester: Manchester University Press. https://doi.org/10.1111/1468-229X.12057_39.

Titl, Julij (1965). *Socialnogeografski problemi na Koprskem podeželju*. Koper: Lipa.

Toplak, Kristina (2008). *"Buenas Artes": Ustvarjalnost Slovencev in njihovih potomcev v Buenos Airesu*. Ljubljana: Založba ZRC. https://doi.org/10.3986/9789610502814.

Toš, Marjan (2010). Migracije in mobilnost judovskega prebivalstva na Slovenskem—od srednjeveških izgonov iz slovenskih dežel 1496–1515 do množičnih deportacij in uničenja v holokavstu 1941–1945. *Migracije in slovenski prostor od antike do danes* (eds. Peter Štih, Bojan Balkovec). Ljubljana: Zveza zgodovinskih društev Slovenije, 634–650.

Tosi, Luciano (2002). La tutela internazionale dell'emigrazione. *Storia dell'emigrazione italiana: Arrivi* (eds. Bevilacqua, Piero, De Clementi, Andreina, Franzina, Emilio). Roma: Donzelli, 439–456.

Trdina, Janez (1951). *Hrvaški spomini: Zbrano delo 3*. Ljubljana: Državna založba Slovenije.

Trebše Štolfa, Milica (1996). Canadian Slovenes in the Oldest Fraternal Organization "BLED" Mutual Benefit Society. *Etnični fraternalizem v priseljenskih državah / Ethnic Fraternalism in Immigrant Countries* (ed. Matjaž Klemenčič). Maribor: Pedagoška fakulteta, 243–257.

Trebše Štolfa, Milica (2001). Slovenci in njihove organizacije v Kanadi. *Slovensko izseljenstvo: Zbornik ob 50-letnici Slovenske izseljenske matice* (ed. Milica Trebše Štolfa). Ljubljana: Združenje Slovenska izseljenska matica, 211–221.

Treves, Anna (1976). *Le migrazioni interne nell'Italia fascista*. Torino: Einaudi.

Tršan, Lojz (2001). Slovenski filmski arhiv in izseljenstvo. *Dve domovini / Two Homelands* 14, 181–183.

Trunk, Jurij Matej (1912). *Amerika in Amerikanci*. Celovec: Self-published. https://www.dlib.si/details/urn:nbn:si:doc-m1c9sgov, November 11, 2022.

Tschinkel, John (2010). *Zvonovi so umolknili*. Ljubljana: Modrijan.

Tschinkel, John (2012). *The Bells Ring No More.* CreateSpace Independent Publishing Platform.

Tumpić, Dušan (1991): *Istarska emigracija – svjedočanstva*. Zagreb: NIP "Alinea".

Turk, Ernest (1938). Naši vojni dobrovoljci. *Misel in delo* 4/12, 56–68. https://www.dlib.si/stream/URN:NBN:SI:DOC-SOWRB7NL/cfa6021d-3fa8-439d-b05a-377698446485/PdF, November 11, 2022.

Turk, Helena (1925). Moja pot po svetu. *Amerikanski Slovenec*, September 23, 24, 30, October 1, 2, 7, 8, 1925.

Umek, Ema (1967). Izseljenska agencija Jožefa Paulina v Ljubljani. *Slovenski izseljenski koledar 1968*, 234–238.

Umek, Ema (1968). Izseljevanje v Brazilijo v zadnjih letih 19. stoletja. *Slovenski izseljenski koledar 1969*, 218–219.

Umek, Ema (1979). Brazilija v "rožnatih barvah". *Slovenski koledar 1980,* 213–214.

Ureditev slovenske izseljenske službe. (1940). *Slovenec*, March 28, 1940, 5. https://www.dlib.si/stream/URN:NBN:SI:DOC-6B8RY0QX/620562f9-75d4-441f-9e58-7b84695ef6dd/PDF, November 11, 2022.

Valenčič, Joseph (2001). Slovenski narodni domovi v ZDA. *Slovensko izseljenstvo: Zbornik ob 50-letnici Slovenske izseljenske matice* (ed. Milica Trebše Štolfa). Ljubljana: Združenje Slovenska izseljenska matica, 187–191.

Valenčič, Joseph (2002). Z delom uma in srca – ustvarjalnost med Slovenci v Severni Ameriki. *Ustvarjalnost Slovencev po svetu: 38. seminar slovenskega jezika, literature in kulture* (ed. Boža Krakar Vogel). Ljubljana: Center za slovenščino kot drugi/tuji jezik pri Oddelku za slovanske jezike in književnosti Filozofske fakultete, 163–176.

Valenčič, Joseph; Milharčič Hladnik, Mirjam (2011). Dosežki ameriških Slovencev: Prispevek kulturam dveh domovin / Slovenian-American Achievers: Cultural Contributions to Two Homelands [exhibition, Ljubljana, Atrij ZRC, June 23, 2011]. http://achievers.zrc-sazu.si/, November 11, 2022.

Valenčič, Vlado (1958). Prebivalstvo kamniškega področja skozi tri stoletja. *Kamniški zbornik.* Kamnik: Občina Kamnik, 41–86.

Valenčič, Vlado (1959). O slamnikarski domači obrti. *Kamniški zbornik*. Kamnik: Občina Kamnik, 165–193.

Valenčič, Vlado (1980). Iz gospodarske preteklosti Bistrice in Trnovega. *Bistriški zapisi*, 67–87.

Valenčič, Vlado (1981). Tovorništvo na Kranjskem. *Zgodovinski časopis* 35/3, 243–277. https://www.dlib.si/stream/URN:NBN:SI:DOC-L48X4EYK/934af83b-954d-44c2-b807-c57cbbf9b501/PDF, November 11, 2022.

Valenčič, Vlado (1989). Gospodarske in družbene razmere Bistriškega okraja v 19. stoletju. *Bistriški zapisi*, 1–35.

Valenčič, Vlado (1990a). Izseljevanje Slovencev v druge dežele habsburške monarhije. *Zgodovinski časopis* 44/1, 49–71. https://www.dlib.si/stream/URN:NBN:SI:DOC-TONGJEQR/0e6dbd01-7866-4f5a-b388-81f8434d468d/PDF, November 11, 2022.

Valenčič, Vlado (1990b). Izseljevanje Slovencev v tujino do druge svetovne vojne. *Dve domovini /Two Homelands* 1, 43–82.

Valenčič, Vlado (n.d.). *Migracije in strukturne spremembe prebivalstva na Slovenskem v dobi gospodarskega liberalizma.* (AS 1996 *Zbirka Inštituta za slovensko izseljenstvo ZRC SAZU*, škatla 113), Arhiv Republike Slovenije, Ljubljana.

Vanello, Lorena (1981). Colonizzazione e snazionalizzazione nelle campagne della Venezia Giulia tra le due guerre. *L'imperialismo italiano e la Jugoslavia: Atti del convegno italo-jugoslavo Ancona 14–16 ottobre 1977* (ed. Massimo Pacetti). Urbino: Argalìa editore, 489–510.

Velikonja, Jože (1996). Slovenska naselja v Ameriki. *Zgodovinski časopis* 50/3, 383–389. https://www.dlib.si/stream/URN:NBN:SI:DOC-IKYW24EE/33ea3014-755b-42e4-bfff-579b4ea9d78e/PDF, November 11, 2022.

Verginella, Marta (1990). Družina v Dolini pri Trstu v 19. stoletju. *Zgodovinski časopis* 44/2, 171–197. https://www.dlib.si/stream/URN:NBN:SI:DOC-1IYB2B63/7c5a8788-98c5-4329-95ed-f6d9c1b45315/PDF, November 11, 2022.

Verginella, Marta (2001). Sloveni a Trieste tra Settecento e Ottocento. Da comunità etnica a minoranza nazionale. *Storia economica e sociale di Trieste. La città dei gruppi* (eds. Roberto Finzi, Giobanni Panjek). Trieste: LINT, 441–481.

Verginella, Marta (2006). *Ženska obrobja: Vpis žensk v zgodovino Slovencev.* Ljubljana: Delta.

Verginella, Marta (2008). Il paradigma cittá/campagna e la rappresentazione dualistica di uno spazio multietnico. *Contemporanea* 4, 779–792. https://doi.org/10.1409/28132.

Verginella, Marta (2010). Velika vojna med spominom in pozabo. *Kultura spominjanja* (ed. Vincenc Rajšp). Dunaj: Slovenski znanstveni inštitut; Ljubljana: Založba ZRC, 79–88.

Verginella, Marta (2011). Nacionalna pripadnost žensk in njihovo delovanje na nacionalnem obrobju. *Jeziki, identitete, pripadnosti. Med središči in obrobji* (eds. Kozma Ahačič, Petra Testen). Ljubljana: Založba ZRC, 253–262.

Verginella, Marta (ed.) (2017). *Slovenka: prvi ženski časopis (1897–1902).* Ljubljana: Znanstvena založba Filozofske fakultete.

Verginella, Marta (2021). Women Teachers in the Whirlwind of Post-War Changes in the Julian March (1918–1926). *Acta Histriae* 29/4, 859–882. https://doi.org/10.19233/AH.2021.34.

Vodopivec, Peter (1979). Odmev industrijske revolucije na Štajerskem v prvi polovici 19. Stoletja. *Časopis za zgodovino in narodopisje* 1–2, 264–277.

Vodopivec, Peter (1999). Velika gospodarska kriza 1873 in Slovenci. *Prispevki za novejšo zgodovino* 2, 25–38.

Vogelnik, Dolfe (1965). *Razvoj prebivalstva Slovenije zadnjih dvesto let z jugoslovanske in evropske perspektive.* Ljubljana: Univerza v Ljubljani, Ekonomska fakulteta.

Voje, Ignacij (1952/53). Kajžarstvo v logaškem gospostvu. *Zgodovinski časopis* 6–7, 650–662. https://www.dlib.si/stream/URN:NBN:SI:DOC-MWREEOXW/a49c42c0-6e04-4ed7-b947-80af6eeec11a/PDF, November 11, 2022.

Voje, Ignacij (1978). Upliv puteva (saobračajnica) na raslojavanje slovenačkog sela u logaškom vlasteništvu. *Jugoslovenski istorijski časopis* 1–4, 149–57.

Vovko, Andrej (1978). Organizacije jugoslovanskih emigrantov iz Julijske krajine do leta 1933. *Zgodovinski časopis* 32/4, 449–473. https://www.dlib.si/stream/URN:NBN:SI:DOC-7NK5QEUD/5f4b931b-25ce-4d90-b819-56987895c7e7/PDF, November 11, 2022.

Vovko, Andrej (1979). Delovanje Zveze jugoslovanskih emigrantov iz Julijske krajine v letih 1933–1940. *Zgodovinski časopis* 33/1, 67–102. https://www.dlib.si/stream/URN:NBN:SI:DOC-SPOVDJXE/e0272819-91ca-4a1b-bc04-b36ff3617b33/PDF, November 11, 2022.

Vovko, Andrej (1990a). Udje Družbe sv. Mohorja v ZDA do leta 1900. *Dve domovini / Two Homelands* 1, 121–135.

Vovko, Andrej (1990b). Izseljenstvo. *Enciklopedija Slovenije*, zv. 4 (ed. Alenka Dermastja). Ljubljana: Mladinska knjiga, 233–234.

Vovko, Andrej (1992). Izseljevanje iz Primorske med obema vojnama. *Zgodovinski časopis* 46/1, 87–92. https://www.dlib.si/stream/URN:NBN:SI:DOC-G5VH212L/e689037b-103c-4d85-b883-f1bad90c8f86/PDF, November 11, 2022.

Wakounig, Marija (2010). Ženske v prvi svetovni vojni: Znanstvenoraziskovalni desiderat slovenske historiografije. *Soška fronta 1915–1917: Kultura spominjanja* (ed. Vincenc Rajšp). Dunaj: Slovenski znanstveni inštitut; Ljubljana: Založba ZRC, 43–50.

Wanderarbeit in Steiermark. (1913). *Der Auswanderer* IV/7–8, 124.

Wanderbewegung und Zahlunsbilanz. (1912). *Der Auswanderer* III/1, 2.

Werner, Erich (1985). *Slovenci v Porurju*. Ljubljana: Slovenska izseljenska matica.

Wickberg, Daniel (2007). What Is the History of Sensibilities? On Cultural Histories, Old and New. *American Historical Review* 112/2, 669. https://doi.org/10.1086/ahr.112.3.661.

Willcox, Walter F.; Ferenczi, Imre (eds.) (1929). *International Migrations. Volume I: Statistics*. National Bureau of Economic Research nr. 14. New York: Arno Press.

Wörsdörfer, Rolf (2004). *Krisenherd Adria: 1915–1955. Konstruktion und Artikulation des Nationalen im italienisch-jugoslawischen Grenzraum*. Paderborn: F. Schöningh.

Wörsdörfer, Rolf (2017). *Vom "Westfälischen Slowenen" zum "Gastarbeiter": Slowenische Deutschland-Migrationen im 19. und 20. Jahrhundert.* Paderborn: F. Schöningh.

Wörsdörfer, Rolf (2021). Gli immigrati storici nella Ruhr sotto il nazionalsocialismo. *Tedeschi contro Hitler? La società tedesca tra nazionalsocialismo e Widerstand* (ed. Federico Trocini). Torino: Rubettino, 179–192.

Zakrajšek, Kazimir (1938). Ustanovitev Slovenske izseljenske zbornice. *Izseljenski zbornik Izseljenske zbornice v Ljubljani.* Ljubljana: Družba sv. Rafaela, 7–33.

Založnik, Saša (2009). Delovanje in organiziranost Slovencev v zagrebški urbani regiji. BA thesis. Ljubljana: Univerza v Ljubljani, Filozofska fakulteta.

Zanini, Alba (2004). "Per la Germania e l'Ongheria": L'emigrazione temporanea dalle valli del Natisone. *Annales: Series historia et sociologia* 1, 73–84.

Zavertnik, Jože (1925). *Ameriški Slovenci: Pregled splošne zgodovine Združenih držav, slovenskega naseljevanja in naselbin in Slovenske narodne podporne jednote.* Chicago: Slovenska narodna podporna jednota.

Zelinsky, Wilbur (1971). The Hypothesis of the Mobility Transition. *Geographical Review* 61/2, 219–249.

Zempel, Solveig (1991). *In Their Own Words: Letters from Norwegian Immigrants.* Minneapolis: University of Minnesota Press.

Zgodovinski arhiv Ljubljana. Reg 1/1004, fol. 308–312.

Zgodovinski arhiv Ljubljana. Reg II/21 (V.).

Zitka, Ivan (1969). Trije Pirci v Ameriki. *Kamniški zbornik 1969*, 129–154. https://www.dlib.si/stream/URN:NBN:SI:DOC-XISJNSFF/064d7718-d798-456e-9329-c87488a92909/PDF, November 11, 2022.

Zobec, Miha (2013). *Nevidni in pozabljeni: Raziskava o izseljevanju v Argentino med svetovnima vojnama na primeru vaške skupnosti Pliskovica.* Pliskovica: Razvojno društvo Pliska.

Zobec, Miha (2019). Surveillance and Persecution of Slovene Antifascists in Argentina: Coordination of Authorities in Combating the Undesired Immigration. *Dve domovini / Two Homelands* 49, 221–238. https://doi.org/10.3986/dd.v0i49.7262.

Zobec, Miha (2021a). Creating the Unbound Yugoslav Nation: The Kingdom of Yugoslavia and Emigrants from the "Unredeemed" Julian March. *Nationalities Papers.* https://doi.org/10.1017/nps.2020.90.

Zobec, Miha (2021b). Prva Jugoslavija in "njena" diaspora v Braziliji. *Prispevki za novejšo zgodovino* 61, 2. https://doi.org/10.51663/pnz.61.2.05.

Zorn, Peter (2012). *Aleksandrinke so prišle domov.* Nova Gorica: Educa.

Zwitter, Fran (1936). *Prebivalstvo na Slovenskem od XVIII. stoletja do današnjih dni*. Ljubljana: Znanstveno društvo.

Žagar, Monika (2015). *Dva brata, trije svetovi: Zgodba o družini Žagar*. Ljubljana: Mladinska knjiga.

Žigon, Zvone (1998). *Otroci dveh domovin: Slovenstvo v Južni Ameriki*. Ljubljana: Založba ZRC.

Žigon, Zvone (1999). Janez Benigar – Araucanian Slovene. *Intelektualci v diaspori: Intellectuals in Diaspora: Zbornik referatov simpozija "100. obletnica rojstva Louisa Adamiča"* (ed. Irena Gantar Godina). Ljubljana: Založba ZRC, 93–101.

Žitnik Serafin, Janja (1992). *Louis Adamič in sodobniki*. Ljubljana: Slovenska akademija znanosti in umetnosti, Razred za filološke in literarne vede.

Žitnik Serafin, Janja (1993). *Pero in politika: Zadnja leta Louisa Adamiča*. Ljubljana: Slovenska matica.

Žitnik Serafin, Janja (1995a). *Pogovori o Louisu Adamiču*. Ljubljana: Prešernova družba.

Žitnik Serafin, Janja (1995b). Progresivne Slovenke Amerike. *Enciklopedija Slovenije*, zv. 9 (ed. Alenka Dermastja). Ljubljana: Mladinska knjiga, 370–371.

Žitnik Serafin, Janja (1998). Slovensko-ameriški narodni svet. *Enciklopedija Slovenije*, zv. 12 (ed. Alenka Dermastja). Ljubljana: Mladinska knjiga, 56.

Žitnik Serafin, Janja (1999). Literarni poskusi Slovencev v drugih deželah Evrope do leta 1945. *Slovenska izseljenska književnost 1: Evropa, Avstralija, Azija* (eds. Janja Žitnik, Helga Glušič). Ljubljana: Založba ZRC, Založba Rokus, 77–116.

Žitnik Serafin, Janja (2011). Pomen Louisa Adamiča v 21. stoletju. *Studia Historica Slovenica* 11/2–3, 763–781. https://doi.org/10.3986/dd.2020.1.01.

Žitnik Serafin, Janja (2014). Prerez zgodovine slovenskih kulturnih društev v jugoslovanskem prostoru. *Priseljevanje in družbeno delovanje Slovencev v drugih delih jugoslovanskega prostora: Zgodovinski oris in sedanjost* (ed. Janja Žitnik Serafin). Ljubljana: Založba ZRC, 137–179.

Žnidaršič, Ivica; Kaplan, Zdenka (2003). *Spomini in pričevanja*. Ljubljana: Društvo izgnancev Slovenije 1941–1945.

Žnidaršič, Ivica; Kaplan, Zdenka (2006). *Spomini in pričevanja, II. knjiga*. Ljubljana: Društvo izgnancev Slovenije 1941–1945.

Žnidaršič, Ivica; Kaplan, Zdenka (2012). *Spomini in pričevanja, III. knjiga*. Ljubljana: Društvo izgnancev Slovenije 1941–1945.

Žnidaršič, Sabina (2015). Ženske na trgu dela: Čas sprememb 1880–1910. *Žensko delo: Delo žensk v zgodovinski perspektivi* (eds. Mojca Šorn, Nina Vodopivec, Žarko Lazarević). Ljubljana: Inštitut za novejšo zgodovino, 119–137.

Žumer, Aleš (1996). Organizacije vestfalskih Slovencev 1900–1933. *Etnični fraternalizem v priseljenskih državah / Ethnic Fraternalism in Immigrant Countries* (ed. Matjaž Klemenčič). Maribor: Pedagoška fakulteta, 395–401.

Župančič, Oton (1932). Adamič in slovenstvo. *Ljubljanski zvon* 52/8, 513–520. https://www.dlib.si/stream/URN:NBN:SI:DOC-K93KZHFS/54d0ad51-2d12-471d-9768-a386ad1772f0/PDF, November 11, 2022.

Index

Thought, Society, Culture
Slovenian and South Eastern European Perspectives

Edited by the Research Centre of the Slovenian Academy of Sciences and Arts

Vol. 1 Kozma Ahačič: The History of Linguistic Thought and Language Use in 16th Century Slovenia. 2014.

Vol. 2 Luka Vidmar: A Slavic Republic of Letters. The Correspondence between Jernej Kopitar and Baron Žiga Zois. 2016.

Vol. 3 Boris Golec: Temporary Croatization of Parts of Eastern Slovenia between the Sixteenth and Nineteenth Century. Changing Identities at the Meeting Point of Related Peoples. 2017.

Vol. 4 Neva Makuc: Border Identities in the Early Modern Period. Venetian Friuli and the Habsburg County of Gorizia Mirrored in Contemporary Historiography. 2021.

Vol. 5 Aleksej Kalc, Mirjam Milharčič Hladnik and Janja Žitnik Serafin: Daring Dreams of the Future. Slovenian Mass Migration 1870–1945. Translated by Barbara Skubic. 2024.

www.peterlang.com

www.ingramcontent.com/pod-product-compliance
Ingram Content Group UK Ltd.
Pitfield, Milton Keynes, MK11 3LW, UK
UKHW020833290926
15470UKWH00004B/1327

9 783631 898970